Movement

Movement
Functional Movement Systems: Screening, Assessment and Corrective Strategies

Gray Cook
MSPT, OCS, CSCS

with
Lee Burton, PhD, ATC, CSCS, Kyle Kiesel, PhD, PT, ATC, CSCS
Dr. Greg Rose & Milo F. Bryant

Foreword
Jon Torine

On Target Publications
Santa Cruz, California

Movement
Functional Movement Systems: Screening, Assessment and Corrective Strategies
Gray Cook

with
Dr. Lee Burton, Dr. Kyle Kiesel, Dr. Greg Rose & Milo F. Bryant

Foreword: Jon Torine

Cover photo: Aaron Stearrett
Interior photos: Paul Liggitt
Athlete models: Brandy Apple, Heidi Brigham, Paul Hylton, Wyle Maddox

Copyright © 2010, E. Grayson Cook
ISBN: 978-1-931046-72-5

Also by Gray Cook
Athletic Body in Balance

15 14 13 12 11 10   3 4 5

On Target Publications
P. O. Box 1335
Aptos, CA 95001 USA
(888) 466-9185, Fax (831) 466-9183
info@ontargetpublications.net
www.ontargetpublications.net

       Library of Congress Cataloging-in-Publication Data
Cook, Gray, 1965-
  Movement : functional movement systems : screening, assessment, and corrective strategies / Gray Cook, with Lee Burton, Kyle Kiesel, Greg Rose and Milo Bryant; foreword, Jon Torine.
     p. ; cm.
  Other title: Functional movement systems
  Includes bibliographical references and index.
  ISBN 978-1-931046-72-5
 1. Movement disorders. 2. Human locomotion. I. Title. II. Title: Functional movement systems.
    [DNLM: 1. Movement--physiology. 2. Exercise Therapy. 3. Motor Activity--physiology.
    4. Physical Fitness--physiology. WE 103 C771m 2010]
RC376.5.C66 2010
616.8'3--dc22
                 2010015256

I have always been a curious and driven person. This, combined with an obvious attention deficit disorder, has made for an interesting ride. Thankfully, I've been blessed with a force in my life that counters many of my dysfunctions. This influence has green eyes, blond hair and can't even make the scales register 120 pounds, but nevertheless I find myself outmatched.

In our house, we call this force of nature Danielle.

She is much of the strength and balance behind this work.
This one is for my wife. Thanks, D.

CONTENTS

APPENDICES

For articles, video, media, downloadable forms and regular updates to support this book, please visit www.movementbook.com.

For 16 years, I have had the privilege of working as a strength and conditioning coach in the National Football League, the past 13 with the Indianapolis Colts. My passion for trying to maximize physical performance and increase durability based on science and practical application has formulated the basis of our system, which has led us to become harsh critics with a tough filter for programs. We have no margin for error, and do not have the luxury of trial and error when that may mean the difference in millions of dollars and in wins and losses.

Years ago, in the late '90s, I'd heard about movement screening. The methodology fascinated me, leading me to study it, and watch and wait and wonder.

A year or two later I met Gray Cook, who put me through the Functional Movement Screen and pointed out my rotary stability pattern was, well... lacking. He taught me proper rolling, made some corrections and asked me to repeat the pattern. I got up and executed the rotary moves almost perfectly, and that's what put me over the top with the system. During the visit, he explained the neural aspects and the role of human motor development components in stark contrast with isolated joint and muscle thinking, and a big light flashed in my head: *This system will work for us!*

And so it did, and so it does. I use the Functional Movement Screen in my work of training professional football players, and you can use it for your work with hearty athletes, personal training clients and rehabilitation patients as well. It's that versatile, that effective and that appealing.

Gray has brilliantly taken complex neurological and anatomical physiology and broken it down into a simple, usable and practical system that can be applied in any setting. Gray's method gives us a baseline and a system to check our work. That's another key I learned from his labor and explorations—how important it is check to my own work. That's huge; it's an objective measure, one I just didn't have with the players before I started using the screen.

In our profession, it's difficult to measure results. We look at power output in meters per second and watts, speed, strength and movement screens. We verify that players can lift respectable loads, that they have power, individual and position-specific body composition, anaerobic endurance, good movement capability and applied sport nutrition. We teach circadian rhythm and sleep patterns—the factors affecting the game are multitudinous. The movement screens outlined in the book you hold in your hands provides a new kind of measurements. These will tell you when you're on the right track, and will tell you when to make adjustments. In my mind, movement screening provides the platform for all our other measures.

Simply put, *it works that way.*

In team settings, the screen opens up lines of communication with the athletes at the center, and the strength and performance coaches working with the physical therapists, athletic trainers, chiropractors, medical doctors and other medical professionals. A medical and performance team is then focused on the athlete's health, rehabilitation and performance, making each transition seamless whether post-operative, painful, dysfunctional or healthy and ready to train and play.

Coaches, trainers and rehab professionals can screen for pre-season physicals whether in school, athletics or the military, and then use the data as part of the return-to-play criteria. In sports, we have to keep the players on the field or court—movement screening is our best tool for predicting injury risk before a player gets sidelined. In our training room, it's used to identify mobility and stability issues and it guides the transition to increased levels of training. When movements aren't clean, it's a big red flag that gains our attention every time.

Screening establishes a movement baseline. If our pre-season screening uncovers pain, the player sees one of our medical pros, who then does a clinical evaluation. The system allows pain to split the decision process into clinical evaluation or simple corrective exercise based on the observed pattern dysfunction. The rehab specialist pinpoints the potential problem and clears it, thus reducing the risk of injury. There's no question that the Functional Movement Screen is a serious biomarker of player durability.

Everything we do at the Indianapolis Colts is built on a Functional Movement Screen base— it's the foundation of our program.

The fact is, full strength and power is not realized or used without movement efficiency. Being strong doesn't mean much without fluid, efficient movement; staying strong and stable while being bombarded is what players need in football. The screen has provided this clarity for me. I now have a huge appreciation for movement efficiency—functional and foundational movement. Understanding human motor development, as you'll learn through your Functional Movement Systems study, clears up cloudy thinking, and healthy, powerful action follows.

You'll gain insight into motor development and human postures and patterns. And you'll understand the reality of the core, of posture and of breathing. It's all in here, and when you *get it,* it'll provide a system upon which your rehabilitation and training programs can be checked for movement.

Do what's best for your people by having a system that creates lines of communication from the medical field to the performance field, one that increases durability, predicts and decreases injury, increases movement efficiency and provides a purpose for exercise while reducing the time spent training.

Enjoy and appreciate the brilliance of a never-ending lifelong journey provided by Gray Cook. I know I—and those I work with—will continue to reap the rewards.

Jon Torine
Strength and Conditioning Coach
Indianapolis Colts

What if you live in a culture obsessed over exercise and diet, a culture where medicine and exercise science are known for world-leading advances, but a culture in a state of severe physical decline where obesity, heart disease, back pain and injuries are ironically becoming more prevalent?

What if you are a personal trainer with a sobering awareness of how the general public is confused by exercise marketing and magic-bullet fitness claims, a confusion made worse by the differing professional opinions within the fitness industry?

What if you are a physical therapist forced to live with the irony that a dentist is routinely compensated for a yearly checkup, but you are not, even though Americans spend more than $50 billion a year on low back pain?

What if you are a sports medicine professional or a strength and conditioning professional with an ethical commitment to prevent athletic injuries and you come to the realization that your professional preparation has not provided a comprehensive tool for the task?

What if you have advanced training in weightlifting, kettlebell lifting and other historical forms of exercise and physical development and that knowledge makes you question most modern prepackaged exercise programs?

What if you observe how the oldest forms of movement training such as the martial arts and yoga fully integrate moving and breathing, but modern forms rarely incorporate those same complementary effects?

What if you work with the best of the best in the fields of exercise and rehabilitation, and plainly see a common perspective among this elite group, a perspective based on something so fundamental it is taken for granted and largely overlooked in education and research?

What if you go behind the scenes of collegiate and professional sports and witness the same fundamental factors that explain performance and durability with greater clarity?

What if you have the opportunity to look for these fundamental factors in seasoned military operators, new recruits, firefighters, junior and senior athletes, industrial workers, dancers, weekend warriors and sports and orthopedic patients?

What if you own that perspective, if you have those opportunities—what opinions would you form? What would you have to say? What would you do to make a difference?

I struggled with these questions because I do this. I see this. I experience this.

I could have added to the confusion by developing new exercise and rehabilitation programs to target the problems.

Instead, I've tried to build a system with the advice and help of the best professionals I know.

We have enough programs and protocols. We need a standard operating system for movement fundamentals. We need a system to help us gauge *movement quality* before we gauge *movement quantity*.

Systems make our programs better. Systems make *us* better.

In this text, I will introduce you to that movement system. As you read, contemplate my foremost philosophy of movement, a simple statement that unfortunately does not work in reverse.

First move well, and then move often.

Gray Cook

INTRODUCTION TO SCREENING AND ASSESSMENT

Movement is at the core of our early growth and development spectacle, and movement remains the central theme throughout life. As exercise professionals, we promote movement, but as a group, we don't have standard screens for risk factors associated with movement-based activities. As rehabilitation professionals, we do not use a grading scale or standard for whole movement patterns. We measure the movement parts and assume that whole movements can be predicted. The screens and assessments now in use are not comprehensive. They are impairment-based, performance-based and activity-specific.

As a profession, we have trained, assessed, treated and profited from movement practices without a proven ability to observe and manage movement fundamentals. This makes us part of the problem, and the problem grows exponentially as we continue with the status quo.

This book represents a standard operating procedure for movement-pattern appraisal in fitness and rehabilitation. Without such a system, we are missing the opportunity for a movement screen and assessment that could improve our identification of risks, and provide for more complete injury diagnoses.

Each of us has developed a preferred way to look at movement issues of specialized interest and arrive at solutions based mostly in our own professional comfort zones. Highly specialized, myopic movement evaluations have been one of our biggest errors, if not the biggest. This is not as much a scientific error as it is a logical and philosophical lapse.

But at its core, the lack of movement understanding is the root of the problem.

The best health and fitness practitioners already look at whole movement patterns. They consider fundamental movement before specific movement.

The best professionals, educators and researchers agree on one wise, simple and time-honored approach: *Fundamentals are always first.*

Healthcare and fitness practices often neglect fundamental movement, paying too much attention to the surface view. Weakness and tightness are often attacked with isolated and focused strengthening and stretching remedies that don't work toward a movement-pattern standard. When one person's hip hurts and another complains of upper back stiffness, we are quick to find solutions. A surgeon, a physician and a physical therapist see these problems through eyes biased by solutions of their training instead of working from a comprehensive baseline. One sees a surgical solution based on structure; one considers which medication to manage pain and inflammation, while the other looks for mechanical issues to rehabilitate. Ultimately, the region of complaint classifies each patient more than that of a comprehensive movement profile.

The initial perspective is the same for the physician, physical therapist and strength and conditioning professional. We study the same anatomy, but regard movement from many different views. We become highly specialized, and in specialization lies the problem. We study specific aspects of biomechanical activities without a baseline movement standard. We each look at tightness, weakness and pain, but we don't all see the same thing. We have the anatomy map, but we can't agree on a movement map.

Fundamental movement isn't brought into the conversation on the same level as other issues that are qualified and quantified in the exercise and rehabilitation professions. That needs to change.

Forward progress cannot be made without a screen or an appraisal of the current state of movement—an appraisal of movement must precede an appraisal of physical fitness or performance.

Whenever possible, we must separate movement dysfunction from fitness and performance. Aggressive physical training cannot change fundamental mobility and stability problems at an effective rate without also introducing a degree of compensation and increased risk of injury.[1]

Whole movement patterns should not only be in the conversation, we should consider and standardize these with scientific and professional attention. We should consider patterns of movement over the appraisal of isolated body segments, range of motion and basic kinesiology.

We can measure flexibility and strength as normal, and still not see natural movement patterns free of deficiency and risk of injury.[2] In fundamental movement, whole movement can be greater or less than the sum of its individual parts. Logically, the appraisal of movement-pattern competency should be the starting point.

Attention to detail regarding exercise is important, but it is only significant after the basics are established and stable. Only then can the foundation efficiently support the rest of the structure and its functions.

This book's proposal is simple: Screen movement patterns before you train them. Training poor movement patterns reinforces poor quality and creates greater risk of injury. If you identify poor movement patterns, you can correct them with simple exercises, later rechecking against the baseline.

Imbalances and limitations within movement patterns are markers we test, because these identify greater risk of injury for our clients involved in exercise and activity. These also identify potential risk when we release our patients, even if they no longer display symptoms. Following a screen of movement, an individual will be placed in one of three groups.

Acceptable Screen—cleared to be active without increased risk

Unacceptable Screen—at risk for injury unless movement patterns are improved

Pain with Screening—currently injured, requiring more advanced movement and physical assessment by a healthcare provider

Screening reveals sobering information about the high percentage of any group who will experience setbacks and injuries with increased activity. Without screening, exercise programs and rehabilitation protocols incorrectly perpetuate the assumption that functional movement patterns, mobility, stability and proprioception are otherwise normal.[3-6]

The number one risk factor for musculoskeletal injury is a previous injury, implying that our rehabilitation process is missing something. Although the current medical and rehabilitation model can manage the pain and symptoms resulting from an initial injury, they have less ability to influence the likelihood of a recurrence. The medical and rehabilitation process should not just manage the painful episode, but also target and contain the recurrence risk factors. By simply screening movement once pain has been managed and resolved, the medical or rehabilitation professional can gauge the chance of recurrence and advise the patient how to reduce risk.

If movement is dysfunctional, all things built on that dysfunction might be flawed, compromised or predisposed to risk even if disguised by acceptable levels of skill or performance. Poor movement patterns demonstrate increased injury risk with activity, but good movement patterns don't guarantee reduced injury risk. Once fundamental movement is managed, other factors like strength, endurance, coordination and acquisition of skill also play a role in prevention. Movement comes first.

Whenever possible, it is important to separate pain with movement from movement dysfunction. It is possible to move poorly and not be in pain, and it's possible to be in pain and move well. A licensed healthcare professional experienced in musculoskeletal evaluation and treatment should address pain with movement regardless of fitness ability. All good exercise professionals can help prevent injuries, but once pain is present, a healthcare practitioner with knowledge of pain evaluation and movement dysfunction should provide a comprehensive assessment.

We must use what we know. We should screen and assess movement dysfunction to reduce risk and promote durability. This we do through the Functional Movement Screen (FMS®). Once we've addressed movement dysfunction, we must be vigilant that it does not return; we need to monitor and address other areas that may indicate increased risk by looking at movement performance and specific movement skills.

The FMS is the *predictive system*. It is a reliable, seven-step screening system with three clearing tests, designed to rank movement patterns basic to the normal function of active people. By screening these patterns, you will be able to identify, rate and rank movement limitations and asymmetries.

Basic movement pattern limitation and asymmetry are thought to reduce the effects and benefits of functional training and physical conditioning. Recent data suggest these factors may be related to injury in sport. One goal of the FMS is to identify movement pattern limitations so professionals can prescribe individualized corrective exercise to normalize movement before increasing physical demands with training. This individualized approach has been shown to be effective in one-on-one situations, as well as in group settings.[7]

The FMS is a screen designed for individuals who do not have a current pain complaint or a known musculoskeletal injury. Clients reporting pain prior to or during an FMS require and deserve the benefit of a medical diagnosis and treatment not provided by movement screening. For this, we use the Selective Functional Movement Assessment (SFMA®).

The SFMA is the *movement-based diagnostic system*, a series of seven full-body movement tests designed to assess fundamental movement patterns in those with known musculoskeletal pain.

When the assessment is initiated from the perspective of a movement pattern, the clinician is able to identify meaningful impairments that may seem unrelated to the main complaint. This concept, known as regional interdependence,[8] is the hallmark of the SFMA. It guides the clinician to the most dysfunctional but non-painful movement pattern, which is then assessed in detail. By attending to this pattern, the application of targeted therapeutic exercise choices are not harmfully affected by pain and can then reduce movement dysfunction. The system focuses corrective exercise at movement dysfunction, not at pain.

The SFMA complements the existing medical exam and efficiently integrates the concepts of posture, muscle balance and fundamental movement patterns into the practice of musculoskeletal rehabilitation. The SFMA provides a systematic process for the best possible therapeutic and corrective exercise intervention.

The two systems, the FMS and SFMA, work together as an effective classification and communication tool between professionals. In these systems, an individual or group is guided to the most basic physical parameter necessary to reduce the risk of injury and promote safe progress toward greater conditioning or rehabilitation goals.

Movement pattern corrective strategy is a form of exercise that focuses more on improving mobility, stability, basic motor control and whole movement patterns than the parameters of physical fitness and performance. Once established, the movement patterns create a platform for the general and specific parameters of fitness, including endurance, strength, speed, agility, power and task specificity.

Movement is what this book is about—screening, assessing, rating and ranking movement. The purpose of this book is to create a perspective concerning movement, and to coach your review of quality in whole movement patterns. Anyone appraising, training or restoring movement will find indispensable information in these pages.

Our bodies are miracles capable of unbelievable durability and resiliency, with an amazing performance and physical capacity. We are made to grow strong and to age gracefully. Reclamation of authentic movement is the starting point. We cannot simply have better fitness, conditioning and sports performance. We must cultivate it.

THE PRACTICE OF MOVEMENT SCREENING AND ASSESSMENT

The science and practice of movement screening and assessment is an organized system for discussing and documenting movement patterns. As you learn this qualitative movement model, you'll be able to implement the system as your starting point in exercise and rehabilitation. You'll gain confidence in your ability to understand movement, and your clients and patients will be the beneficiaries.

As a practical guide for the professional, we will discuss how a qualitative movement model serves each discipline, and will address the need for a functional movement appraisal. You'll review a detailed presentation of the Functional Movement Screen (FMS), a tool used for understanding movement quality in healthy populations, and will work in depth with the Selective Functional Movement Assessment (SFMA), which is used with patients who have pain with movement.

One of the primary intentions of this system is to help frame a professional standard for movement-pattern observation and documentation through screening and assessment. Space is dedicated to develop a paradigm shift in movement science, as movement patterns become an additional biomarker for movement dysfunction, musculoskeletal problems and injury prediction. Our initial observation should move away from the detailed and quantitative analysis of selected motion and anatomical structures, and toward fundamental movement quality. Once a movement map has been created, the quantitative analysis of motion and structures will have greater relevance. This simple change in focus helps create new strategies for screening, assessment and corrective measures.

If this book fails as a conventional text, consider that the objective. Textbooks are not designed to change practice; evidence and new ideas change practice. As exercise and rehabilitation professionals, we are doing our clients and patients a serious disservice if we work with or on them while failing to understand why their bodies move the way they move. Our sincere presentation in this book should provide the knowledge needed to support that statement, as well as demonstrate the need for a new model—new movement logic founded on the working of the central nervous system together with the musculoskeletal system.

Motion is Life

~Hippocrates, Greek physician, 460–377 BC

Motion, specifically movement, symbolizes life. We move with conscious intention, and we move in automatic response. Our activities combine both reflex and purposeful movement behavior. In most cases, the first supports the second, and the second triggers the first. Like two sides of a coin, we cannot separate these easily or practically.

When we move, we think about the intended movement, but we dismiss the subtle adjustments our bodies and minds make to support the initial intention. To some degree, our movements represent our physical strengths as well as our limitations, and our movements and body language can also forecast an emotional state.

We often communicate about movement in pure and clean mechanical terms, but human movement surpasses simple angles, vectors, forces and directions. Human movement is a behavior, and we should think of it within behavioral parameters. In general fitness, conditioning, rehabilitation and medicine, movement measurements set a baseline. We measure our attempts to improve some facet of movement with this baseline, whether enhancing performance or restoring a previous level of function.

If we can agree on movement's behavioral and mechanical aspects, we must understand measurement applications and interpretation strategies that respect both disciplines equally. To that end, we temper the mechanical movement facts with the subtleties of human behavior to create comprehensive movement management. Developing that idea with a system of implementation is the goal of this book.

BODY PARTS VERSUS MOVEMENT PATTERNS

In mechanical science, one big item gets broken into manageable parts. That breakdown—called reductionism in science—often creates one perspective while destroying another. Just as bodies are destroyed by dissection, movement patterns are destroyed by reductionism. We see this in our modern perspective of foods as we focus on calories. This single perspective might allow one to assume that highly processed foods are the same as whole foods of equal calories. Most people don't examine enzymes, micronutrients, fiber and glycemic index—measured by calories alone, two meals might seem similar, but may produce an opposite metabolic response. The discovery of calories and their subsequent counting has not made us a leaner society. In fact, bodyfat percentages have for the most part gotten worse.

Likewise, breaking down movement into isolated segments has not reduced our musculoskeletal injuries[1] or made us fitter or leaner. The problem is reductionism without equal parts of reality and practicality, moderated by a comprehensive view. As movement measurement sensitivity improved, movement scientists followed other sciences by embracing reductionism. At a dissected level, movement observation and categorization became organized and manageable, while consideration of fundamental whole movement patterns died.

Conversely, patterns and sequences remain the preferred mode of operation in biological organisms. Patterns are groups of singular movements linked in the brain like a single chunk of information. This chunk essentially resembles a mental motor program, the software that governs movement patterns. A pattern represents multiple single movements used together for a specific function. Storage of a pattern creates efficiency and reduces processing time in the brain, much as a computer stores multiple documents of related content in one file to better organize and manage information.

Fundamental movements get stored in basic patterns, as do frequently reproduced movements. Although a scientist may want to look at a pattern's parts to enhance understanding, we as exercise and rehabilitation professionals must understand that the brain recognizes sequences and uses them to generate true function and realistic movement.

Viewing the parts can give clarity, but viewing the patterns will produce a global understanding. Studying the details imparts movement intelligence, but understanding the patterns creates movement wisdom. For academic study, dissection is appropriate for terminal understanding. However, if the goal is to affect realistic and functional movement in a practical way, we can't stop at simple dissection, but instead must focus on reconstruction and reinforcement of whole movement patterns.

THE MOVEMENT VERSUS MOTION PARADOX

In some cases, a semantic problem exists where terms become interchangeable. In other cases, professionals who routinely measure motion make subtle assumptions. This builds a paradox between movement and motion: There's an assumption that if each joint involved in a movement has normal motion or range of mobility, a movement involving all of these joints will also be normal.

The definitions of movement and motion are similar, but imply different things when dealing with exercise and rehabilitation. Movement often denotes the act of a functioning body as it changes position under its own power. Motion might represent the available range of flexibility within a single body segment or group of segments. Movement is associated with basic and advanced full-body activities such as crawling or running, or perhaps a golf swing. Motion, such as 180 degrees of shoulder flexion, often relates to a specific amount of directional freedom.

With this view, we might say normal motion is necessary for normal movement. However, normal motion does not *guarantee* normal movement. Motion is a component of movement, but movement also requires motor control, which includes stability, balance, postural control, coordination and perception.

This book introduces systems designed to reduce professional subjectivity and semantic problems such as these to simplify issues and circumstances where assumptions cloud focus and misdirect attention.

A PROBLEM WITH PERSPECTIVE

We see examples of problems with perspective in many healthcare evaluation practices. Some practices operate under the assumption that managing a dysfunctional system's defective aspect will correct the entire system. Our highly specialized approach is often set up to evaluate parts, not patterns. This corrective system typically results in only the appearance of successful management of a single, isolated part. However, focused piecework ignores synergistic power and the integrated patterns that produce true function.

Side effects emerge, causing secondary problems in other systems, and at some point, we realize we must consider the whole pattern. If we normalize the movement impairment into elements of isolated weakness or stiffness, the pattern or function may be unchanged—the problem remains present at a functional level. The isolated measurement indicates a positive change, but the problem is present at a practical level.

Dealing only in parts is safe and comfortable; after fixing a small part, but not fixing the problem, we might say, "That's not my specialty" or "I did all I can do." I can make this cutting statement because I am part of the medical and rehabilitation profession. I have witnessed and been a part of the small-mindedness of an incomplete movement management system.

This problem is not limited to the medical profession—it simply serves as a common and obvious example of a problem of perspective that exists mainly in established science-based activities. Today, the holistic approach is akin to popular appeal and marketing instead of authentic practice, when authenticity is what we want people to believe about our services. Yet, with the right system in place, a comprehensive approach can become the mode of operation and the mission, and not just another marketing ploy.

THE WHOLE IS GREATER THAN THE SUM OF ITS PARTS

Most of us know and agree with the saying *the whole is greater than the sum of its parts,* but do we act on that belief? Consistent with that concept, a whole movement pattern is greater than the sum of its moving parts. Do those in exercise and rehabilitation practice this, or just agree with the statement without practical execution? When the pressure is on, do we manage moving parts or take responsibility for entire movement patterns? Do we choose to focus on a movement pattern's part and correct that, hoping the whole movement pattern will adjust on its own? Alternatively, do we look at the whole pattern and first address the problem as a whole, letting the fundamental forces within the pattern normalize naturally?

In some instances, managing the small parts of a movement pattern is beneficial. That does not endorse the practice of neglecting the whole pattern once the single aspect is successfully controlled. A great example of the *pattern versus the part* phenomenon happens when a person sprains an ankle severely or breaks a leg, and then continues to limp long after healing. The brain writes new software—the limp—to work around a temporary problem.

This system does not always reset to a primary or normal operating scheme once healed. The body repaired the damage, but the now-unnecessary dysfunctional pattern remains unchanged—the whole is less than the sum of its parts. All of the parts check out with no particular rationale to support the dysfunctional pattern, giving us a behavioral curve ball that separates pure movement mechanics from reality. Residual habits and habitual tendencies often remain in place after we remove the quandary that created them.

These movement tendencies demonstrate illogical behavior. We need a logical system to monitor this behavioral departure—we need to look at movement patterns as closely as their supportive elements. Movement patterns should be at our work's beginning and end. At the outset, we screen and assess patterns before parts, and only then do we judge the parts within the most limited patterns as primary problems. All other components are not a concern at the beginning, but should be documented, referenced and monitored.

Let's look at an example of a car with a flat front tire: The relative pressure in the other three tires does not really matter, even though we can measure and perhaps discover sub-optimal air pressure. The first order of business is to recognize the flat, fix it and then finish by testing the methods used for other corrections. Once we've attended to the flat front tire, the other information becomes significant and appropriate measures can be taken. If we tested the tires with a pressure gauge, they might have all shown varying degrees of low pressure, but thinking of the flat as just another low-pressure problem would direct us to add air instead of recognizing the need to repair the flat. This simple example demonstrates priority.

In the body, if we work on a specific part of the most limited pattern, we need to reconsider the pattern at the end of the intervention. If the most limited pattern is unchanged, the results are minimal at best. Something may have changed, but the central nervous system did not recognize it.

In basic terms, the most beneficial facet of whole patterns is that they will direct the focus initially, and confirm the intervention in the end.

Use functional movement patterns in the beginning to decide what to investigate. Use them in the end to see if the brain recognizes the changes made to the parts within the pattern. Intelligent and well-trained people ignore the "flat tire" every day, and mistakenly measure the air in the four tires with equal attention. Focused on systematically measuring imperfections and departures from the norm, they fail to rank the information and identify the primary problem. The obvious distinction between low air pressure and none is not recognized or prioritized. Low air pressure is a problem, but *no pressure* changes the weight of the initial problem altogether. The issue goes from caution—low pressure—to complete immobility, a flat tire. At times, the modern clinician looks more like a motorist frantically re-measuring three low tires and a flat than a confident mechanic with tire-changing skills.

This basic logic does not tell us to ignore every problem observed, just to use an objective system to rate and rank the problematic information efficiently and effectively. We waste valuable time precisely measuring problems we do not need to act on. Set baselines if necessary, and use other information to mark progress, but understand that some information always takes priority. Once you observe change in the initial variable, another variable may come to the forefront. If you do not find the flat tire, nothing else matters; the flat is the variable that stops forward progress.

Always rate and rank information before measuring any detail or variable. We must objectify, organize and start with prioritization, not with precise measurement. *This rule cannot be overstated.*

MOVEMENT IS OUR BUSINESS

Since movement is our business, we should push to look beyond the current practices and understanding within the professions of human movement. From athletic performance enhancement to rehabilitation of the severely disabled, we must not just talk about a new comprehensive approach to movement, we need to build one.

Attempts to improve our profession sometimes embody more enthusiasm than objectivity. Many successful professionals develop effective pro-

grams and protocols, and then devise evaluation techniques toward their old familiar programs. Sometimes seasoned professionals fall victim to their own subjectivity. The most objective professionals are those who know they are not objective and therefore defer to systems to recheck the work.

BIAS: MANY THINGS LOOK LIKE NAILS ONCE YOU HOLD A HAMMER

This bias is a pitfall of human nature. Games lose meaning and merit without rules and officials. We should not make rules as we play—we should set rules based on inductive and deductive logic, and then use our skills to affect the variables we rate, rank and measure.

We do not have a qualitative movement standard that gauges exercise and rehabilitation effectiveness. Professionals in our field recognize and discuss movement quality. We have varying degrees of quality in the exercises we supervise, but most do not practice qualitative movement testing with the same degree of precision that quantitative standards display in the current research.

Researchers have a natural bias to study movement quantities because they are clean, and fit into numeric models. Studying quality can be messy and difficult to measure. The evolution of modern exercise and rehabilitation practices has gone unchecked for quality because of this bias. Without a qualitative standard, we are doomed to observe only the quantitative features of movement.

This missing variable may partly explain how major movement issues such as adult low back pain and scholastic athletic injuries continue to escalate in the presence of our so-called new and improved movement sciences. In the scientific undertaking, qualitative and quantitative characteristics need to be represented and considered continually. When equal weight and consideration is not present, perspective is often lost.

We must devise user-friendly systems that protect us from the subjectivity that often emerges because of our expertise and experience, and because of the natural bias and affinity researchers have of quantities over qualities.

SYSTEMS VERSUS PROGRAMS

Program—a predetermined action plan for achieving a result

System—a method of achieving results where one procedure's outcome determines the next

Many of us were taught to relate problems to the programs and protocols designed to fix them. The critical thinking needed to address movement-based problems is not so much about assigning a program to a particular problem. It's about taking a problem to its source and systematically managing the primary issue, as well as detecting the secondary problems that cloud judgment and clarity.

Programs and protocols are limited when they depend on a preset plan without a set of procedures to modify it. Systems are always more effective because they possess a constant reappraisal method. The system will continue to refine itself with built-in procedures that match progress to baselines.

This text will help develop measured responses and to avoid quick reactions to movement problems. It will introduce an objective system to assist and direct you, but it is not a program to govern. Instead of a program without consistent constant checks and balances, it's a system built on a reassurance of continuous review, followed by a measured and calculated response.

The intended role is to assist with information gathering and management of movement screening, assessment and corrective exercise strategies.

In reading this material, you will consider a qualitative approach in parallel with a quantitative approach. This material may appear to have a qualitative bias, but that is not the intent; one approach is not more important than the other. In our society, we more clearly understand, define and discuss the quantitative perspective, and exercise and rehabilitation professionals are often rewarded for quantity over quality. This text will show the importance of using both forms equally.

To gain the clearest perspective, we must first use quality to rate and rank information. Then we can use measures and quantities to focus on the steps responsible for achieving the primary goals.

Some personal fitness clients have a greater concern for rapid weight loss than for total fitness. Many athletes focus on superhuman feats of speed, power and endurance, instead of athletic fundamentals and a balanced training approach. They do not consider it productive to build a superior foundation, even though most consistent physical accomplishments center on that platform.

Patients with musculoskeletal problems often focus so much on pain relief, they fail to see the disparity between basic symptom management and true healing or resolution of a problem. In their distress, they mistake pain as the problem without understanding it is merely a signal, and that the problem may remain even after the signal is gone.

We can blame our consumers, or, to better explain our work, we can give the public qualitative standards in addition to the quantitative standards they know. If we move first, insurance companies and consumers may eventually follow. When the clients and patients focus on quantity, our professional responsibility is to educate them to produce awareness and understanding. We, as a group, must first prize a qualitative standard before we can pursue a qualitative approach. Professionally balancing quality and quantity will help us map the most efficient and effective path toward the goals we set.

However, before discussing a new system, it is necessary to expose a few current issues in our profession.

THE NEED FOR A FUNCTIONAL MOVEMENT SYSTEM

The scrutiny and appraisal of movement is not new. We have been interested in movement since the beginning of human observation. Movement is how we survive, communicate, recreate and thrive. Our aptitude to measure and examine movement has evolved into an enormous modern science. The information provided by this science can increase our knowledge about movement, as well as perpetuate our confusion.

Movement screening and movement assessment are important because these two systems bridge the gap between real life activities, and medical or performance testing and advanced biomechanical analysis. The importance of bridging the gap is to establish a qualitative movement-pattern standard.

Human movement arises from patterns designed to protect and feed us, or in a different view, to help us receive pleasure and avoid pain. We are hardwired with reflex movement behaviors that develop into a blend of conscious and subconscious movements as we grow. Movement patterns are purposeful combinations of mobile and stable segments working in coordinated harmony to produce efficient and effective movement sequences. These linked sequences give us a command of physical posture during activity and movement through our surroundings.

True science does not exist solely to dissect a natural behavior to its smallest segment. It must also demonstrate the interrelationship between problematic segments and movement patterns. If we aggressively pursue the segmental dissection of movement without mapping the pattern matrix needed for functional activity, we show our ignorance of nature and the supporting system behind the structures and energies that move us.

Pattern recognition is at the heart of movement screening and movement assessment. If we can first recognize the functional movement's supporting patterns, we can successfully dissect the segments of each pattern or choose to manage the pattern as a whole. This provides insight into rebuilding primary movement patterns. Once the patterns are established, function can rest on the resulting foundation.

Screening and assessing these patterns creates clarity and understanding when we want to change, improve or rehabilitate a particular movement. The subtle degradation of patterns may also provide predictive information about increased injury risk during activity.

In other words, our ability to measure portions of movement with greater precision will only help us collect information. The way we use that information depends on the logic we bring to our study.

MOVEMENT SCREENING MODELS

To increase comprehension and promote professional guidelines, this book is broken into two distinct models, using two tools—the Functional Movement Screen (FMS) and the Selective Functional Movement Assessment (SFMA).

The original team, whom you'll meet in the appendix on page 357, and I used the letters FMS to represent the Functional Movement Screen. We soon recognized the need to develop a clinical counterpart because of confusion about where to draw line between movement screening and clinical movement assessment. The clinical model became the Selective Functional Movement Assessment, and now both systems fall under the umbrella of Functional Movement Systems and both are associated with the letters FMS.

Our intent is to develop yet another concept represented by these same three letters—*Functional Movement Standards*. If we can collectively use functional movement standards alongside our current standards of physical capacity, fitness, performance, sports skill and rehabilitation, we can reintroduce a quality of movement and begin to reverse the current decline.

We cover the first model, the FMS, beginning in Chapter 5. The FMS targets professionals who work with movement as it relates to exercise, recreation, fitness and athletics. It also has applications for the military, fire service, public safety, industries and other highly active occupations.

The FMS is not intended for those displaying pain in basic movement patterns. Painful movement is covered in the SFMA. The FMS is for healthy, active people and for healthy and inactive people who want to increase physical activity. The following professionals will benefit greatly from using the FMS.

Recreational Activity Instructors
Tennis and Golf Professionals
Outdoor Activity Instructors
Sports and Conditioning Coaches
Physical Educators
Health and Safety Instructors
Dance Instructors

Yoga Instructors
Pilates Instructors
Personal Trainers
Massage Therapists
Strength Coaches
Athletic Trainers
Physical Therapists
Chiropractic Physicians
Medical Physicians

The second model, the SFMA, targets professionals working with patients experiencing movement pain. We refined the SFMA to help the healthcare professional in musculoskeletal evaluation, diagnosis and treatment geared toward choosing the best possible rehabilitative and therapeutic exercises. The SFMA will enhance the work of the following licensed and certified medical and rehabilitation professionals.

Athletic Trainers
Physical Therapists
Chiropractic Physicians
Medical Physicians

Our teams and instructors teach the FMS to all fitness and healthcare professionals, and the SFMA to those licensed and certified in the medical field. The SFMA, which is specifically designed to address pain, is beyond the scope of practice of those in the fitness and coaching fields, but it is important for these professionals to be aware of the system. It is equally necessary for the healthcare professional to understand proactive movement screening.

THE LINE BETWEEN THE MODELS

We draw the line between the functional movement systems at pain. Pain changes everything, and it therefore nullifies the FMS results. If an FMS demonstrates that an active and healthy person has pain in a given movement pattern, the screen has done its job. It is not intended to classify painful movements; it is simply designed to capture them prior to exercise and activity.

We do the initial FMS screening for injury risk and program design, not for musculoskeletal evaluation. Once risk is determined, we gain greater clarity through an organized professional assessment in areas commensurate with the physical activity to be pursued. In contrast, the SFMA is done as part of a complete musculoskeletal evaluation when pain complicates movement.

A disconnect exists between our professions. Strength and conditioning experts need a better understanding of medical and rehabilitation systems. Likewise, those involved in medicine and rehabilitation need a better understanding of fitness and conditioning systems. To bridge that divide, healthcare professionals must understand screening, and exercise professionals must understand assessment. There is no getting around this if the goal is to provide the best service for our clients and patients.

We need to create an understanding and an active dialog between the professions. Our team does not advocate, not for a second, that any of us work outside of our particular specialties. This is merely a call to understand how to interact and communicate with others in or around the profession. A true paradigm shift requires better communication and new semantics may be required.

We all work with movement, and each of us needs to understand the fundamental rules or principles of human movement. In all actions, fundamentals and principles take precedent over the methods and detailed complexities characteristic of our particular expertise. The best teachers, coaches, trainers, therapists and doctors have always had a clear perspective of fundamental principles even in the presence of advanced concepts and of changing scientific circumstances.

Our attention to the basics has not impaired creativity or intellect. Instead, it has helped turn intelligence into professional wisdom. Clear fundamentals create a philosophical foundation and a perspective that supports other information without discrimination or bias unless the information challenges a fundamental.

This book goes beyond being a quick reference or a recipe for a speedy movement solution or a corrective plan. You will learn how to understand functional movement through screening, assessment and corrective approaches, and to recognize the interrelationship of the three.

However, let's not be naïve. Many readers will skip what they consider philosophical mumbo jumbo to get to the discussion about screening, assessment and corrective strategies—after all, tools are the cool stuff. Nevertheless, skipping forward without understanding the basics would be the equivalent of studying the medical remedy for a perceived problem before having the skill to diagnose the cause.

The first three chapters of this book force us all to look at the perspectives, opinions and attitudes surrounding movement. This introductory material is not intended to point fingers or call out professional mistakes. If anything, it's an admission of guilt on my part. My professional education trained me to look at movement in a particular way, but it also provided many cautionary hints I did not heed. I can speak about poor logic because I've used it, and can discuss reductionism and isolated approaches because I've practiced them. I can identify with every mistake that can be made in exercise and rehabilitation after owning them all. That is the perspective under which we'll discuss some of the philosophical, practical and social mistakes surrounding movement.

The single purpose in providing all this information is to help you root the ideas in a strong accessible foundation. Many intelligent exercise and rehabilitation professionals have attended our screening and assessment workshops. They have enjoyed the experience and the fresh perspective, excited to return to their jobs to apply what they have learned. But when questioned by clients, patients and co-workers, they cannot readily defend or explain the ideas they are ready to implement with enthusiasm. They often call to request help in the explanations. Although honored they are so ready to use our tools, I'm equally disappointed they have not prepared themselves to explain and intelligently discuss the logic behind the functional movement systems.

The systems are not complicated; they're actually quite simple and that's the beginning of the problem. With all the technological advancements at our fingertips, why do we need to look at movement patterns? How can these systems benefit us when we have force plates, high-speed video and other highly sensitive measurement devices? Don't we already have tools to screen and assess movement?

We review and discuss all of this book's information in our workshops, and although most attendees are entertained and agree with the information, perhaps they don't think they will be called upon to defend and explain the concepts. Take the time now to read the introductory material, or at least come back to it later. You may agree with most of the information, but if you can't explain and defend it, you don't own it.

FUNCTION VERSUS ANATOMY

This is a functional approach to movement rather than an anatomical approach. The anatomical approach follows basic kinesiology and is often complicated by assumptions in isolation.

We understand the assumptions. For instance, if knee extension is less than optimal in some movement patterns, we can target the knee extensors with exercise, giving isolated attention to restore the knee extension function. Once restored, it will be incorporated automatically into the lacking movement pattern. That is basic kinesiology, and it is clean and logical.

But what is basic about movement? Movement is varied and complex. A basic kinesiology approach to the knee extension problem fails on many levels as demonstrated by science and growth and development. Our simplistic observations may mislead us to handy solutions.

- What we view as weakness may be muscle inhibition.

- The weakness in a prime mover might be the result of a dysfunctional stabilizer.

- Poor function in an agonist may actually be problems with the antagonist.

- What we view as tightness may be protective muscle tone, guarding and inadequate muscle coordination.

- What we see as bad technique might be the only option for the individual performing poorly selected exercises.

- What we see as low general fitness may be the extra metabolic demand produced by inferior neuromuscular coordination and compensation behavior.

Strengthening, stretching, extra coaching and more exercise will not correct these problems. Making decisions on surface observations is the medical equivalent of treating the symptom and not the cause.

Many professionals appreciate function and yet insist on an anatomical approach to exercise, training bodyparts instead of movement patterns. By the end of this book, you'll know how to focus on movement patterns, letting the bodyparts develop naturally, instead of zeroing in on bodyparts and expecting natural movement patterns to spontaneously emerge.

DYSFUNCTION, PAIN AND REHABILITATION

During my physical therapy schooling, post-professional training and advanced training in orthopedic and sports rehabilitation, I was largely unimpressed with the standardized prepackaged exercises that often followed brilliant manual treatment. Corrective exercises were not correcting anything. These exercises just rehearsed movements that were awkward or faulty in the hope that arbitrary resistance loads would somehow create strength, integrity and competency.

Most of the corrective exercise targeted tissue physiology and not motor control. It was all highly coached, verbal and visual two-dimensional movement. This did not fit the definition of function my investigation and experience had revealed. We didn't make anyone react to or perceive anything. We didn't challenge the sensory motor system. We just rehearsed exercises that fit the simplest application of local kinesiology. We just exercised areas of pain and dysfunction and hoped for positive changes and less complaining, didn't we?

Many physicians and physical therapists assumed if we provided activity at or around the dysfunctional region, motor control would spontaneously reset. Yet, we were not so much causing a reset as we were creating greater opportunity for compensation behavior.

Pain affects motor control in unpredictable and inconsistent ways. This, coupled with poorly planned and poorly reproduced exercises, gave the average patient little chance of reestablishing authentic motor control. We treated people until the pain was gone or diminished to a tolerable level, assuming we had done something. We didn't thoroughly check function. We had no idea how much compensation the patients had developed on the road to recovery. We concerned ourselves with removing pain, not restoring function against a movement-pattern standard. When I reviewed my discharge notes and those of my peers, I saw the truth of the matter—far more was written in the physical therapy discharge notes about pain and impairment resolution than functional restoration.

As I started to refine my evaluation skills, I also started experimenting with drills that fit my definition of reactive neuromuscular training (RNT), developed more in the section beginning on page 199 and further discussed beginning on page 294, drills that used a light load to exaggerate a movement mistake. Seeing valgus collapse in a lunge, I put an elastic band on the knee and pulled the knee inward even more. When pulled too hard, the move would be too difficult to complete. If I didn't pull hard enough, the pattern would not change. But if pulled just enough, there would be a reactive countermeasure. The knee caving in would reset itself in a more functional position. This credit for this concept goes mostly to proprioceptive neuromuscular facilitation (PNF)—these drills were simply an extension of the tenets of PNF.

The best resistance is the one that causes the problem to correct itself without verbal or visual feedback, like giving the simple command to *lunge and don't let me pull you off balance*. These techniques worked, and my colleagues began to copy the exercises. They observed a drill I performed with a knee patient, often trying the same drill unsuccessfully with their knee patients. I eventually realized they were classifying problems by a patient's diagnosis or by the site of the pain.

Meanwhile, I was on a completely different path, choosing corrective exercise based on movement dysfunction, not pain or diagnosis. My peers provided treatments presumably appropriate for the pain and dysfunction, but that had no bearing on movement pattern correction. In many cases, I found myself working on regions of the body far from the site of pain. In this new model, two patients with low back pain could have completely different exercise programs. They might receive the same pain control treatments, but their movement dysfunctions could require completely different corrective exercise approaches.

When I started teaching functional exercise courses, I became even more convinced we needed a functional standard. Sixty rehabilitation professionals would show up for a weekend workshop on functional exercise with 60 different functional baselines. How could we standardize a functional treatment if we couldn't standardize a functional diagnosis?

As I continued to teach and practice, this approach to corrective exercise worked well and seemed to accelerate progress. However, the new approach had rules. Two major rules became clear determinants of effectiveness.

The first rule required the consideration of movement patterns alongside other parameters such as physical performance and diagnosis. These considerations soon became the basis for the FMS and the SFMA.

The second rule was an acknowledgment of a natural law: *Mobility must precede stability.*

The reactive drills are only effective if mobility is not compromised. This means we must address mobility before expecting a new level of motor control. Stated differently, if we change perception, we can change behavior. If there is no mobility problem, the RNT drills and exercises can be expected to improve motor control and improve movement patterns if sensory and motor pathways are in working order. If mobility is limited, we need to address the mobility first.

Of course, it is unrealistic to expect to normalize mobility in all cases. However, do not assume that since it cannot be made normal, no attempt should be made. In most cases, mobility has the potential to improve. With each measurable improvement, it is also likely motor control can be addressed with a basic stabilization exercise or an RNT drill.

Mobility problems are movement dysfunctions. They are probably the byproduct of inappropriate movement, or they could be the result of a poorly managed injury, physical stress, emotional stress, postural stress or inefficient stabilization. All these issues alone or in combination can reduce mobility in the body's attempt to provide function at some level. Loss of mobility is sometimes the only way the body can achieve a point of stability, but that stability is not authentic. It is often seen or observed as stiffness or inflexibility, but on a sensory motor level, it is part of a system with no other available choice. It is basically engineered dysfunction at a local level to allow continued physical performance at a global level.

Those with a weak core might develop tightness in the shoulder girdle or neck musculature as a secondary attempt to continue functioning. Those with chronic low back pain and stability problems may develop tightness in the hip flexors and hamstrings as secondary braces even if it reduces mobility. The body has worked out a solution and although the solution might compromise mobility in some regions, it affords functional survival.

We often assume that hip tightness causes back problems, but the back problems may very well create the hip tightness. The point is to not assume that tightness is the central problem. Improving hip mobility may allow core control to reset spontaneously, but exercise may be needed to facilitate the process. Problems usually occur in layers, with a mobility problem and stability problem residing as neighbors. Both problems need to be monitored and addressed, but intervention starts with mobility. You'll find a long discussion of mobility and stability in the joint-by-joint appendix beginning on page 319.

As long as mobility is compromised, the stiffness and increased muscle tone are providing the requisite stability needed for function. If mobility is not addressed in any way, the system will not need a new level of motor control; it will use its own creation.

However, if mobility is improved, a window of opportunity appears in which the body cannot rely on stiffness and inappropriate muscle tone. Within this window, motor control exercises that engage both the sensory and motor systems will call on primary stabilizers to work, while tightness and stiffness are temporarily not options. Dosage is everything in this window. If exercise is too stressful, the individual will default to old patterns, and if exercise does not challenge the primary stabilizers, they will not reintegrate into posture and movement.

The system requires us to improve mobility in a region where a limitation has been identified. We put the person into a challenging posture or activity such as rolling, quadruped, kneeling or half-kneeling. He or she might perform a movement or simply be challenged to hold the position from a stable posture, progressing to less stable postures and into dynamic movement patterns.

Babies enter the world with uncompromised mobility and follow this progression naturally. I gained perspective by studying the movement patterns used in the sequence of growth and development: rolling, creeping, crawling, kneeling and walking. I studied the way one movement pattern could serve as a stepping stone, an actual foundation for the next. Although my professional training provided me with this background knowledge, I did not embrace the concept until I watched my daughters as toddlers work it out for themselves. Our best efforts in exercise and rehabilitation attempt to replicate this gold standard when movement patterns are dysfunctional.

This is what we need to do to change movement.

We map the dysfunctional movement patterns, and note asymmetry, limitation and inabilities. We address the most fundamental movement-pattern problem with specific attention to mobility issues. This is the reset button—all new programming means nothing without the reset. When measurable mobility improvement is noted, we challenge the system without its crutch of stiffness and tightness.

We tap into natural reactions that maintain posture, balance or alignment at a level of stability our clients and patients can handle—a level where they can demonstrate success and receive positive feedback. We avoid fatigue at all costs, and

minimize verbal instruction and visual feedback, attempting to challenge each individual to respond through feel.

Balance is automatic. Balance is natural. Encourage your clients and patients not to over-think or try too hard, and make sure they are not stress breathing. If stress breathing is noted, stop the drill and try to get a laugh or use breathing drills. As they develop control, progress them, but always be mindful to not to overload or turn a motor-control drill into a conventional exercise.

End each session with a reappraisal of the dysfunctional pattern. If you're successful, you know where to start next time, and if not, you know exactly where not to start. If successful, recommend a small amount of corrective activity at home to maintain the gains. If not successful, suggest only mobility exercises and maybe some breathing drills until the next session, since you have not yet established the best motor-control exercise.

When you look at it, the entire system is very simple: Identify the primary movement dysfunction, verify appropriate mobility and reprogram the movement pattern.

THE FMS HISTORY

My physical therapy education at the University of Miami prepared me to ponder movement and exercise from many different perspectives. My orthopedic education was straightforward, and it applied the basic principles of kinesiology and biomechanics. My neurological education broadened my scope of understanding and reasoning as I considered movement and its many unique problems. As I studied PNF and started to see movement as interconnected patterns, I realized conventional orthopedic rehabilitation did not incorporate neurological principles with the same weight it gave to basic biomechanics. You can read more about this in the Jump Study appendix beginning on page page 359.

At the time, general principles in fitness and athletic conditioning did not give neurological principles the same weight as those of exercise physiology. Neither general exercises nor orthopedic rehabilitation made effective use of neurological training perspectives. These neurological techniques were designed to improve all types of movement, and are extremely effective for neurological problems. These techniques allow the therapist to make muscle tone more appropriate and to facilitate the way the muscle creates movement.

Neurological techniques tap into the sensory motor system and use forms of stimulation to create more optimal environments for movement. PNF and other techniques designed to facilitate movement use passive movement, assistive movement, tactile stimulation, body position, light resistance, breath control and other forms of subtle stimulation. They are all based on natural perspectives of movement and movement control.

Many of these perspectives are so common we ignore them. We watch babies go through the progressive postures of growth and development during which they develop command of one mode of movement, and then tinker with a more challenging pattern. We watch them use different parts of their bodies for locomotion, not realizing they are stimulating better support and movement with every point of weight bearing and contact.

We watch sports movement without considering the many spiral and diagonal movements that go into each athletic form or fitness activity. We fail to note the subtle torso rotation or reciprocal arm action of an elite runner, but when these movements are absent in the less-polished runner, we immediately sense the awkwardness. Many times we note the awkwardness and yet cannot identify what is missing. Since they cannot comment on the defect, they ignore the obvious awkwardness and subtle dysfunction, and awkwardness slowly becomes the norm.

This is why I proposed the FMS in a non-diagnostic way. As my colleagues and I developed the screen, we witnessed how easy it was to fall into physical performance or diagnostic testing. We wanted neither; we simply wanted to standardize movement. The FMS identifies movement problems in a rating and ranking system and first seeks agreement on what should and should not be acceptable before it attempts to suggest remedies or corrections. We based everything on strength, range of motion and performance standards, but we were not basing our movement opinions on standards because there were none.

I first introduced the FMS in formal print in 2001 in a chapter in *High Performance Sports Conditioning,* a book edited by Bill Foran. The screen had been in print since we started teaching screening workshops in 1998, and got more exposure in 1999, when it was presented in many regional athletic training and strength and conditioning events.[9] That same year, the screen gained national exposure at the NATA[10] and NSCA national conferences.

In the first book to describe movement screening connected directly to corrective strategies and approaches, *Athletic Body in Balance*, Human Kinetics 2003, I established a user-friendly system of self-screening and movement-pattern correction. That book is a practical training manual, and although it is a no-nonsense guide for athletes, trainers and coaches, it is also popular with exercise and rehabilitation professionals, and gave teeth to a shift in the way we view comprehensive functional exercise and conditioning.

Athletic Body in Balance has appeal with progressive professionals because it serves as an example of how we should use movement screening in the beginning of any exercise or training activity. Exercise professionals are starting to understand that screening provides insight to movement problems, as well as a logical path to exercise choices and program design. Rehabilitation professionals are beginning to understand that screening at discharge provides insight to dysfunction even when pain has been managed.

The screens and assessments presented in this book are about fundamental movement patterns that support most activity-specific movement patterns. This text will help you redevelop a general functional movement base in your clients and patients before specialization in movement, regardless of population or activity.

Humans all have the same developmental milestones for movement as we grow. We start with head and neck control and progressively move to rolling, creeping, crawling, kneeling, squatting, standing, stepping, walking, climbing and running. Missing a primary movement milestone inevitably results in obvious movement limitations or dysfunctions and causes delay in the complete movement-system maturation. As we age, grow and become self-sufficient—and then as we decline and lose some capabilities—we must always maintain some degree of our original functional movement patterns or we will be disadvantaged.

This original function represents basic mobility and stability in the body's moving segments working together to create movement patterns. This is not about strength, power, endurance or agility. The basic function is what supports these attributes; the basic movement patterns lay the foundation for higher movement skills.

This is a case for the observation and consideration of quality before quantity. This is the general blueprint that comes before the specific needs of a particular movement, exercise or rehabilitation specialty. Only after developing a command of general movement patterns should we strive to screen, assess, train or control specific and skilled movements.

A true movement management system develops a general movement base to support the specific. In most cases, a general functional movement system can remain the same for numerous populations and activities. We may, if necessary, weight or change the implications of the system's data while the basic system remains the same.

Once we establish general movement-quality minimums, performance and skill considerations become relevant between different populations.

RECOGNIZING PATTERNS

The screening and assessment systems are set up to rate and rank functional and dysfunctional movement patterns, respectively. Rating and ranking information helps identify problems within certain movement patterns, and, after identifying the problems, we can focus on the weakest link or greatest limitation within the group of patterns. Working on any link in the chain other than the weakest link will not adequately change the strength of the chain.

For example, a 400-meter relay team may have the four fastest people on the planet, yet they display difficulty exchanging the baton. The start is great, their stride frequency and stride length are phenomenal. Their strength and grace are unmatched, but they consistently fail to finish or are disqualified after dropping the baton, exchanging outside the zone or running out of their lane. Would it help them to get stronger, faster or more powerful? No, they will only fail quicker. Cleaning up their exchanges—working the weakest link—will be the best path to perfect their racing.

In working the weakest link, once we've rated and ranked the different patterns of movement and have identified a key pattern of weakness we can then investigate the specific details. We do not dismiss details outside the most limited movement pattern, but these are of lesser importance even as they help refine the problem.

Pattern identification is important because the human brain uses patterns instead of isolated muscle and joint activity to create practical perception and behavior regarding movement. These patterns create harmony, efficiency and economy, or during pain and dysfunction, create movement that compensates to maintain a degree of function regardless of quality. That compensation is a survival mechanism overshadowing any specific information gathered at a single joint or muscle group, or any preset movement patterns.

The human brain also has an affinity toward habits. Repetitive behaviors become patterns, and these patterns require reprogramming when they become problematic. The body and its parts are like the hardware of a computer, and the motor programs that produce movement patterns are like the software. When the hardware of a computer is changed, the software does not update automatically, but most of the new hardware's benefits cannot be fully realized using old software. Likewise, changing the strength or flexibility of a particular bodypart will not likely change the movement quality unless motor programs are also addressed.

The attempted management of the body's stiff and weak regions describes the focus of most older exercise and rehabilitation programs. In this new model of movement science, if the movement pattern does not improve, we have accomplished nothing, even if we have positively affected stiffness and weakness as demonstrated by isolated measurements. This does not suggest movement patterns are the only part of movement science, but we do propose it is the first part of the information to be gathered, and the last thing considered.

Modern technology creates the illusion of advancement as measurement sensitivity is increased. However, if innovation in logic does not improve along with measurement sensitivity, nothing really changes. We need new movement logistics. Ponder these definitions as you scrutinize the Functional Movement System against current standards of practice.

Note that measurement is a detail, a data point collected at the request of a logic system. A logic system is organized to rate and rank data generated by measurement. It is common to have increased focus on the most detailed measurements, the easiest to collect or most readily available, but data collection should not influence data importance. Science is often thought to be improved by greater levels of focus and technical precision of measurement. We must remember the word *focus* does not mean zoom—it means a state or quality producing clear, uncomplicated definition.

Measurement—meas·ure·ment—n
1. the size, length, quantity or rate of something that has been measured

2. the size of a part of somebody's body

Logic—log·ic n
1. the branch of philosophy that deals with the theory of deductive and inductive arguments and aims to distinguish good from bad reasoning

2. any system of or an instance of reasoning and inference

Logistics—lo·gis·tics n
(takes a singular or plural verb)
1. the planning and beginning of a complex task

Encarta World English Dictionary 1999

THE FIVE BASIC PRINCIPLES OF FUNCTIONAL MOVEMENT SYSTEMS LOGISTICS

• Basic bodyweight movement patterns should not provoke pain. If pain is present in movement, activity and exercise should be modified, interrupted or stopped as additional information is gathered. If not, compensation and altered movement patterns can result, and these are likely to aggravate the problem and cause secondary movement problems if left unaddressed.

• Gross limitation of fundamental movement patterns, even if pain-free, can cause compensation and substitution leading to poor efficiency, secondary problems and increased injury risk in active populations.

• Fundamental movement patterns involving the body's left and right sides should be mostly symmetrical. These movement patterns are not skill-based and are present before handedness becomes specific. Although activities and unilateral dominance are commonplace in a normal active lifespan where skills are involved, a significant amount of symmetrical ability in basic patterns should be present across the lifespan.

• Fundamental movement capability should precede performance-based capability. To assure a performance measure is evaluating only performance, we must first establish a sound functional movement base. Otherwise, tests revealing poor performance can also capture a fundamental movement problem that cannot be corrected or remedied with performance training.

- Fundamental movement capability should mostly precede complex movement activity or complex skill training. Basic movement patterns are part of growth and development before complex or specialized patterns, and should remain present throughout life. These movements form a neuromuscular foundation for advanced activities and reduce the need for compensation and substitution often observed when fundamental movement is limited.

The following two statements can summarize practical application of the five principles.

Pain produced by movement should be reported, managed, diagnosed and treated by a medical professional.

Manage movement pattern limitations and asymmetries—mobility and stability problems—before applying a significant volume of fitness, performance and sports training. This is appropriate, ethical and justified, because we can link these to increased risk of injury.[3, 5, 6, 11-17]

We should make sure our methods always reflect our principles. It is easy to get caught up in methods, but those will change, improve or be replaced. Innovation, research, experience and expertise will always move us along to better methods, but we must always judge them against our principles. That is how we make sure the glitter is actually gold.

OVERVIEW OF THE FUNCTIONAL MOVEMENT SYSTEM GOALS

Research has already shown that reliable risk factors exist within the neuro-musculoskeletal system, but we need to gather these risk indicators with objective and consistent documentation and communication. We can do this now, because through basic screening and statistical analysis we can track people participating at different activity levels in different populations.

Used properly, the Functional Management Systems—this umbrella we use for both the FMS and the SFMA—enhances communication. If applied reliably, it creates usable data across general and specific populations. Those in direct contact with athletic and exercising populations will be able to make informed decisions and recommendations, decisions that can potentially reduce the musculoskeletal risks associated with increased activity. If the risks are present, the professional can recommend corrective measures as part of the progress toward increased activity.

Early detection of musculoskeletal problems gives us a greater advantage in reducing dysfunction and disability that can accumulate with poorly managed or unmanaged problems. Screening may not produce symptoms, but may show faulty movement patterns associated with elevated risk. This creates a second opportunity to prevent injury.

Unmanaged problems force the neuromuscular system to compensate in the presence of both pain and dysfunction. This compensation often hides the primary problem and creates secondary problems, which complicate matters and prolong activity limitations.

As more fitness professionals participate in the gatekeeper role of musculoskeletal treatment, we must look for consistent methods and screening processes that enhance communication and increase the reliability of detecting undiagnosed or potential musculoskeletal problems. This calls for the highest level of professional responsibility and objectivity.

Medical science has screening processes that help detect disease and dysfunction and allow us to observe potential risks in the early stages. But often we do not manage the musculoskeletal system appropriately until injury, dysfunction or disease cause symptoms. We are not proactive with the largest functioning system in the body, while we are constantly improving screens created for early disease detection in other body systems.

Potentially, extensive problems lie ahead. We have specialists from different backgrounds functioning as guides for musculoskeletal problems, yet we have no consistent screening tool to communicate effectively. We also have not properly identified risks in the musculoskeletal system to the extent we have in other organ systems.

The goal of Functional Movement Systems is to pull together those now participating in all facets of medicine, rehabilitation, athletics, fitness, wellness and performance enhancement. These disciplines have opportunities to prevent musculoskeletal problems and injuries, rather than to respond and react when these occur.

By instituting safe and reliable screenings, our clients and patients will enjoy a reduction in injuries and musculoskeletal problems. We will also be able to increase our skill at early detection of potential problems in those unaware they possess injury risks with increased physical activity.

The information collected from screening and assessment can create research data and real-world practical data to improve our professional development.

Please see www.movementbook.com/chapter1 for more information, videos and updates.

ANATOMICAL SCIENCE VERSUS FUNCTIONAL SCIENCE

There is an inseparable interplay between structure and function as it pertains to movement—a movement matrix. Certainly, we can discuss functional movement without discussing the internal framework that supports it. Likewise, we can discuss anatomical structures without considering that each small segment somehow contributes to every movement we perform. In functional movement screening and assessment, we don't focus on the structure, but we still need to be aware of its properties. We will first review structural interconnections and then dive deeper into functional movement.

The body's tissues complement and support each other in a multitude of movements and positions, and its system of muscles perfectly trusses three-dimensional spiral and diagonal movement. Skilled dissectors appreciate how the nature of one structure is completely dependent on its supporting and opposing structures. The muscles and joints the system sustains are practical representations of first-, second- and third-class levers.

Every joint benefits from the relationship of its supporting and movement-producing muscles. Noting the attachment points of muscles, early engineers referred to muscles based on their pull point on the bone and the distance from the joint axis as shunt or spurt muscles.

- Shunt muscles compress or produce structural integrity to the joint because the muscle's distal attachment is far from the moving joint.

- A spurt muscle has the mechanical advantage to produce movement since it has a distal attachment close to the axis of rotation.

Kinesiology is the study of the muscles, their attachment points and the basic action of each muscle, but this superficial study only scratches the surface. In this view, the brachialis is considered a spurt muscle, and the brachioradialis a shunt muscle. Of course, this assumes that the dumbbell curl demonstrates the mechanical role of each. The singular perspective is not representative of reality.

If we review the chinup, the rule is broken and the situation inverted. The muscles change roles responding both mechanically and with neuromuscular accommodation as they perform the task, unaware of the academic classifications. In one example, the hand moves to the fixed body and in the other, the body moves to the fixed hand.

Shunt and spurt muscles are present, but the roles are an interplay between their structures and the situations in which they function. The result is muscles that complement each other regardless of the action. They perform the necessary task whether or not they carry the correct moniker.

Through motor learning and development, the brain has learned to organize muscle synergy and contribution to familiar activities. The conscious brain does not act alone. It is supported by an automatic system of reflex activity with involuntary adjustments occurring in the background of every intended movement. This is possible because the sensory system constantly monitors our real-time movement to the intended movement pattern. We don't really think about our muscles, we think about movement and our muscles act in accordance with our intensions and automatic support system. This is the single most important reason that movement training does not intentionally try to direct a client's or patient's focus onto a particular muscle group. When we focus on a single muscle group, we demonstrate that we do not understand the supporting matrix behind superficial muscle action.

Many people would like to have a spreadsheet of stabilizer and prime mover muscles, but what most fail to realize is the role of the muscles often change depending on the body's position and the joint in action.

All muscles move and stabilize to some degree, but we have muscles that only cross or span one joint. These muscles are dedicated muscles. The influence of other joints has little or no affect on the function of the muscles at or about the joints these span. Other muscles cross two or more joints where the influence of the position of one joint greatly affects the function of the muscle on another joint. This gives rise to the terms *active and passive insufficiency.*

Active insufficiency—the inability of a bi-articulate or multi-articulate muscle to exert adequate tension to *shorten enough* to complete full range of motion in both joints simultaneously

Passive insufficiency—the inability of a bi-articulate or multi-articulate to *stretch enough* to complete full range of motion in both joints simultaneously

The deepest layers of muscles seem to be dedicated muscles; they're close to the bone, close to the joint and we can visualize their stabilization contribution easily. Now picture adding more and more muscle layers onto the skeleton until it takes on the human shape. Note with each layer how the body is tied together crossing multiple joints. The layering system provides support and movement in multiple patterns and purposes.

Ironically, a central objective of many fitness and conditioning programs has been to focus on the development of the superficial muscles trained as prime movers, assuming these muscles play a more important role in performance than the supporting stabilizing muscles.

The muscular system supports and moves the skeletal system. The skeleton is supported against gravity and through movement by the constant and coordinated work of the stabilizing movers. These smaller, deeper muscles enhance the efficiency and power of the prime movers by creating resistance, stability and support of movement at one movable segment, and allowing freedom of movement at another. This interaction happens in milliseconds and occurs without conscious control.

The quadriceps and hamstring muscles work in opposition of each other. There are basically four muscles in each group. Only one muscle in the quadriceps group crosses both the hip and knee, while three muscles in the hamstring group cross both. This leaves three muscles in the quadriceps group dedicated exclusively to the knee, and only one hamstring muscle dedicated to the knee. People often assume that long multi-joint muscles are movers and short dedicated muscles are stabilizers, but this is an interesting example of the opposite situation.

EXAMPLE

As you rise to standing from sitting, the quadriceps and hamstring muscle groups fire simultaneously, even though the rectus femoris and three of the hamstrings are completely antagonistic.

The rectus femoris flexes the hip and extends the knee. The biceps femoris (long head), semi-tendonosis and semi-membranosis extend the hip and flex the knee.

For many years, this very real co-activation of two antagonistic muscles during the activity of standing from a sitting position has been known as Lombard's paradox.[18]

Both the rectus femoris and the three hamstrings are active, and neither change length from sitting to standing position. The rectus femoris lengthens at the hip and shortens at the knee. The three hamstrings shorten at the hip and lengthen at the knee. No net change in length is noted for either between the start and finish position. Ironically, two active antagonists display tension, but no change in length over a large movement occurs—sounds a lot like a stabilizer. These muscles actually cancel each other out, which may sound inefficient, but the body is too wise to waste energy.

In reality, the muscular tug of war creates supportive joint and tissue compression and serves as a global stabilizer and proprioceptor. The three local quadriceps muscles most likely serve as local movers and proprioceptors at the knee, while the single local hamstring muscle serves as a local stabilizer and proprioceptor. The three hamstring muscles probably assist the glute maximus in hip extension. The glute maximus has both local and global influence since it attaches onto the hip as a local mover and to the iliotibial band as a global mover.

These muscles might take on completely different roles during other activities involving different movement patterns. The example simply demonstrates that muscle roles and contributions are task specific and not necessarily anatomically specific. These terms should describe the roles of muscles within particular patterns, and not an absolute definitive category of anatomical classification.

Global Stabilizers—Larger and longer superficial muscles spanning two or more joints, contracting primarily to create tension to produce stability. Their roles are stabilization and static proprioceptive feedback.

Global Movers—Larger and longer superficial muscles spanning two or more joints, contracting primarily to produce movement within a specific movement pattern. Their roles are movement and dynamic proprioceptive feedback.

Local Stabilizers—Shorter and smaller deep muscles mostly spanning a single peripheral joint or few spinal segments, contracting to primarily create tension to produce stability. Their roles are stabilization and static proprioceptive feedback.

Local Movers—Shorter and smaller deep muscles mostly spanning a single peripheral joint, contracting to primarily produce movement within a specific movement pattern. Their roles are movement and dynamic proprioceptive feedback.

TRAINING STABILIZERS VERSUS TRAINING MOVERS

It's common to see stabilization programs that attempt to train the stabilizers like primary movers by using concentric and eccentric movements. Unfortunately, this assumes that strengthening a stabilizer will cause it to stabilize more effectively. Common strengthening programs applied to muscles with a stabilization role will likely increase concentric strength but have little effect on timing and recruitment, which are the essence of stabilization.

Stabilizers control movement in one local segment while movement occurs in another, or they create supportive tension within multiple global joints. Their role is to *not move* in the presence of movement. They should therefore be trained to produce integrity, alignment and control in both static and dynamic situations.

A static situation would be a near-isometric contraction in one segment while movement occurs at another. A dynamic situation would require adjustments with timing and tension to stabilize a joint in one or more planes, while primary movement is produced within the confines of a different plane. The contribution of stabilizers can change throughout the range as well, performing a static role in one phase of movement and a dynamic role within another phase.

Stabilizer training goes far beyond the simplistic isometrics found in popular stability exercises such as the side plank. In this isometric exercise model, conscious rigidity and stiffness are the goal, but true authentic stability is about effortless timing and the ability to go from soft to hard to soft in a blink. Stability is also confused with strength, where concentric and eccentric contractions build mass and endurance. The muscles do become stronger at shortening and lengthening, but again they lack the timing and control needed for true functional stabilization. We should train muscles in the way we use them. Stabilizers need to respond quicker than any other muscle group to hold position and control joint movement during loading and movement.

This may be controversial, but here it is—*Train stability with exercises that are more dynamic with the highest movement quality possible, and when dynamic quality cannot be achieved, revert back to static postures where alignment can be challenged.*

At their best, stabilizers are the complementary dampeners that refine and control the explosive energy of the prime movers as they protect our joints, align our segments and balance our bodies in space. Calling stabilizers *multitaskers* is an understatement.

To take this a step further, it would not even be necessary to train stability if quality and functional patterns had not at some point been neglected. The neglect occurred the minute we started to train partial movement patterns instead of whole movement patterns, the minute we focused on quantity maximums and did not set a quality minimum. One might argue we need progressions, but breaking down movement patterns into isolated muscle training is not as effective as following a developmental progression. Toddlers develop a command of movement patterns without isolation. They work patterns of movement in stages of progressing difficulty.

The way to reinforce stability is to train fundamental and functional movement and reject any conditioning or fitness endeavor that causes a departure from this platform. Some sports and activities can actually compromise stability over time, so measures must be taken in these situations to check and reinforce authentic stabilization whenever possible.

MUSCLE FUNCTION— MOVEMENT AND SENSATION

Muscles are made of motor units, small contractile elements within each muscle. Motor units are identified by a nerve ending that provides information for contraction and sensory information for muscular tension and feedback. The fibers within each motor unit have been referred to as fast twitch or Type II, and slow twitch or Type I. Fast-twitch fibers are often associated with explosive power and with limited capacity for endurance. Slow-twitch fibers have been coupled with continuous contractile ability and resistance to fatigue. Other non-specific fibers develop over time, depending on which activities dominate, as they become helpers to the specialized fibers getting the most action. Since genetic predisposition is the predominant factor, many researchers have moved past fiber types.

On a larger scale, the central nervous system has systems that complement the different muscle-fiber characteristics. The phasic and tonic systems manage different activities within the body. Phasic systems control explosive and robust movement patterns and are mostly associated with prime movers. The tonic system is dedicated to postural control and to maintenance of alignment and integrity throughout the skeletal system. The tonic system supports the body's structure and provides appropriate stabilization for prime movers to function efficiently.

Muscles have proprioceptive roles as well as moving roles. Muscle spindles gauge tension and contraction to provide feedback to the brain by creating a three-dimensional moving map for holding tension and postural tension against gravity.

Information from the joints, vestibular system, visual system and muscles play an important role in the way the mind perceives movement. If any of these systems are compromised, the others must compensate. This compensation is a great strategy for survival, but over time can cause the proprioceptive system to erode.

The body's proprioceptive capacity is an important aspect of human movement. With respect to movement, proprioception can be defined as a specialized variation of the sensory modality of touch that encompasses the sensation of joint movement and the sense of joint position. This awareness in each segment of the kinetic chain must be functioning properly for appropriate motor control, which is the foundation of efficient patterns. Likewise, when movements are limited, stiff or sloppy, proprioceptive awareness cannot provide normal feedback. Movement influences proprioception, and proprioception influences movement.

JOINTS AND LIGAMENTS

When muscles fail to function well, it places unnatural stress on joints. The unnatural stresses cause micro-trauma and wear, and the resulting stiff joints give poor feedback and create a greater demand on the muscular system. Damaged joints can cause muscle inhibition, muscle guarding and muscle imbalances.

Joint stiffness can be a byproduct of injury or lack of activity, and can be a secondary attempt at stability. Much like abused hands develop calluses and abused joints become stiff, when postures and activities put stress on connective tissues, they often thicken and become rigid as protection. The stiffness can produce muscle fatigue and muscular strain, and we blame the degenerated joint in our failure to recognize that lifestyle and activities compromised the body's resilient nature.

Other tissues complement the muscle functions. Ligaments provide basic integrity to the joint and the joint capsule. Both ligaments and joint capsules surround and encase the joint to isolate it as a single unit, separate from the rest of the body. The joint encasement involves a fluid-filled space where cartilage surfaces come into contact. These cartilage surfaces are near frictionless and remain bathed in fluid that nourishes and lubricates them.

Certain areas of the joint have greater stress than others, causing thicker capsules. This thickening in some cases is closely related to a ligament. Some ligaments are identified closely with the joint capsule, but other ligaments could be visualized as straps creating joint integrity and helping maintain pivot points to foster mechanical movement throughout the joint.

Ligaments do not only serve a mechanical role; they provide feedback, too. They don't have the contractile abilities of muscle tissue, but they can greatly influence the contractile abilities of the muscles surrounding them. The ligament receptors can facilitate and inhibit muscles as part of reflex behavior, which happens below the level of conscious control.

Typically, when a ligament is stressed, it automatically and immediately generates signals to reduce the stress. It does this by facilitating agonist muscle action (muscle arranged or aligned to potentially reduce ligamentous stress) to protect it and by inhibiting antagonist muscle action (muscle arranged or aligned to potentially increase ligamentous stress) that could damage it. Ligaments are arranged at all the potential stress vectors on the joints. By protecting themselves, ligaments ultimately protect and help maintain joint integrity.

When a ligament fails, it is sometimes from contact, collision or when a stress is too great and a tear cannot be prevented. Other times, the ligaments tear without outside trauma—in sports called non-contact injuries. Fatigue can play a role, but poor movement patterns also contribute to many ligament failures.

Poor mobility and stability place unnatural stresses on ligaments with what looks to be natural movement. Normal exercises and activities appear on the surface to be uncompromised, but when mobility and stability is limited, subtle compensa-

tion occurs. The compensation can cause sheer forces and poor alignment, resulting in stress to ligaments and joints. The stress can compromise muscle activity, which can further compromise mobility and stability. The compensation can also increase energy expenditure and reduce muscular control as activity progresses.

The ligaments and joint capsules don't simply protect and stabilize a joint; they interact with our neurological system giving awareness of joint position, direction and speed of movement.

Though the current state of medical science allows for the reconstruction and replacement of ligaments, surgeons could never recreate the fine dexterity of the neurological interplay between the natural ligamentous tissue and its contribution to the movement matrix. While some serious injuries are unavoidable and need surgical repair, we should do everything possible to build an injury buffer zone by training healthy movement. It is always better to bend than break—and strong agile bodies bend better than weak, stiff bodies.

THE FASCIAL MATRIX

Fascial tissues support and connect the arrangement of moving parts throughout the body. Unlike ligaments, fascial tissues do not simply tie together a single joint, crossing the small space from one bone to another. They weave throughout the body as a three-dimensional web-like structure, with long lines spanning from head to toe and index finger to index finger. The web runs deep, superficial and crisscrosses the body. Arranged in a matrix that complements whole-body movement patterns, they redirect stress, and provide dynamic structure. The fascial web works intimately with the muscles and enhances muscles' contractile quality. The web also links the muscles common to particular movements together in a biomechanical chain that creates automatic synergy and support. Author Thomas Myers calls the matrix *Anatomy Trains*, which is also the title of his unique book.

Part of the fascial web contains muscle within a sheath. This sheath provides support and pressure against the muscle, creating a near hydraulic effect as the muscle contracts and bulges, pulling the fascial lines into greater tension. These structures ultimately provide tension and support in all lines of natural and functional movement and stress.

Since the lines extend throughout the body, the influence on one muscle contraction can be effective in supporting another part of the body. The central nervous system recognizes muscles that are complementary in action as synergistic partners within motor programs. The fascia linking these same muscles builds a mechanical relationship that connects them as well. In a nutshell, the mental framework and the mechanical framework seem designed for each other—why wouldn't they be?

The fascial system creates as much dynamic structural support as the skeletal system creates static structural support. The static support of the skeletal system is consistent and is the most rigid tissue by functional movement standards. The fascial system can become stiff and rigid in some patterns and allow movement freedom in others.

Each line of fascia has a role within a movement pattern. Sometimes the role is supportive and sometimes the role is yielding. Its inherent dynamics complement the skeleton by providing that something extra where it's needed—and it's needed in different areas within different movement patterns at different times.

THE BREATH

Breathing connects all parts of the movement matrix, but it remains the most neglected aspect of the Western approach to exercise, athletic conditioning and rehabilitation. When we acknowledge and discuss the breath, we quickly migrate to the measurements of breathing mechanics or discuss VO$_2$ maximums, but neglect the qualitative aspects of authentic breathing. We neglect the potential power and rhythm of breathing as we jog, lift weights, play sports and do our back rehabilitation. Instead, we use shallow and disconnected breathing patterns that would appear obviously inefficient to a yoga or martial arts master.

The typical, excessive mouth breathing seen as exercisers and athletes take rest breaks is related to superficial upper-chest breathing patterns. Dominant nose breathing couples with deep diaphragmatic breathing patterns. The complete boxer, the powerful discus thrower and

the elite ultra-marathoner all tap the power of correct breathing to fuel violent explosion and near-Herculean strength and stamina. Correct breathing provides power through a central drive of energy supported by the matrix.

Control of correct breathing can help relax and reset the system in the presence of unmanaged stress breathing related to dysfunction, anxiety and tension. Deep, slow breathing has been connected to parasympathetic nervous system stimulation and the production of alpha brain waves. The return to slow, controlled breathing between bouts of exertion is a hallmark of the supreme athlete and elite warrior. Command of the slow and controlled breath tends to increase heart-rate variability (HRV), a measurement of the fluctuation of heartbeat activity during exertion. HRV is actually a favorable quality; reduced HRV has been shown to be a predictor of mortality following heart attack.[19] The lack of HRV and breathing quality represents a rigid system that cannot adapt or respond to stresses at appropriately managed physiological levels. See the appendix on page 355 for an introduction to HRV.

Exercise and healthcare professionals who do not know how to perceive and respond to fluctuations in breathing quality are missing an access to the mechanical and physiological goals of training and rehabilitation. While one exercise can stimulate the correct and automatic postural and breathing response, another can distort both. If yoga, the martial arts and the greatest feats of human strength and stamina are built on a respect for the breath, we must acknowledge this in our professional discussions, and more importantly, in our actions.

Although the primary focus of this book is dedicated to movement appraisal, be aware that a qualitative movement dysfunction is a subtle and parallel indication of qualitative breathing dysfunction. The breath and breathing rhythm is essentially stuck in the middle regions of function, in most cases not reaching its potential at either end of the range—whether it be total relaxation or superior fatigue management.

We see a mirror of this in movement where a pattern is available and possible, but does not rise to its potential. A good example is poor squatting quality in which a person will move up and down through the squatting pattern, but never reach the potential squat depth or rise to the starting position with correct alignment, erectness, neutral pelvis and complete hip extension. The breathing dysfunction equivalent involves breathing patterns that cover the middle spectrum of breathing, but not the quality of extremes where tremendous relaxation and extreme exertion are addressed efficiently by coordinated efforts of the breathing structure.

Using the Functional Movement Screen (FMS®) and Selective Functional Movement Assessments (SFMA®), you will push people to movement pattern extremes where their limitations and asymmetries show clearly. You should also note changes in breathing at these extremes of movement. Obviously, some postures and position changes can reduce natural breathing mechanics, but it is common for people to breathe in shallow succession, hold their breath or brace simply to move though a pattern. This is not normal, nor is it authentic.

People are often unaware of breathing incorrectly in certain positions. These positions should not be considered normal. The FMS has scoring criteria that usually pushes these patterns to a lower score. Those trained at advanced levels with the SFMA are instructed to look specifically for breathing changes in each pattern.

The movement limit of a pattern is called the end range. If the end range causes unnecessary bracing, breath holding or breathing difficulty, the end range is not authentic. The individual is not doing the movement—more like surviving the movement.

The ability to cycle a complete, undistracted breath at the end range of a movement pattern is called the breathing end range. Breathing end range and movement end range may differ. If these differ, you should assume that only the ranges of movement within mostly normal breathing cycles are part of the functional movement map. All else is outside the region of natural functional movement and proper reflex stabilization.

A good indicator of compromised or shallow breathing is the anterior neck musculature. The individual who moves in a pattern outside the range of confidence or comfort will attempt to access extra stability and control by shallow breathing and using excessive bracing with the neck muscles. The inappropriate muscular activity in and around the neck is obvious by movements in any direction other than neutral, and by increased bands of tension representing muscles such as the sternocleidomastoid and scalenes. Note this behavior and do not create exercise situations that reinforce it. It is an indication of overload or unconstructive stress.

Modern society is stuck in the midpoint of its movement and breathing potential. The movement matrix is responding to what it is fed—a diet of movement and breathing opportunities based on incomplete remnants of authentic activity. The remnants are synthetic singular movements and one-dimensional training performed with just enough breath quality to cover the activity, but not enough to make it authentic.

You'll read more about breathing quality in the appendix on page 353.

THE NEUROMUSCULAR NETWORK

The muscular system is not just the contracting tissues that move us; it is also the arrangement of muscles within the matrix that creates tension in opposition to gravity. The natural tendency for humans is to keep the head upright and the eyes level. As balance and posture are changed or disturbed, the sensory and movement systems work to right us. One goal of the sensory motor systems is uprightness between the forces of gravity and our surrounding environment.

Three sensory systems provide information regarding uprightness, changes in uprightness, or the lack of uprightness. The three systems are the vestibular, proprioceptive and visual systems.

- The vestibular system provides information regarding the position of the head in relation to gravity and movements of the head.

- The proprioceptors, especially those associated with spinal and core joints and muscles, provide information about movement of the body segments on and around each other.

- The visual system provides information about

the body's postures and positions in relation to the surrounding environment.

The systems are intimately connected and involved in the production of functional postures. However, don't think of posture as rigid, manikin-like poses. Postures are very dynamic when all systems are functioning in a complementary way. They are constantly changing and adjusting to address changes in the internal and external environment.

The counter-rotation in walking and running is a perfect example. The alternate movement of the arms and legs provide a counter-balance that requires minimal spinal movement, and minimal spinal movement provides the anchor for the alternate extremity movement. In this example, the extremities are in constant motion, while the posture of the spinal column and the core function with little movement but constant adaptation. The spine and core continually redistribute and transfer energy to efficiently create this rhythm and movement. In fact, our eyes are immediately drawn to situations in which the natural rhythms of walking and running are out of sync or the balance of posture and movement seem irregular. Our eyes cannot help but focus on a limp, reduced natural arm swing or a flexed spine in a runner. We do not notice the natural movement, but instantly detect compromised function.

Consider this example of a client asked to perform a double-arms raise to 90 degrees of forward flexion. As the straight arms lift upward and forward to 90 degrees, the body slightly shifts backward and away. In normal situations, the spine and core do not move; most of the movement will occur at the ankle. This maneuver is an automatic posterior weight shift to maintain the center of mass over the base of support. As the small mass of the arms moves away from the center, the large mass of the body shifts in the opposite direction to maintain balance. The muscles of the spine and core activate but do not produce movement—they sense the movement and allow adjustments to occur at the ankle and foot.

We can see that muscles contract to create support, maintain posture and to transfer energy. The muscles aren't merely arranged levers that move us independently on a straight plane. They are organized in a spiral and diagonal fashion to create efficient three-dimensional motion and to complement the most productive patterns of movement.

The central and peripheral neurological systems drive the moving matrix through sensory, motor and reflex behavior. The sensory system takes in information as the body moves through space, changing position, posture, terrain and speed. It responds to load, gravity and different forms of tactile feedback.

The motor system creates and controls tension in stabilizing and moving muscle. It responds to sensory feedback with gross and fine motor control. Reflex behavior functions below the level of conscious attention, making subtle adjustments in muscle tension and contraction. This allows focus of attention more on the task at hand than on the thousands of automatic adjustments that support it.

The neurological system develops in a head-to-tail fashion, which means we gain visual motor control and then motor control of the head and neck. We then progressively gain the same control through the shoulders and torso, then into the shoulder girdle and hips and finally out to the extremities. We also develop in a proximal to distal fashion—from the spine out to the feet and hands. Gross motor qualities of the trunk, shoulders and hips precede the fine dexterity that later develops in the hands and feet.

The hands and feet, more than any other part of the moving body, maintain a sensory feedback system as they interact with the environment. Significant portions of the brain are dedicated to the sensory and motor control of the hands and feet. As we manipulate objects and move through space, that control keeps us in constant contact with objects around us, with the environment and with movement.

As we first start to move, reflex behavior rules the show. Many reflex behaviors serve to not only protect us, but also help develop neurological pathways and movement patterns to assist us with locomotion and in the manipulation of objects.

Some of us develop better hand-to-eye coordination, such as is used in throwing and swinging, while others migrate toward activities of mass movement such as gymnastics and running. Some excel at feats of strength; others enjoy speed, endurance, quickness and agility. As we develop, we migrate towards certain activities, and over time, we develop movement patterns that intrigue us and patterns we try to avoid.

The neurological system gives us options when conditions are not optimal. Most people sustain injuries and painful situations, and as we continue to move, we sometimes adopt poor movement patterns as a result of the pain. Part of the survival system allows movement around problems we cannot move through. We use some segments and movements in excess to avoid movement in other segments. These patterns are behaviors modified to fit situations. They often remain in place even after the situation that caused their emergence is gone. If we use these compensations long enough, they can become our primary movement method, triggering long-term problems in other regions of the body. Compensations are temporary solutions, not effective long-term options.

The current best evidence suggests that movement changes after an injury, and that these changes occur at multiple joints away from the injury site. Pain adversely affects motor control and the results of pain-related motor-control changes are somewhat unpredictable and highly individualized.[20-22] This simply means the absence of pain doesn't suggest normal movement or the absence of risk factors.

Movements are not only a result of mechanical and neurological systems; they also represent the emotional system. We use movements to forecast body language, and we often store unnecessary tension in muscles when under emotional stress. The state of the muscles and the body's posture represent mood and emotion—from a relaxed and comfortable condition to an extremely threatened and tense one. We respond and act by posturing the body and moving muscles to fit the perceptions of the activities ahead.

Movement problems do not always follow the rules of anatomy. It is entirely possible for muscle function and coordination to perform adequately in one movement pattern and poorly in another. The core and hips might perform well in squatting but poorly in single-leg stance on the right or lunging on the left. The same muscles are at work but each movement pattern is distinctly different with its own timing and coordination signature.

Poor core stability in that example cannot be remedied by selective focus on a single group of core muscles since they perform normally in one movement pattern but poorly in another. The solution can only be found by addressing the pattern, not by attempting to create isolated strength or performance in a faulty muscle group.

Movement appraisal must first establish levels of acceptable function and dysfunction. If multiple movement patterns have a common anatomical hindrance, it must be identified and remedied. Likewise, if an anatomical region provides adequate contribution in one movement pattern but appears inadequate in another, the faulty pattern should be investigated since the anatomical region is not behaving consistently. Problems within specific movement patterns may be a result of poor pattern-specific motor control and organization.

AUTHENTIC MOVEMENT

The observation and appreciation of whole movement performed without restriction or dysfunction demonstrates systems working in harmony to create functional movement patterns. For this reason, we map other contrasting whole movement patterns before rushing to judgment. We don't dissect a solitary movement once we identify a pattern as limited or dysfunctional. Avoid breaking down or dissecting any single movement pattern if you have not completed the full appraisal of basic movement patterns. By first viewing movement at a whole pattern level, we can see if multiple patterns are dysfunctional, or if it's a single dysfunctional movement.

If a single pattern is limited, we can systematically break it down. The remedy is found by reconstructing the pattern, not by exercising the problematic parts. If multiple patterns are dysfunctional, we look for common threads that could compromise movement within each pattern.

We must also be aware that the things we cannot measure remain as important as those we can. All these systems function with one complementing the other. One system may take a supportive role, while another seems to take a primary one. If an activity or body posture changes, the supportive system may move to the foreground and become the activity's frontrunner while the other takes on a secondary role. Using simple kinesiology or basic anatomy to describe this moving matrix does a disservice to the miracle that allows authentic human movement.

There's great difficulty in defining every aspect the integrated relationship between the neurological systems, muscles, fascial network, ligaments, joints and bones plays in movement. Highlighting one neglects the contribution of the other, but whole movement patterns can provide a starting point for moving matrix observation.

As exercise and rehabilitation professionals, we should be thankful for the introduction of science, because it has given us the ability to quantify the contributions of these structures. However, we must not excessively discuss these systems in isolation, lest we forget each part's inter-connectiveness and qualitative contribution.

As we map the human body, we must continually be aware that the map is not the territory. The map is a vehicle to improve communication, navigation and understanding. There is much more to human movement than a simple movement screen or assessment could ever reveal. It is, however, the best starting point. These tools simply capture fundamental patterns, and these fundamental patterns are what function is based upon.

Once appropriate levels of movement pattern function have been established, performance and skill can be investigated. If these are prematurely investigated without an appropriate movement-pattern baseline, poor performance and skill tests may actually be attributed to a faulty fundamental movement pattern.

THE MOVEMENT MATRIX

The interplay between structure and function is a central theme among professionals when discussing movement and dysfunction. Taking a purely mechanical approach to human movement, researchers and clinicians often migrate toward examples of how structure drives function. For example, those who mold orthotics or build braces or who perform orthopedic surgery rely on structural support or change to enhance function.

Structure governing function is an obvious and bold statement we can all understand. However, we must also understand the less obvious but equally valid statement that function can govern structure.

We can look at infant and childhood development and witness how certain functions and stresses enhance structural integrity. Some activities promote greater bone density because they place stress onto and across the bones. Hardened bones give up structural integrity and begin to develop demineralization and osteoporosis in the absence of weight-bearing and fundamental activity. The factors of stress and function are also at work in muscle development. Function influences structure through the SAID principal, *Specific Adaptation to Imposed Demands.* The body tissues adapt or remold continuously based on the activity or lack of activity that occupies most of our time and attention.

Think of a tennis player who displays greater bone density in the racquet arm. Was this athlete destined to be a tennis player with the genetic gift of superb dominant-arm bone density, or was that extra bone density developed through years of functional adaptation?

Is a great athlete prewired to superior balance and control, or did the athlete develop these attributes on the path to greatness? Elite athletes often display superior balance and neuromuscular control. Many might argue it is in the genes, but another group exhibits this great balance in laboratory testing as well: flight attendants. Are we to conclude flight attendants were born to fly the friendly skies while superbly balancing our drinks?

Genetics are an incredibly important facet of development—it is foolish to disregard what this means to our maturation. However, it is equally foolish to think we were born with all the ability we will ever have. In the well-researched book *Talent is Overrated*, author Geoff Colvin lays out a great argument that extraordinary performance is not innate. Indianapolis Colts quarterback Peyton Manning was not born holding a football playbook; Wolfgang Amadeus Mozart didn't enter this world with a conductor's baton in hand. Colvin argues with tremendous scientific and anecdotal evidence that their immense successes came because of a term researchers call *deliberate practice.*

Both received expert teaching early in their lives; both learned early what skills were required for greatness. Moreover, both practiced those skills for years before reaching international acclaim. The quantity of practice is not Colvin's point. He emphasizes that the specific way practice is executed is what defines the deliberate practice common to those deemed talented. There is a certain quality among the best of the best. They set themselves up for specific feedback. They apply uncanny objectivity to the investment of practice time and do not rely on gross repetition to build skill.

As professionals, we practice the art of rehabilitation and exercise to enhance the movement of those who seek our skills and guidance. If we are to set up deliberate practice within our specialties, we must hold movement to a baseline and continually check our work. Without a movement baseline, how can we be sure we have improved movement-pattern quality? Sure, we may improve physical capacity or exercise tolerance, but did we use an objective scale for movement-pattern quality against our work? An objective scale will provide us with positive or negative reinforcement for our exercise choices. Both types of reinforcement will help us grow, and that feedback will improve our professional development if we are willing to accept it.

The body's movement success is no different. It is also driven by positive and negative reinforcement. We started learning how to move before we were born—call it movement training. Our bodies were born with mobility, and some movement, but we did not have posture, stability or motor control. We had twitches, gestures and responses to stimulus, and reflexes drove much of our movement. We had to earn our stability, alignment, balance and control.

Our bodies were born, but we built our posture and movement patterns. We combined mobility, stability and posture to create movement. Though we did not know it, much of the waking day was spent practicing movement, limited only by what our maturing musculoskeletal system could handle, and, of course, our huge eating and sleeping demands. Mother Nature taught that movement, and it was expert teaching: basic, pure and unmolested by the interpretation of professional instructors. The practice was so pure, we didn't know we were practicing. The rules and goals were clear: Here's gravity; explore your world with your senses, and, by the way, an added benefit—your gift—will be movement.

DISSECTION AND RECONSTRUCTION

We need both dissection and reconstruction skills, but unfortunately, more focused education is placed on dissection with the assumption that reconstruction skills will emerge spontaneously. Reconstruction is significantly more than dissection in reverse, but it's not a bad basic model. In dissection, we work back to the fundamentals. Following this logic, reconstruction would force us to learn those fundamentals and to start with them. We should consider these as the fundamental supporting structure and always have a way to check the status.

Many rehabilitation approaches do not revisit the fundamental movement patterns that walking is built upon when older individuals lose balance or have difficulty walking. Instead, many seniors are placed on recumbent bicycles or given resistance exercises for their thighs, under the assumption that weakness is the only problem. However, we must consider that coordination, patterning, reflex stabilization and timing also play a role, and these will not be reconstructed with generalized strengthening or cardio exercises.

If we consider reconstruction as a process of starting with fundamentals, we will take a different approach, an essential approach. These essentials are evident to anyone who takes the time to observe the rolling patterns of an infant and understand how these become the gait patterns of a toddler. The simple disassociation of the head, shoulders and pelvis in rolling are natural, fundamental building blocks for the coordination needed for successful ambulation. We must consider reacquisition of rolling as a step to ambulation, and not expect successful ambulation if rolling symmetry and proficiency is not present.

Ultimately, we need to discuss anatomy. We will need to consider structure. The point of the following list is to demonstrate that we must consider the contrasting and complementary relationship of structures. All these structures intersect at movement.

They can contribute to superior function and can also present the major limiting factor. Ultimately, a single structure may be discovered as the primary limiting factor to a movement pattern. However, the wise professional will be constantly vigilant of the multiple secondary problems that will also be present. Some problems will be resolved with correction of the primary problem, and some will require more focused attention.

A global perspective to dissection and reconstruction demonstrates the dance of contrasting and complementary opposites. With each example, consider the near-automatic interplay between the examples, and remember that separating them provides substance for academic conversation, but undermines the substance of functional movement. These systems are uniquely different and totally dependent on each other.

Some examples of contrasting and complementary opposites—

- Static and rigid support by the skeletal system versus dynamic and flexible support by the fascial system

- Tonic muscle behavior for postural control versus phasic muscle behavior for movement production

- The dedicated control of single-joint muscles versus the regional influence of two- or multi-joint muscles

- Sympathetic nervous system that controls fight or flight versus a parasympathetic nervous system that controls resting and digesting

- The oxygen-dominant aerobic system versus the non–oxygen-dependent anaerobic system

- Movable segments created by joints versus rigid segments created by bones

- Static non-contractile control of ligaments versus dynamic contractile control of muscles and tendons

- Sense of movement and balance for guidance and information gathered by the kinesthetic system versus image-based feedback gathered by the visual system

- Touch, pressure, vibration, position and movement sense regulated by joint mechanoreceptors versus gauging muscle tension regulated by muscle spindles

- Intentional movement action and response determined by conscious movement control versus automatic movement action and response determined by reflex movement control

- The relaxed connection between the brain and the body connected with deep, slow breathing versus the energized unification of the brain and body with vigorous controlled deep breathing

In reductionist science, we often dissect the body's structures and functions to understand their importance and purpose within the movement matrix. Problems arise when we identify a particular focus area; we unwittingly impose importance on the singular part or function. Doing that, we might fail to recognize the supporting network that allowed the structure or function to do its work.

Using football as an example, the quarterback's effectiveness would change greatly without the offensive line's protection, but we focus on the quarterback. However, the quarterback's achievements cannot be considered in isolation as long as other players are connected to the accomplishments. In reviewing the productivity stats, the novice observer rarely recognizes the relationship between a quarterback and his protection.

We know Peyton Manning's exploits and prowess—four NFL MVPs 10 Pro Bowls and statistically on pace to become the most prolific passer in NFL history. How many people know the linemen who blocked for him? Manning's stats would not be as great as they are without these men. The veteran observer understands the entire offense's interconnectedness and how it progresses.

So it is with us as we reduce and dissect the body. We must not forget to reconstruct it in a manner consistent with the supporting matrix.

MOVEMENT DEFICIENCY AND DYSFUNCTION

The source of movement deficiency or dysfunction is rarely a singular event or has a single cause in any given person. To understand the factors and foundations that drive deficiency and dysfunction, it is best to designate general categories. The categories establish a clear platform for how to avoid insult to the movement matrix across a lifespan.

The movement dysfunction categories are—

Developmental

Traumatic

Acquired

DEVELOPMENTAL MOVEMENT DYSFUNCTION

First, don't confuse developmental movement problems with developmental problems that cause lifelong disability. Developmental movement problems arise when movement opportunities are denied or modified, or inappropriate activities are introduced in an otherwise normal system.

At birth, normal structures and systems require a sequence of movement challenges and opportunities to develop and function. Researchers and medical professionals mark movement advancement episodes as developmental milestones. An infant will develop head and neck control and then progress to rolling. The infant will work its way up to standing with various postures and strategies, each of which must occur within a certain age range or there will be a delay. This delay may or may not have long-term implications, but it can be associated with other lifelong problems.

The body and brain have numerous periods of accelerated growth between infancy and adulthood. Growth is not linear, and neither is the command of movement. A child may become competent with movement, and later have a setback due to a growth spurt or a long infirmary.

During puberty, we can grow at seemingly astronomical rates. For example, a 14-year-old golfer plays a tournament on June 1st, plays another a month later, and might have grown an inch taller in that month, or his body could have grown taller while his arms remained the same length. Then again, the opposite could have been true, with the arms growing longer while the body stayed the same size. Competency in movement suffers because he has yet to become comfortable with his maturing body.

Other problems can arise when children practice higher-level skills with compromised or undeveloped fundamental movement abilities. Pitching, throwing, kicking and swinging activities practiced repetitively without a sound athletic base can slow or alter complete and balanced functional development. As these children become teens and then enter adulthood with incomplete functional movement patterns, poor physical performance and elevated injury risk can result.

Screening is one way to determine the presence of movement-pattern problems carried into the present from the developmental past. As with other physical problems, early detection provides the best opportunity for appropriate correction. In times of change and fluctuating activity, repeatedly check fundamentals, as these are the foundation of everything specific.

TRAUMATIC MOVEMENT DYSFUNCTION

Trauma can cause obvious movement problems. Pain can alter movement, but movement might remain altered even after the pain is resolved. Tissue damage from the initial injury is compounded when movement patterns remain abnormal after healing. Inflammation, swelling, joint effusion and immobility can all compromise neuromuscular coordination, timing and control.

Movement compensation is a primitive survival behavior. These compensations and alternate movement patterns cause stress to other regions and are far less efficient than authentic movement patterns. However, they do allow us to temporarily move and function. Long ago, this option allowed us to continue moving out of harm's way following injury. However, these compensations are not our best long-term options and will themselves cause problems if not identified and addressed.

Coaching and execution of proper exercise technique is not likely to change a problem below the level of conscious control. We must identify the movement pattern that demonstrates dysfunction and reconstruct it, not coach or hope that general exercise will train the problem away.

Modern science offers many opportunities to artificially reduce or cover up pain, allowing us to move into patterns and postures we would otherwise instinctively avoid. Poorly informed athletes and fitness diehards commonly take painkillers to get back to competition or training. When they do, they give up the natural alignment and stabilizing reactions that occur as normal reflexes to support movement when moving into or around painful patterns. Even when the pain is covered up by synthetic means, motor control, reflex stabilization and reaction times are less than authentic.

Although there are immediate tangible benefits to getting back to the action, early return initiates simultaneous, sometimes imperceptible erosion. In athletes, these are debilitating effects and dysfunctions that might not be observed until the next season when a minor injury becomes catastrophic.

People often consider themselves recovered or rehabilitated because the pain has subsided, but have they completely reestablished the previous functional level? What was that level in the first place?

Determination and fortitude are assets when overcoming traumatic insult to the body, but those attributes should not cloud appraisal or judgment of post-injurious abilities or limitations. There is a motor control adaptation that occurs following injury—partially pain-driven—that many times does not normalize with rehabilitation efforts. We are trying to measure this with body-relative movement screening and testing.

Absence of pain is not a sign of complete recovery. Baggage follows injuries; rehabilitation professionals should use screening to demonstrate recovery and minimal risk of recurrence nearing the end of the rehabilitation process. Proper diagnosis is the initial responsibility of a healthcare provider, but equally important is the final responsibility of proper prognosis. This can only be done with an objective tool designed to appraise potential future risk.

In the old system, we measured impairments such as the return of isolated strength, joint mobil-

ity, muscle flexibility or balance. We also might consider performance parameters such as job or sport-specific movements.

In the new system, we know these measurements are possible without complete reacquisition of functional movement patterns. Impairments can be normal, but movement patterns can still represent dysfunction. Likewise, it is possible for some performance parameters to fall within normal limits even when acceptable movement patterns are not present. The initial focus should be normalization of the impairment, followed by movement-pattern quality and once that is established, performance quantities can be the secondary goals.

ACQUIRED MOVEMENT DYSFUNCTION

Acquired movement dysfunction generally occurs in two ways.

Unnatural activity repeated on a natural movement base

Natural activity repeated on an unnatural movement base

Unnatural Activity on a Natural Movement Base

Movement dysfunction can result from activities that require special skills, training or are against a natural movement pattern.

For instance, the pitching and throwing motions are within the boundaries of normal movement patterns. However, when they are repeated for large volumes over long periods, imbalances can occur. This is because pitching and throwing is a skill not balanced with equal activity from the non-skilled side. The unnatural pattern, and movement dysfunction occurs because of its repetitiveness. Players often throw hundreds of times a day and thousands of times a week. They throw standing up, sideways, jumping and twisting, from their knees, sitting, while running and from a variety of other positions.

Frequency, intensity and duration are separate ways to gauge the cumulative effects of these activities and patterns. These movements could be part of daily activity, work duties, hobbies or sport skills. Even on a good movement base, these could impose a significant bias to one pattern or toward certain specific movements. Habitual activities often take on a life of their own, causing overuse, compensation, postural changes and musculoskeletal imbalances, and, if left unchecked, symptoms such as pain and inflammation will arise.

We can manage the pain and inflammation quickly and easily. This is often mistaken for the source when it is just the outcome, a normal response to an abnormal situation. The best way to avoid an accumulation of problems is to impose balancing and contrasting activities to counteract the effect of high-volume unnatural patterning. Screening tracks and curtails movement-quality erosion in situations where certain specialized activities are performed.

Unilateral movements are normal in many sports from golf to swimming, where breathing to only one side may be preferred. Opportunities do not always present themselves to create balance within sport-specific training, but if we continually screen fundamental movements, we can be proactive with corrective approaches before problems present themselves.

Habitual postures can also present the same type of imbalance. We are more sedentary than ever before, and we have a predominance of flexion through the spine and hips as compared to extension. Over time, the predominant flexion behavior can compromise natural extension. Original humans were on their feet for a large part of the day without leisure or entertainment opportunities designed around sitting in one place.

We still own the same bodies; we simply don't own the full array of movements that serve to maintain a body's good working order. We can supplement extension activities to offset the predominate flexion postures, or we can look for opportunities to sit less, or both. The best way to gauge the sedentary affect of flexion is to screen movement patterns. This simple act will quickly indicate if we already have the need to correct movement or if we need to be aware of potential future unfavorable adaptation.

NATURAL ACTIVITY ON AN UNNATURAL MOVEMENT BASE

Movement dysfunction can result from activities that would appear to be natural and within functional limits. Problems arise when preexisting fundamental movement limitations and asymmetries cause people to compensate when performing basic tasks. It's not the activity; it's the lack of mobility and stability giving rise to the problem. Repeated functional activities on a faulty base can cause overuse, compensation, postural changes and musculoskeletal imbalances. Many times, the activity gets the blame when the blame should be placed on the poor foundation the innocent activity was placed upon.

As in the aforementioned scenario, if left unchecked, symptoms such as pain and inflammation will develop. Some people break down, perform poorly and with more injuries than others. It is common to blame activities when problems occur, but this singular-cause view casts a shadow over the pre-activity movement quality, which can usually account for the problem.

Of all the categories of movement dysfunction, this category may be the largest and least understood. Every day, out-of-shape people attempt to regain fitness, lose weight and become more active. They assume if they just move more, they will start to move well. Unfortunately, they will just get better at moving poorly for longer periods of time or with larger amounts of weight or at greater speeds. As problems arise, some will change equipment and some will modify the workouts. Some will simply take a daily anti-inflammatory and some will just quit, only to try again the following year.

RECAPPING THE CATEGORIES

Each of the three categories—*developmental, traumatic and acquired*—can compound one of the others, but in our jobs as exercise and rehab professionals, we can only act on one. The first two are usually parts of the past; the third is a product of the past and the present. The acquired movement dysfunction is the most manageable since it involves lifestyle choices. After the fact, it may be hard to ascertain if unnatural movement patterns preceded the problems, but we need to address them regardless.

Those with the most risk of injury will have hallmark signs of movement-pattern limitation, deficiency and dysfunction. Routinely screening as a precursor to exercise, activity and competition is one way to establish a proactive model for risk management and to develop better programming when deficiencies are noted. Screening is not a one-time thing. Activity levels change; fatigue, strain and tension levels fluctuate. Movement patterns can be biomarkers of the effect lifestyle stresses impose on movement behavior.

The need to identify a current problem is paramount to speculation of future risk. Screening can detect painful movement patterns in people who are unaware of them or are naturally avoiding certain patterns because of pain. It is important to view a cross-section of movement-pattern extremes before increasing exercise and activity in order to avoid compounding a painful situation. These individuals need a healthcare professional, not just a fitness professional.

Ultimately, the team approach is best. Some people will not have pain with screening, but will display dysfunction that places them in a high-risk category. Fitness professionals who understand the difference between corrective exercise and conditioning exercise can help them safely navigate out of risk and into movement patterns not considered high-risk. Once risk is removed with corrective exercise, conditioning programs can be designed that help maintain movement patterns while improving energy systems and physical capacity.

PAIN CHANGES EVERYTHING

The presence of pain in the movement matrix changes the rules of fitness and rehabilitative exercise. Because of pain, we cannot use physiological principals to attack movement dysfunction, nor can we rely on consistent outcomes of strength, endurance and flexibility.

Pain changes the way we move. It is unpredictable and highly individualized. We do not know how a body will move when pain dictates the movement; we just know it is altered from situations that are pain-free. Dysfunctional movement and movement complicated by pain do not allow the authentic and sustainable creation of strength, endurance and flexibility in a consistent and reproducible manner.

Pain is a part of the evaluation when present in movement dysfunction, and must be managed in a methodical and reliable way, which is different from the way we work with pain-free movement dysfunction. Modern technology allows us to compete with pain, but temporary solutions have become standard training practices and permanent fixtures. Wraps, braces, drugs and tape were developed in the athletic arena to allow athletes to complete a single game or event. Now these provisional measures have become common training aids as we routinely play and train with topical creams and therapeutic devices.

Television ads proclaim that pain shouldn't sideline us—take a pill or gel cap and get on with life. The ads frame pain as an inconvenience and declare our schedules more important than the messages sent by our bodies. We act surprised, irritated and inconvenienced when the subtle pain message becomes an alarm, as if it is the first we have heard it.

Pain is a warning sign. Long before pain represents a chronic problem, it can alert us to poor alignment, overuse, imbalance and inflammation. We embrace all the other warning signs in our lives—computer virus alerts or the oil light on the dashboard—but when it comes to the body, we act as if the warning sign of pain is an inconvenience. We cover it up so we can keep moving. If we ignore pain's natural self-limiting nature, we are ignorant to the lessons its ancient design provides.

*Please see www.movementbook.com/chapter2
for more information, videos and updates.*

UNDERSTANDING MOVEMENT

To address the needs of a culture that needs to move better, we have to understand its intentions. We must also look into the assumptions, misconceptions and mistakes that define the perception of the collective population.

As fitness and rehabilitation professionals, we often assume universal agreement on definitions of movement quality, return of function, fitness and athleticism. Review a list of the top fitness and healthcare professionals. Each might have multiple credentials, 20,000 hours of practice and unparalleled success. These authorities have written books and produced videos, and newspapers and magazines call on them as experts. Yet prepare for an argument when putting them together to define the best method to increase any aspect of a client's fitness, an athlete's performance or the most effective path for patient rehabilitation. These are technical professionals and artists, and naturally we would expect some disagreement, but most don't even agree on an entry-level standard operating procedure (SOP).

Pilots and surgeons defer to a SOP before each flight or as they start a new case. On the other end of the spectrum, you could say exceptional artists rarely agree, but the actual physics and chemistry of the mediums they use are standard. Sculptors and painters alike use tools suited to the media of their work.

Physical therapy, chiropractic, sports medicine, formal physical education, personal training and strength coaching are very new professions—most formalized standard education is less than 100 years old. These all work with same medium of movement, but lack the consistency and SOPs of pilots, surgeons and artists. Our disagreements beyond the SOPs are expected and make us unique, but our disagreements about the SOP make us seem less professional. Without an SOP, we often fall victim to personal perspectives and subjectivity. We will always debate methods and that is no problem—it's professionally healthy—but SOPs should guard our principles and guide our methods.

Our experiences lead us to methods and methods can alter our professional opinions. Personal trainers with bodybuilding experience might focus on muscle development and fat loss and may de-emphasize components of flexibility and posture. A physical therapist with a love of yoga might concentrate more on the flexibility and postural aspects of rehabilitation and neglect the basic strength and power necessary for safe return to work, sport or daily activities.

These are simple examples of how personal preference and lifestyle influence each of our fitness and rehabilitation views. If we accept and allow this diversity and subjectivity in our own professions, what must the public think? The public has the impression of us as the gurus, but at times, we do not even seem agree on the fundamentals and principles. The sad truth is that media and advertising interests have greater influence on the fitness culture than the professionals dedicated to fitness, athletic development and rehabilitation.

We need to investigate the opinions and misconceptions, and understand the confusion the public must endure to simply become fit or to recover from a debilitating experience.

They need—

- basic logic and sound advice

- us to understand their assumptions

- practical examples and confident professionals

- us to communicate what they cannot

They do not need fads and quick fixes, nor do they need negativity and professional peer critique.

Understanding their perspectives reduces communication problems. Understanding your subject matter is intelligence; understanding the way your clients and patients see your subject matter is wisdom.

MOVEMENT KNOWLEDGE VERSUS EXERCISE KNOWLEDGE

We need to understand how the people who depend on us view movement. Many of us look at fitness and training as an exercise culture, but not necessarily as a culture of movement. Our clients and patients are even less aware of movement, and they need that education.

The perception is if we simply exercise, we will automatically move better, but without a movement baseline, this debate can never end. It is likely the ability to perform the exercises practiced most often will improve, but that is a myopic view of the movement spectrum. That is practicing the test.

No single exercise can represent the full spectrum of human movement. A pre-exercise movement baseline will show that sometimes exercise helps us move better and sometimes it contributes to greater levels of dysfunction.

Current exercise programming has two inherent problems: Some movements are performed too frequently or with too much intensity, and some movements are used too infrequently or with too little intensity.

The magic recipe is not universal; it is unique to each person's movement map. Completely clean movement maps are possible but rare today. Movement maps were probably all good when we hunted our own food and tilled our own soil, but from the time we entered factories and offices, we started down a slippery slope of movement dysfunction, and each of us has responded in unique but predictable ways.

Knowledge of exercise cannot help movement until we have exposed a baseline.

THE TWO FUNCTIONAL MOVEMENT SYSTEMS

The two systems presented in this text are basic and logical, intended to reduce professional errors and assumptions. They are designed to capture tightness, weakness, poor mobility and poor stability within the pattern that represents the most significant movement pattern dysfunction. Both systems consider basic information before specific information—these systems are high logic, but not high tech.

The concept behind these movement appraisals is not new—this is how the best of the best do it. It simply has not been outlined in this specific way. When we look at our mentors, we often focus on the complexities of their expertise instead of the fundamentals to which they adhere. Our tendency to make ideas more complicated than necessary is the main reason movement quality was not standardized and simplified early in our professions.

The mistake is simple: Specificity and special interests kill basic objectivity and logic.

Focus must become narrow to observe the specific details that interest us, but a narrow focus reduces the broad view needed to always consider and incorporate fundamentals. There's blindness in the exercise profession, where movement exists only to serve exercise research and programming.

Research and exercise programming should first support and promote a comprehensive approach to movement. Exercise research often elevates one program over another. Research may study some microscopic part of a movement or a singular event that may or may not affect whole movement patterns. If we do not establish goals for movement quality, continued research into movement quantities offers a diminished value both for knowledge promotion and for practical application.

People assume that knowledge of an exercise science that favors a heavy metabolic focus over the mechanical is equivalent to knowledge of movement science. This is backwards; instead, thorough movement knowledge should precede a specific knowledge of exercise science.

Do we believe and act on the premise that the laws of activity govern movement, or do we believe the laws of movement govern our activities?

Think seriously about that for a moment. The laws of exercise and skills such as athletics and recreational activities are our inventions. Humans developed explosive lifting, tackling, passing, shooting, inverted skiing tricks, roundhouse kicks and cartwheels, jumping jacks and leg extensions. These are not nature's inventions.

But our movement grows out of nature. We develop activities, games, sports and exercises that apply a movement array. This is far from a *did the chicken come before the egg?* question. Authentic movement absolutely precedes evolved activity. The opposite would go against the rules of nature.

We enjoy the activity, competition, challenge and benefits of specific skills and movements. The problem is, we often get too specific.

Experts do not agree on the methods of instructing and teaching exercises and sports skills. From football conditioning to a golf swing and from endurance training to yoga, we develop opinions and methods to support specific movements. This focus often perpetuates neglect of the basics we have in common.

The above-mentioned activities require strength, flexibility, endurance, coordination, mobility, stability and balance. Nevertheless, it's common for a golfer to ask for golf-specific flexibility exercises when generally inflexible. It would be more valuable to become flexible overall, and then pursue golf-specific flexibility only if needed. Likewise, many parents look for pitching-specific drills for their young baseball players, without recognizing that the aspiring athletes cannot do a single quality pullup or pushup, and have little or no physical activity other than baseball.

We often build specific fitness on poor general fitness and think nothing of it, but sport skills are built on general athleticism. Perhaps because of our impatience, lack of commitment or lack of understanding, many skip the steps of general fitness or basic athleticism, and advance directly into a movement specialty or a specific sport. The movement specialty should be the goal, not the starting point.

Nature demands that we crawl before we walk. Failure to notice the common attributes between activities demonstrates a failure to see the common foundations on which we build specific skills.

Humans go through the same movement stages in growth and development. We cannot determine in an infant who will grow to be a runner or who will grow to be a boxer. As we mature, our interests and movement activities go in many directions, but foundations key to early development support those varied pathways.

That variety might beg the question: If our movement foundation is so similar and so basic, is it ever okay to lose it, even when our chosen sports and activities differ? The only way to answer in the affirmative is if inefficient movement is acceptable.

To enjoy specific activities, we need to develop special skills on top of basic function, not in its place. If basic function erodes beyond minimal acceptable levels, some movement efficiency and durability will be lost and will reduce an advanced movement skill. The gap between the fundamentals and the elite skills sometimes makes these seem unrelated, but this is faulty thinking.

TYPICAL EXAMPLE

You start to develop knee pain after your training run, and relate the pain to age, shoes or mileage, not realizing flexibility and core strength have been on the decline. You have been pressed for time, but not wanting to miss your running, you've been skipping the pre-run stretching and the twice-weekly strength training. You haven't done either activity for months, and do not relate the current knee pain to the neglected aspects that once created a well-rounded fitness program. Your life consists of sitting at a desk and forcing a few ill-prepared runs each week.

Most assuredly, someone will recommend an analgesic or anti-inflammatory medication, and while those will reduce the discomfort, they will not fix the underlying issue. It is the logical equivalent of mopping a wet floor and thinking you've addressed a leak in a pipe. The leak is the primary problem, not the wet floor. The painful knee is not the problem—the supporting mechanics are the problem, and the pain is the result.

In our highly specialized and activity-specific society, we forget that a simple movement problem will limit specific physical improvements. Establishing perfect basic movement is not necessary, but fundamental minimums are mandatory.

Our foundational human goal is to survive. We can only achieve perfect movement if we are alive to do it. The brain must consider technical perfection and specialized movement secondary within a system created to ensure survival.

Remember the laws of human movement, and value these in every specific endeavor. These laws are common tendencies; they are simple and straightforward, and become strong in the presence of poor movement patterns. Stiffness, weakness, injury, asymmetry, fatigue, pain and unfamiliar quandaries can magnify these tendencies, and the brain triggers these to ensure survival. The brain offers the temporary solution to survive, but these survival techniques shouldn't become the norm.

Imagine the brain following commands in survival or stress mode as defaults. These commands are instructions that help conserve energy and avoid further stress. Now imagine that someone is enduring physical stress to lose weight or push through physical therapy. These are some of the suggestions offered by a brain in survival mode—

- Avoid positions that produce restriction and stiffness
- Avoid unfamiliar movement patterns
- Avoid pain and stress
- Compensate and substitute if needed and whenever possible
- Compromise movement quality to gain movement quantity when needed
- Conserve energy whenever possible
- Do not rely on positions and patterns of weakness or instability
- Take the path of least resistance
- Seek comfort and pleasure

Now imagine trying to teach exercises to someone with all this going on behind the scenes. People with basic movement-pattern dysfunction will respond this way whenever asked to perform exercise within a dysfunctional movement pattern. Most are not aware of all that goes into avoiding dysfunctional movement patterns. This is why over-coaching a movement does not produce efficient motor learning.

For example, imagine a female client who has poor hip stability and core control. You subsequently remember that every time she performs a lunge, her knee caves into valgus collapse. More lunges will only rehearse this dysfunction, not correct it. Non-threatening exercises must be created and corrective exercises will be required. The entire lunge exercise problem might have been avoided by screening movement first.

Overriding automatic tendencies is completely possible, but automatic tendencies are a default mode during instruction, education, training and rehabilitation if underlying problems are present. These tendencies will commonly occur without a consistent tracking system to target movement dysfunction and the associated behaviors.

This can happen when ignoring movement with a focus only on exercise, training, sport skills and rehabilitation protocols. Often the body will avoid a movement when something is wrong. Compensation will occur and exercise and rehabilitation professionals will try to instruct or coach away the problem. However, verbal instruction is not appropriate because the problem is not a conscious issue. All the proper instruction in the world is of little help when nature disagrees with the suggestion.

We can manage human movement tendencies easily and successfully. To create the potential for clean correct movement patterns and good movement basics, we—

- Remove pain
- Reduce or resolve movement and movement-pattern and asymmetries

Once we've achieved this, we build on the foundation by providing repetition and basic pattern reinforcement. This will create familiarity and improve mobility and reflex stabilization. Movement patterns that were once faulty but are no longer painful, limited or asymmetrical should receive specific attention to reinforce the corrected patterns.

The moves we have our clients and patients perform should use functional flexibility and should naturally engage the core's reflex stabilization. If they need mobility and stability work, they must do it, but once mobility and stability reaches a respectable level, we should provide programs to make them fit and simultaneously maintain that foundation of mobility and stability.

Because of today's training programs, some sculpted people with beautiful, aesthetically pleasing bodies cannot touch their toes or perform a full backward bend. Others compete in triathlons, yet cannot squat into a relaxed resting position with their heels flat on the ground.

Many patients have successfully completed cardiac rehabilitation programs, meeting the cardio-respiratory goals while still displaying severely compromised functional movement patterns. These patterns may have originally compromised movement efficiency and contributed to the cardiovascular stress in the first place.

Peddling a recumbent bicycle while watching a heart rate monitor and a TV program can create a cardio-respiratory demand, but it does nothing to improve the multisystem relationships required to truly increase authentic functional movement capacity. Modified versions of tai chi or yoga may very well achieve the same cardio-respiratory goals with parallel improvements in breathing management, endurance, balance, coordination, posture, self-image and physical confidence. It is unlikely the next heart attack will occur on a recumbent bicycle—it will occur when physical and emotional stresses combine to overload systems. Most would agree that control of breathing and movement patterns instead of recumbent bicycle competency would be better insurance against cardiac stress and a future cardiac episode.

Nevertheless, here we are. Modern fitness equipment allows training while sitting and even slouching comfortably. This equipment accommodates pushing and pulling with the arms, and flexing and pressing with the legs. The equipment also furnishes torso flexion, extension and rotation without forcing users to balance on their feet or naturally engage the stabilizing musculature.

People move muscles without the burden of controlling bodyweight, maintaining balance or managing alignment, but that is not life. When Grandpa lifts the carry-on into the plane's overhead compartment, he does not have a backrest to lean against; he has his inner and outer core. He has muscles in his feet, legs and back to sense his position and load and hold him upright. His body is in a state of dynamic movement, one that

requires coordination many modern machines do not provide or even allow.

A dogmatic focus on fitness goals and sports skills without considering basic movement patterns is not the individual's mistake. That inflexible focus comes from a society that does not include fundamental movement quality in its definition of fitness.

The tendencies of human movement exist in significant contrast to the rules of activities, exercise and athletics. The postures, alignments and movements we teach can be the correct ways to perform specific sports or activities. The irony is we try to teach and coach a movement skill without first checking basic movement patterns. This makes us exercise and rehabilitation professionals, but should not infer that we possess the ability to appraise and teach authentic movement unless we have a system in place to do just that.

BASIC MOVEMENT BEFORE SPECIFIC

When basic movement is limited or compromised, it follows the natural laws of energy conservation, compensation and avoidance of pain, avoidance of the unfamiliar, and the essential tendencies of survival. Often the poor technique we observe is the body trying to survive a predicament it is not ready to address. Only when the movement patterns are present and functioning at a basic quality is it time to add volume or intensity or to work on specific skills.

Most professionals agree we need a foundation before we enter specific activities, but actions speak louder than words. Many do not take consistent and organized action to enforce this belief. This would take time, organization, training, structure and a reliable system. Unfortunately, someone else is always ready to sell consumers the shortcut to performance, fitness or wellness. None of our clients or patients want to hear about fundamentals when excited to play, train or recover from a problem, and it's hard to stand strong against the tide of quick-fix advertising.

Specific activities can serve to undo a basic functional level, forcing the body to work in only certain patterns, and this is okay if we take countermeasures. Examine sport activities that feed one particular movement more than the others, like the golf swing that usually has a one-side focus, swinging left or right. Alternatively, those who concentrate only on running or field or court sports have a tendency to overdevelop lower body muscles and movement patterns and neglect others. Specialized activities will always lead toward a degree fundamental compromise. These can elevate strength, endurance and power in some movement patterns, but reverse basic mobility and stability in others.

Maintenance of the foundation is a constant battle, more a journey than destination. The more specific, complicated and extreme the activity, the greater the maintenance schedule must be. In our profession, we need to routinely monitor fundamental movement. There is no situation where it is advantageous to overlook the fundamentals.

Even though specialized movements can promote movement and some degree of fitness, they may have limited holistic effect or long-term benefit. High levels of fitness and activity will often disguise basic dysfunction.

Modern conveniences and protocols allow us to put fitness on dysfunction, but the resulting fitness is not well supported—it will be short-lived and hard to maintain. It will act as an exterior cover, placed over internal dysfunction and an unbalanced system. This dysfunction will slowly erode the attempts at conditioning by causing compensation and reduced efficiency.

The body will have the physical capacity to exceed the limits of basic movement patterns and supporting parameters of fundamental mobility and stability. Muscle capability can surpass joint integrity; strength can exceed stability; flexibility problems can compromise postural control; and muscle imbalances can cause premature fatigue of some muscles and poor engagement of others.

An energy system that surpasses the basic movement framework is a welcome mat for injury.

UNDERSTANDING PERSPECTIVES ACTIVITIES, EXERCISES AND ATHLETICS

Activities, exercises and athletics require the physical body to move and behave in more vigorous ways. These words engage each of us differently.

Activities

Exercises

Athletics

People identify with one term more than the others, but each of these in some way represents a more dynamic life. We derive the importance of the words from our own preferences and experiences. If we hate to practice and love to play, competition, athletics and sports drive us. If we love the peace of mind and internal confidence a good workout provides, we may enjoy training and exercise more than the formality of public sporting display or the intensity of competition. Some may enjoy the arts, where dance and music might require significant physical demand, but are not considered sport.

People from a different generation or culture may have limited recreational opportunities that allow for exercise, the arts and athletics. However, these people may have a great appreciation for physical activity, physical labor and productive physical accomplishment. These enthusiasts can enjoy an active lifestyle without competing or training. They don't see the point in sets and reps, but can still move as well as if they were at the gym two or three days a week. They labor and work against physical tasks or against nature, expending the same energy others reserve for exercise and recreation.

Most healthy human beings migrate toward some form of enhanced movement and enjoy the resulting mind-body experience. Life moves.

However, perspectives get distorted. The culture that gave us the martial arts did not intend yet another competitive activity. The martial arts sprang from a basic need for self-defense. Early martial artists did not do the training drills and exercises for calorie expenditure, exercise or competition, but to gain competency and efficiency with offensive and defensive movements.

Purposeful movement perfection was the focus, and physical conditioning happened as a natural side effect.

The culture that gave us yoga did not intend to provide trendy flexibility exercises, but for many Westerners, that is what yoga represents. We fail to see that yoga does not serve as a workout, but as a daily moving meditation where breathing and movement become one, creating mobility, stability, endurance, strength, patience and focus that can increase that day's quality.

Instead, we hear our client's say, "The instructor really picked it up in yoga today, and it kicked my butt." How did the increased intensity help the remaining part of the day? Perhaps a good butt-kicking is the best way to start the morning.

We often spotlight the physical side effects of calorie burning, and neglect the primary purpose, the chance to gain competency and efficiency in a movement. As we focus on a qualitative standard, it will produce both increased movement quality and the secondary conditioning benefits.

Training, conditioning and rehabilitation should have purpose, precision and progression.

Our sedentary society instinctively knows we need to be more physical. We even punish our bodies with activities, exercise and athletics as a penance for our sloth. Nevertheless, the assumption that difficulty will produce fitness is misguided at best. Exercise and rehabilitation professionals should focus on activities that accelerate practical achievement, not just on difficult activities.

Most people make the incorrect assumption that any activity will yield achievement. That assumption is part of a more-is-better philosophy. We assume difficulty is the measure of athletics, exercise and activity, but it is not. The challenge is the measure.

DIFFICULTY VERSUS CHALLENGE

Difficulty—a thing that is hard to accomplish, deal with or understand
Difficulty = Tearing down, struggle

Challenge—a task or situation that tests abilities
Challenge = Training, preparation

Any fitness or conditioning professional can create difficulty in athletics, exercise and activity. It takes wisdom and a higher purpose to design challenges on the razor's edge of possibility for a person or a group. The perfect training situation should challenge constantly changing physical fitness, but also require the trainee to use experience and knowledge to overcome an obstacle such as weight, time, distance, position and specific drills and tasks.

The challenge should stimulate the mind-body connection, movement efficiency, management of emotion and presence of mind under physical stress and fatigue. The challenge should strive for intuitive and instinctual movement behavior blended with a movement skill. These should not be over-taught and over-coached mechanical routines designed to increase exertion.

Physical difficulties may make us tougher, but intelligently devised physical challenges make us stronger. Activities, exercises and athletics should present challenges that foster the logical and instinctual management of physical, mental and emotional difficulties when they arise. Professionals with clear goals and defined standards can convert a difficult circumstance into opportunities for constructive challenge.

In conditioning and rehabilitation, the purpose of training drills is to provide challenge. Soon the recipients of this training will convert difficulty to challenge on their own. That ability is the purpose of training and rehabilitation, and is the true confirmation of learning.

A PROFESSIONAL CROSSROADS

We have established that we migrate to more dynamic movement if not limited by pain or disability. Now we must consider the guide—Who is the teacher? Supervision is prudent and necessary in new activities to promote safety and to accelerate the learning opportunities involving movement.

If we train or teach movement, we have a responsibility to screen functional movements before we teach or start a conditioning program. We have the responsibility to separate movements that require correction from those that are safe to condition. If we evaluate movement in a medical or rehabilitation capacity, we have a responsibility to assess movement patterns before we develop corrective exercises and progressions for patients' with movement dysfunction and associated pain.

The concepts of difficulty and challenge apply to all disciplines as they govern specific movement activity. With physical education, athletics, elite performance and rehabilitation, we are charged to design challenges directed toward achievement and self-reliance. We are not the distributors of difficulty, in the hope that triumph and autonomy will occur spontaneously, nor are we here to instill unnecessary dependence.

Many times in training and rehabilitation, monetary compensation rewards professionals who intentionally make ideas complicated. Our clients and patients might want and even pay for difficult and complicated, but that does not mean they should get it. After all, we're the professionals; we have the insight, skill and understanding. The public has a long-standing belief that difficult and complicated are somehow better for the body—no pain no gain. It is up to us to disclose the truth about fitness and it's up to us to resist the self-importance and perhaps even the greed that encourages client and patient dependency.

The clients and patients will always be there when we help them enjoy life instead of merely surviving it. Health and fitness professionals in authority positions who perpetuate difficulty, either consciously or unconsciously, bring about dependence and frustration, and that is a poor long-term business model.

We must continually look for systems to help generate the most appropriate challenges for those reliant on us for safe guidance and education. Together we make up the landscape forming opinions that influence the active public. These opinions stretch from obstinate to indifferent, but professional responsibility dictates persistent self-appraisal.

DISEASES, INJURIES AND AILMENTS

Diseases, injuries and ailments related to the musculoskeletal system comprise a significant number of medical visits. For example, the National Center for Injury Prevention and Control has estimated that more than 10,000 Americans daily seek medical treatment for sports, recreational activities and exercise-related injuries[23] alone, with lower-extremity injuries comprising most of these injuries. This excludes visits for spinal pain, which is well known as the second or third most common reason to visit a primary care provider.

The medical doctor has long been our source for diagnostics and medical management of musculoskeletal problems. For the most part, we use medications to manage the symptoms associated with musculoskeletal injury, disease, ailment and pain, but these are not corrective in nature, especially with respect to movement.

While using medication can make movement less painful, the prescription does not correct the problem or reset motor control. In most cases, the quick fix placates the person temporarily while the medical professional tries to devise a cure or corrective measure. Unfortunately, both the professional and the patient often perceive the placation as the solution.

In the past decade, we've seen a new shift in the way people seek treatment for musculoskeletal problems. The gatekeeper's role is no longer completely in the medical doctors' hands. Physical therapists and chiropractors operate with direct access to patients and in many cases function independently without a medical referral.

Likewise, strength coaches and athletic trainers serve in positions of direct contact with athletic populations, dispensing sports medicine in a emergency, rehabilitative and preventative capacities. The roles of the strength coach and athletic trainer often overlap in the areas of early detection of risk and in the transition from rehabilitation to full return to activity.

Sports coaches, physical educators and group exercise instructors are often the first professionals to observe musculoskeletal problems, and by making a professional referral, they initiate the treatment process.

Personal trainers have an opportunity to educate aspiring fitness enthusiasts. This is a perfect position from which to screen movement and discuss the differences between corrective exercise for movement dysfunction and conditioning exercise to attain fitness goals.

If pain is present, trainers can refer the client for assessment by a healthcare professional. If no pain is present, but significant limitations or asymmetries are noted, corrective measures can be implemented before conventional exercise programming or the trainer can seek a healthcare professional's guidance for corrective exercise suggestions for the client.

Strength coaches at the high school, collegiate and professional levels are in key positions to screen for injury risk associated with conditioning and athletic competition. The physical screens conducted before initiating exercise often display undiagnosed problems that could be aggravated by physical activity. In each of these cases, quick identification and appropriate referral can prevent the development of a potential problem caused by increased activity.

THIN-SLICING

In the book *Blink*, Malcolm Gladwell discusses the phenomenon of thin-slicing. Thin-slicing is the ability of experts in any field to slice through the information milieu to make observations and decisions. The best of the best do this without the regimented systematic process we would expect them to use. This ability takes on the appearance of professional intuition, but it is much more. Gladwell defines expertise as advanced pattern recognition, and explains that expertise and experience are not the same. Experience may only demonstrate high levels of technical skill, whereas expertise is defined by advanced problem-solving and innovation.

Technical skill is important, but only if it is used in the right situations and at the right time. Experts in any field are able to identify patterns accurately and quickly. Novices stumble through the process and may not recognize the most important features. Details and irrelevant differences distract the beginner. This clouds the decision-making and problem-solving abilities that are the hallmark of the expert.

Gladwell is actually promoting the concept of using movement patterns as initial guides when we observe and discuss other forms of human movement. He describes how successful professionals from different fields all use patterns as their primary directives and then investigate further in a particular direction indicated by the initial pattern.

Movement screening and assessment forces the professional to develop pattern-specific observation skills. By blending these skills with the technical aspects of more narrow and specialized investigation, the exercise and rehabilitation professional embarks on expertise. Some suggest that movement screening and assessment should be mechanized or digitized to increase data and reduce errors, but that would be counterproductive to professional development. We do not need more detachment from movement—we need to lean in and become reacquainted.

Professional intuition is intense familiarity, blended with systematic objectivity. Consider these suggestions and concepts as you look at the definitions and usage of screening, testing and assessment.

A GLANCE AT SCREENING, TESTING AND ASSESSMENT

On the surface, specific testing procedures and involved assessments appear more thorough than simple screens, but screens give us the best starting point since they provide basic groupings and fundamental categorizations. The lack of specificity is intentional, because it is paramount to define a general group by the most reliable and appropriate screening procedures for the classification. Once people are placed in appropriate groups, they can be tested and evaluated with a greater specificity particular and appropriate to their needs.

Specificity may appear more valuable, but perspective is often lost therein. A microscope is great for biological slide investigation, but one must move away from the magnification to find the appropriate slide to investigate. Screening provides reliable perspective in order for specific testing and assessment to refine the most pertinent information.

Specific testing without pre-qualification can yield what is referred to in medicine as false positive information. A common example is X-ray investigation of the lower spine region of a random group. The X-ray will note degenerative changes in the spines of people with no complaint of pain or dysfunction, while normal findings can be seen in individuals who complain of both pain and dysfunction.

To demonstrate clarity and promote clear communication, the terms screen, test and assess are defined as the following.

SCREEN

A system for selecting suitable people

To protect from something unpleasant or dangerous

Meaning—to create grouping and classification; to check risk

TEST

A series of questions, problems or practical tasks to gauge knowledge or experience ability

Measurement with no interpretation needed

Meaning—to gauge ability

ASSESS

To examine something; to judge or evaluate it

To calculate a value based on various factors

Meaning—to estimate inability

Neither the 12-minute Functional Movement Screen (FMS®) nor the five-minute Selective Functional Movement Assessments (SFMA®) replace other forms of fitness performance or skill assessment. The Functional Movement Systems—

- Demonstrate if movement patterns produce pain within excepted ranges of movement

- Identify those individuals with non-painful movement patterns that demonstrate a higher risk of injury with exercise and activity

- Identify specific exercises and activities to avoid until achieving the required movement competency

- Identify the most effective and efficient corrective exercise path to restore movement competency

- Create a baseline of standardized movement patterns for future reference

Movement screening is valuable specifically because it is not a conventional measuring system. Movement screening only appears to be a measurement system because it uses an ordinal scale. This ordinal scale provides a grouping and classification of similar movement-pattern proficiency or deficiency across seven tests and three clearing exams.

Physical performance testing done on a person with significant movement deficiency might yield poor performance values. Without a movement perspective, the tester might incorrectly recommend additional performance work to normalize the performance testing values.

Likewise, involved movement assessment on a person who demonstrates average screening values might uncover specific findings that statistically have little predictive significance. Remember, perfection is not the goal of movement screening—screening is designed to identify deficiency. The significant finding is in the displayed deficiency.

We use screens for grouping and predictive value, and to forecast risk. Screening can also provide indications associated with increased performance, but the first order of business is risk. Screens can also assist with initial program design by directing corrective exercise choices and conditioning exercise choices. When we need extra information, testing should refine the data. This would include situations where more specialized performance or skills are required and demonstrated to have predictive value.

Assessments are more suited to diagnoses than prediction, and are used when screens and tests indicate risk with pain and severe dysfunction. Assessments require unbiased judgment expertise and are more prone to subjective mistakes. Frontloading assessment with screening is the best way to make sure the assessment information is accurate and appropriate. This means it would be better to screen a group for risk than to attempt to assess asymptomatic individuals for potential problems.

In addition to conventional testing, by practicing the screening and assessment concepts appropriately, you can advance your problem-solving and decision-making abilities. You'll become a better thin-slicer in areas involving human movement appraisal and correction.

OVERVIEW OF THE SYSTEMS

It is time to outline movement screening and movement assessment as tools for your use as an exercise or healthcare professional. The screening and assessment systems introduced are not complete evaluations, but are the qualitative component intended to direct the remaining information collection. They work within the conditioning or healthcare intake plan, and should be used at the beginning of your client, athlete or patient relationships. These systems give direction, clarity and priority to the cause and severity of movement limitations, asymmetries and movements that induce pain. The systems will automatically execute a consistent movement appraisal in a logical manner if implemented correctly. You can also use these throughout the training or therapy relationships to demonstrate progress and revisit remaining limitations.

MOVEMENT SCREENING

Pre-exercise screening system

The movement screen—the FMS—will deal with screening the movement risk as it pertains to exercise, sport and increased activity. The screen can also provide information regarding exercise program design based on movement patterns. Other facets of exercise risk are at the complete discretion and responsibility of each fitness professional. You'll still perform quantitative measurements, performance measurements and skill measurements. You are still responsible for looking at contra indications and risk factors for exertion before exercise separately from the screening.

MOVEMENT ASSESSMENT

Clinical movement assessment used prior to corrective exercise in a rehabilitation setting

The movement assessment—the SFMA—will help separate pain-provoking movement patterns from dysfunctional patterns. Once the movement assessment is complete, clinicians must collect appropriate impairment measures such as strength, range of motion and balance before making a diagnosis, or suggesting treatment and corrective exercise. You as a clinician are still obligated to conduct a complete appraisal of the patient's medical history, current patient status, neurological and vascular clearing and any special tests, as well as appropriateness for movement assessment.

The resulting information will expedite the evaluation and treatment of movement dysfunction related to orthopedic and musculoskeletal issues. As the healthcare professional, you are still responsible for identifying relationships between functional movement assessments and impairments as they relate to pain and movement dysfunction.

The SFMA is not necessary for patients with acute debilitating pain, post-surgical patients with movement restrictions or patients who will not participate in corrective exercises. The SFMA can be performed on intake of a new patient or performed prior to corrective exercise intervention.

OUTLINE OF THE FUNCTIONAL MOVEMENT SYSTEM

This overview may not be completely clear until you are familiar with the screening and assessment systems, but it will help you understand each system's purposes. Once you've reviewed the forthcoming detailed screening and assessment section, revisit this outline. The second time you read it, you will have a greater understanding of, and confidence in, the systems.

THE BASIC MOVEMENT SYSTEM— FUNCTIONAL MOVEMENT SCREEN: FMS

We first screen functional movement patterns using the seven tests of the FMS as well as three clearing tests. If you're a fitness professional and you note pain in one or more of the tests, refer the client for a clinical assessment by a healthcare professional. When referring, look for a clinician who understands the FMS. If they also have a clear understanding of the SFMA, this will provide even greater benefit by systematically breaking down the movement pattern that produced pain.

- Rate and rank movement patterns for those without pain based on limitations and asymmetries using the grading system you'll learn in Chapter 6.

- Identify or uncover painful movement patterns prior to increasing activity level.

- Identify the lowest ranking or most asymmetrical movement pattern within the group. If more than one is present, pick the most primitive pattern.

- Look for activities and training habits that could perpetuate the movement problems identified, and suggest a temporary break from the offending activities.

- Initiate a corrective strategy linked to the chosen pattern and make sure the client can understand, tolerate and execute it.

- After the initial application, or a series of applications as needed based on the situation and response, revisit the limited test in the screen and compare it to the previous baseline.

- If the baseline is positively affected, continue with the chosen strategy. If not, recheck the FMS score and the most-limited or lowest-ranking test. Repeat tests if necessary.

- Pay attention to detail during corrective exercise to guarantee you're properly coaching the movement.

- Make sure the progression is effective. Progressing too fast can lead to additional compensation, but progression too slow might not allow for the necessary correction.

- If you observe a change in the lowest-ranking test, perform a second FMS to establish normal parameters or to establish a new pattern-correction priority.

Once the screen's score reaches an acceptable level, formulate an exercise and training plan that minimizes the need for corrective strategy, yet maintains acceptable movement and performance.

THE CLINICAL SYSTEM— SELECTIVE FUNCTIONAL MOVEMENT ASSESSMENT: SFMA

Clinicians use this system in the presence of pain to assess the basic movement patterns of the SFMA. This is outside the fitness professional's scope of practice, since the findings of the SFMA will require some form of professional rehabilitation and treatment. For this reason, we use the word *client* when discussing the FMS, and we use *patient* when the presence of pain indicates the need for use of the SFMA.

- Break down the dysfunctional, non-painful movements—what we call *find the path*—and the functional but painful movements—what we call *find the markers*. Break down painful movements after dysfunctional movements to reduce unnecessary pain provocation.

- Do not break down the normal, non-painful movements.

- Only break down dysfunctional, painful movements if the first two breakouts are not available for review, are not possible or are uninformative.

- The dysfunctional, non-painful breakout should show mobility or stability problems. Observe limitations and asymmetries with changes in load and unilateral inspection.

- Perform impairment measures within the anatomical region to clarify and confirm findings—manual muscle testing, range of motion testing, ligamentous stress testing, neural tension signs, joint mobility assessment and soft-tissue assessments.

- Break down the functional but painful movement pattern to observe the behavior of pain and movement with changes in load and unilateral inspection. Note the lowest level where pain is present or where pain is no longer provoked.

- Formulate a working diagnosis based on the information collected from the dysfunctional, non-painful movement breakout and impairment measurements.

- Uncover daily activities, work activities and exercise habits that could perpetuate the identified movement pattern dysfunctions. Suggest temporary breaks from the offending activities.

- Perform treatment and corrective strategies based on the functional diagnosis.

- Recheck your information, and recheck the impairment measures looking for changes.

- Re-evaluate pain-provoking movement breakouts. If the lowest breakout has changed, move back up through the breakouts toward the basic pattern or patterns.

- Re-evaluate dysfunctional movement breakouts. If the lowest breakout has changed, again move back up through the breakouts toward the basic pattern or patterns.

- If the baseline is positively affected, continue with the chosen strategy. If not, recheck the data and breakouts.

Once the SFMA performance is acceptable and pain is resolved, use the FMS to gauge risk with increased activity and potential injury recurrence. By performing the FMS at or near discharge, the clinician can formulate exercise and training plans that minimize the need for corrective strategies, yet maintain acceptable movement and performance. If the FMS findings suggest continued risk, a plan can be devised to efficiently minimize risk.

CREATING FUNCTIONAL MOVEMENT STANDARDS

Today's exercise professionals should be familiar with current rehabilitation standards. Fitness professionals must also consider restoration, system balance and corrective exercises as a precursor to fitness, and not assume general activity will correct movement problems. Likewise, rehabilitation professionals must not believe that the absence of pain is the primary criteria for patient progress and the discontinuation of service. Clinicians should consider pain and function together. It is also advisable to update your general exercise and fitness knowledge.

Functional restoration is equally important and is a predictor of long-term success. The best way to set functional standards is to resolve strength and range-of-motion issues, and also to understand functional movement pattern standards and agree on acceptable minimums.

In refining theory related to the muscle testing, Florence Peterson Kendall and Elizabeth Kendall McCreary, the developers of graded manual muscle testing in the United States, did not look at muscle weakness or pathological muscle contraction to create protocols for strengthening. They first looked at normal contractile qualities and muscular actions and mapped out as many individual muscles as possible for testing, treatment and exercise purposes. Goniometric studies of asymptomatic subjects suggested isolated joint range-of-motion values. Their observations of normal contractile quality and joint movement provided a baseline for establishing sound muscle-function goals. This also allowed for assessment and grading of dysfunction noted in an isolated muscle group.

Throughout the history of rehabilitation, clinicians have drawn upon the data provided by normal populations. Movement-pattern information from normal populations greatly influences the way clinicians manage patients. As healthcare professionals, we standardize impairments in strength, range of motion and balance measures with those of normal populations. But we should also consider whole movement patterns as a standard as well, because it is possible to demonstrate strength, range of motion and balance, and still display basic movement pattern limitation and asymmetry.

Testing and observing normal subjects defined the information we now employ regarding balance, proprioception and equilibrium. Despite our reliance on data derived from normal populations, there is a lack of information related to normal functional movement patterns. Why not go to normal asymptomatic populations for movement-pattern information and baselines?

The past 20 years have seen the rehabilitation profession move from a traditional isolated strengthening approach toward an integrated functional approach that incorporates the principles of PNF, muscle synergy and motor learning. Exercise professionals have also migrated toward functional approaches and away from muscle isolation.

To further the trend, we have to describe optimal movement in normal individuals in order to develop functional exercise programs and corrective exercise strategies. It is difficult to develop and refer to protocols and programs as functional when a functional movement standard does not exist. Most protocols are established on isolated, objective evaluation techniques such as muscle testing, joint-integrity testing and range-of-motion measurements.

Strength coaches and personal trainers have made substantial efforts to develop functional exercise models that improve pushing, pulling, squatting, lunging and single-leg stance movements without understanding normal movement values. This is great progress, but before proceeding with conditioning programs, functional patterns should be reinforced and dysfunctional patterns should be corrected and improved.

An understanding of movement baselines for an individual or a group will demonstrate which patterns are functional and which are dysfunctional. By documenting the functional movement patterns of active, highly functioning and injury-free people, we gain a greater understanding of ideal movement.

It is common for exercise and rehabilitation professionals to perform specific testing and training for sport movement and occupational tasks without first looking at basic functional movement. Without investigating movement patterns in a systematic manner, we are assuming these are normal, but our experience and initial research[4,5,7] demonstrates they are not, even in fit and active populations. It is important to examine and understand basic aspects and common denominators of human movement and to realize these are common throughout many activities in a variety of applications.

The central goal of this book is to push you to look at movement, to trust your eyes and to look at basic patterns of human movement, not so much to ponder movement within the confines of a chosen field of study or specialized discipline, but to just lean in and look.

Unfortunately, movement is often considered with abstract and isolated measurements or even by computer representation. Some of us have developed scientific detachment from authentic movement, while others have become so dedicated to a singular exercise or rehabilitation perspective that they only see movement within the limits and methods of the preferred discipline.

The two systems in this book are tools designed to force your eyes to capture movement patterns and to imprint your brain with a new perspective, a complementary perspective provided by removing a filter you may not even know is there. Our decisions about exercise and rehabilitation do not currently consider movement pattern behaviors against a standard. Most of us have developed trust in our conventional movement appraisal systems, but we must admit they are incomplete and could be improved.

We convert dynamic three-dimensional movement into language that comfortably fits our perspectives, but in this conversion, we unfortunately lose the bigger picture. Coincidentally, the two systems in this book are also dependent on language, but they force us to face discrepancies and inconsistencies between the many ways we view and describe human movement.

Used correctly, they will help create balance by blending our technological and specialized measurement tools with practical and behavioral tools to achieve a superior perspective. When discrepancies arise between isolated measurements and whole movement patterns, it forces us to dig, investigate and explain them.

We will become experts in the process.

Please see www.movementbook.com/chapter3 for more information, videos and updates.

MOVEMENT SCREENING

The Functional Movement Screen (FMS®) serves as a tool for risk management. Its role is to pinpoint areas of movement-pattern limitation and asymmetry. Fitness trainers and active populations still largely overlook these deficiencies until problems arise, even when these are often associated with risk. The ability to predict injury risk is equally as important as the ability to evaluate and treat injuries. The information gained by screening different groups will provide much needed data toward what is acceptable and not acceptable to the development or decline of functional movement patterns.

Screening can also provide information of cultural- and activity-specific influences on functional movement patterns. The development of norms can alert us to people who deviate from the average. Movement screening can track qualitative deviations within patterns that pertain to limitations and asymmetries, while performance testing can track quantitative deficiencies relating to athletic parameters.

Used together, movement screening and performance testing create a more complete human function perspective than either alone can provide. This union offers the age-old balance of quality and quantity where one without the other might undermine efficiency or effectiveness. Scientific investigation has a bias toward performance and quantities, so we have a professional responsibility to consider quality regarding teaching and training human movement.

For example, instead of developing conditioning programs and balance exercises for the elderly, we should first consider those who have deficiencies in movement quality compared with those who have quantitative deficiencies in strength, flexibility and endurance. The corrective programs for each will differ greatly.

As another example, poor landing mechanics can be one explanation of the increased prevalence of young female ACL injuries. Jump training clinics and programs would seem to be the obvious remedy if we only consider the performance issues of jumping. However, if we separated the group of young women with poor jumping mechanics into two groups, we could devise a more specific corrective strategy.

Movement-pattern problems—basic deficiencies in mobility and stability causing limitations and asymmetries in one or more basic movement pattern or patterns

Athletic-performance problems—fundamental movement patterns are free of deficiencies, but deficiencies are noted in athletic fitness parameters and movement skills related to a specific sport

It is not necessary to identify a third group who possesses both movement-pattern and athletic-performance problems because those athletes are part of the movement-pattern problem group.

- Any negative results with athletic performance testing would include the movement deficiency.

- Any positive results with athletic performance testing would demonstrate undocumented compensation, since basic mobility and stability deficiency are present but undetectable by performance testing.

Beginning with a movement baseline is the best way to explain and investigate movement-related problems. Looking at performance when presented with movement problems creates the assumption that basic supporting patterns are within normal limits. We investigate strength, flexibility, endurance or other parameters that lend themselves to obvious solutions, but although these demonstrate movement-quantity applications, they do not represent the quality of movement, which is the base of all movement function. These represent forms of performance on top of a movement platform.

The obvious reason we gravitate to the performance investigation first is that it fits nicely into a performance-based solution we've been trained to provide. For example, if we observe weakness, we add strengthening activities. If we observe tightness, we recommend stretching. We are so busy measuring the movement quantity, we neglect to consider and standardized its qualitative aspects.

We address specific measurable quantities and hope that whole movement patterns spontaneously correct. Yet movement patterns are the performance bedrock we must maintain at a reasonable and functional level before performance analysis and training.

The appraisal of movement patterns as a starting point can help human movement investigators fit the scientific principle attributed to the 14th Century logician and Franciscan friar, William of Occam: *Entities should not be multiplied unnecessarily.*

Called Occam's Razor, the statement can be expanded as a principle for modern scientists and professionals striving for clarity and perspective. Occam is telling us that solutions to problems should not be unnecessarily complicated—the simpler the better.

The movement baseline concept is simple and yet complex in the responsibility it undertakes. To set the movement baseline, we focus only on movement patterns and not on performance. The patterns observed must represent a large portion of available movement relative to practical and probable functional expectation. They should display patterns currently used, patterns considered key to growth and development. The baseline must be reproducible and lend itself to both communication and grading.

If we are to take the Occam advice and keep things simple, we should let the final word fall to Einstein, who warned us that Occam's Razor could cut both ways: *Everything should be made as simple as possible, but not simpler.*

THE FMS RAZOR

A baseline for functional movement must be set and used to gauge basic movement pattern deficiency to apply Occam's Razor effectively to movement-related problems. If a deficiency is present, we have the simplest reasonable explanation for the movement problem, whether that be balance in the elderly or increased ACL injury prevalence in young women. If movement deficiency is not present compared against the baseline, further investigation into the parameters of fitness, performance and neuromuscular function is necessary. We can develop a razor for Functional Movement Systems that states—

Entities of minimum movement-pattern quality should be considered before entities of movement quantity and physical capacity.

This means—

Movement-related problems and deficiencies should first be weighted against a minimum qualitative standard. If a minimum level of quality is acceptable, then—and only then—the quantities and specifics of movement should be considered.

If movement quality does not meet minimal standards, it should be the primary focus against other physical parameters.

The above statement should serve as the Functional Movement Systems Razor. The statement means we should not attempt to improve movement quantities until a minimum level of movement quality is achieved. Functional Movement Systems are simply a method to uphold a qualitative standard. It is our intent that the Functional Movement Screen (FMS®) and Selective Functional Movement Assessments (SFMA®) serve as methods and place markers for fundamental movement quality until they are no longer needed or are replaced by something more comprehensive, efficient and effective. For now, these systems prove a valuable and reproducible perspective that can be implemented with efficiency and effectiveness. They also help connect current exercise and training practices to the developmental roots of human movement.

During growth, a young person develops perceptions through reflexive movements that perform basic motor tasks. The progression occurs because of maturation and learning, and the development occurs from proximal to distal. Infants first learn to stabilize the joints in the spine and torso, and eventually develop the same controls in the extremities. They learn the fundamental movements by responding to a variety of stimuli through numerous degrees of motor control.

The head-to-tail and trunk-to-extremities development process is operational throughout life and has a tendency to reverse itself as we age. The most recently learned activities in the lower body and upper extremities are the first to exhibit regression signs. Movement evolution also occurs as people gravitate toward specific skills and movements through habit, lifestyle, training or all three.

Most of us migrate toward and routinely repeat specific movement skills. That skill training may perform a secondary role by maintaining general fitness; however, those *specific* movement patterns may not maintain a balance of the *basic* movement patterns.

The fire service industry illustrates this well. Firefighters constantly train certain movements for improved performance. They initially train through voluntary movements, and as they repeat the movements, these become stored as central commands leading to subconscious performance of the tasks.

Subconscious performance involves cognitive programming, the highest level of central nervous system function. However, problems arise with the inefficient or asymmetrical performance of the training. Even if movement patterns are adequate, if the skill predominates one pattern or is asymmetrical, fundamental movement dysfunction can occur. Likewise, practicing movement patterns in the presence of fatigue as is often done during this type of training can also compromise basic movement.

Training skills using compromised movement patterns compounds the effect. An example would be the firefighter who lacks the appropriate balance of mobility and stability to perform tasks such as the hose drag, stair climb or fireman carry. This firefighter will perform the tasks using compensatory movements to overcome these stability and mobility deficiencies. The compensatory pattern develops during the training and as this happens, the individual creates a poor movement pattern used subconsciously whenever executing the future task. This can lead to greater mobility and

stability imbalances and deficiencies, all risk factors for injury.[24]

These tendencies are present in all highly active or exercising populations. There is an underlying assumption that general activity, exercises and athletic endeavors will improve movement, but this cannot be observed objectively until we develop a movement standard. Although the exercise or sport task might create a conditioning effect on the body's energy systems, the initial movement compensations are reinforced, rather than reduced or removed. In this situation, it is possible to add generalized or specific fitness to basic movement dysfunction.

In many cases, movement patterns are lost due to muscle imbalances, habitual asymmetrical movements, improper training methods and incomplete recovery from an injury where a compensatory movement pattern continues dysfunctional activity. Correcting these issues rarely results in spontaneous pattern reconstruction. It is a general human tendency to migrate toward one or two preferred patterns instead of an equal balance of patterns.

People who have suffered from an injury will have a decrease in proprioceptive input if the injury is left untreated or is treated inappropriately. A disruption in proprioceptive performance will have a negative effect on movement-pattern behaviors. This results in altered mobility, stability and asymmetric influences, eventually leading to compensatory movement patterns. This may be a reason prior injuries are one of the more significant risk factors predisposing people to injuries.[25-32]

One issue of rehabilitation may be that a full functional movement appraisal is not part of the discharge criteria. The initial focus in management of an injury is the reduction of symptoms and pain control, but the absence of pain and the resolution of symptoms do not indicate functional movement restoration. It is entirely possible to enjoy pain-free dysfunctional movement, which leaves the patient at risk of re-injury.

Standardized functional movement screening is one way to rate and rank limitation and asymmetry in the asymptomatic patient before discharge from rehabilitation. We could then take measures to resolve movement dysfunctions that indicate increased injury risk.

Possible solutions—

- Continue in rehabilitation with an alternate or secondary diagnosis

- Progress into a post-rehabilitation program

- Continue with supervised corrective exercise in a fitness facility

- Work in-home toward movement goals with a qualified personal trainer

- Work on a self-administered corrective exercise program with follow-ups and rechecks

Insurance companies will not reimburse for functional movement restoration when the patient is pain-free until researchers can provide overpowering evidence that movement dysfunction is an appreciable risk factor for injury. Other healthcare professionals screen for and treat risk factors as commonplace, but this is not yet true with musculoskeletal issues and rehabilitation.

We should create alternate reasonable solutions such as post-rehabilitation programs. Fitness and wellness centers have a bias toward cardiovascular health goals, but with education and training, the industry would embrace corrective exercises that focus on functional movement goals once standards are developed.

When reviewing previous injuries or strength and flexibility imbalances, it is difficult to determine which risk factor has a larger influence on injury. In either case, both lead to deficiencies in functional mobility and stability, and both lead to pain, injury and decreased performance.

Researchers Cholewicki and Panjabi[33] found that limitations in stability in the spine led to muscular compensations, fatigue and pain. It was also determined that spinal instabilities resulted in degenerative changes due to muscle-activation strategies, which can be disrupted because of previous injury, stiffness or fatigue.

Additionally, people with previous low back pain episodes performed timed shuttle runs at a significantly slower pace than those who did not have back pain history.[34]

We can see that an important factor in both preventing injuries and in improving performance is to identify deficits in mobility and stability, both of which influence the creation of altered motor programs throughout the kinetic chain. The movement pattern complexity makes it difficult to evaluate weaknesses using conventional static methods, so we use functional tests that incorporate the entire kinetic chain to first identify movement deficiencies.

THE EFFECT OF INJURY ON MOVEMENT

The study described in the next paragraphs is one of many that direct our observation to patterns and functional relationships and keep us from developing professional tunnel vision.

"The likely influence of a localized injury in a distal joint on the function of proximal muscles is an important consideration in assessment and treatment of musculoskeletal injuries. However, little experimental evidence in humans exists in this area. Accordingly, a controlled study was carried out in which the function of muscles at the hip was compared between subjects who had suffered previous severe unilateral ankle sprain and matched control subjects. The pattern of activation of the glute maximus, the hamstring muscles and the ipsilateral and contra lateral erector spinae muscles was monitored through the use of surface electromyography during hip extension from prone lying.

"Analyses revealed that the pattern of muscle activation in subjects with previous injury differed markedly from normal control subjects, and that changes appeared to occur on both the uninjured and the injured sides of the body. A significant difference between the two groups was the delay in onset of activation of the glute maximus in previously injured subjects. The existence of remote changes in muscle function following injury found in this study emphasize the importance of extending assessment beyond the side and site of injury."[35]

The resulting proximal change could be an inhibitory protective response as a reflex measure to reduce further injury. It could also be reduced or altered proprioception at the distal injury site. The missing information could result in lowered coordination of the proximal musculature during functional activity. Lastly, the change could come from pain associated with the original injury; even though the injury is resolved, a residual inhibitory effect remains as part of the pain response.

Any of the above three could be a plausible explanation of what Vladimir Janda and his partners observed in that study. However, the fact that he discovered it is of more relevance than the reason for the discovery. He expanded the criteria for complete rehabilitation to include the proximal secondary effects of a primary distal injury.

I developed the FMS to identify people who have compensatory movement patterns in the kinetic chain. During screening, we accomplish this by observing right- and left-side imbalances, and mobility and stability deficiencies. The seven movements in the FMS challenge the body's ability to facilitate movement through the proximal-to-distal sequence. This course of movement allows the body to produce movement patterns more efficiently. Once the FMS isolates the most dysfunctional, asymmetrical or inefficient movement pattern, you as either the trainer or clinician can institute the corrective exercise strategies to circumvent problems such as imbalance, compensation, micro-traumatic breakdown and injury.

WHERE DOES A MOVEMENT SCREEN BELONG?

Does the movement screen replace the physical examination by a physician, or is it a physical performance test designed to gauge fitness and conditioning? Where does it go in the hierarchy of examinations and physical testing that responsibly precede physical activity?

The logistics of where to place and how to use the movement screen are actually simple once you step back to review a program to increase, modify or maintain activity. The movement-screening tool fills the void between the pre-participation medical examination and the performance-based tests done on those intending to participate in physical activities. The performance test and skill test are grouped together as physical capacity tests.

Some examples—

Personal fitness or wellness client—The client will get a physical examination by a physician and probably undergo some type of endurance testing to appraise cardiovascular fitness.

High school football player—The athlete will receive a physical examination by the team physician and then undergo strength and speed testing, as well as position-specific drills and tests.

A firefighter—The firefighter has a physical examination by a designated physician and will then undergo job-related physical and obstacle-course testing.

The examples can continue, and all perpetuate the assumption that medical clearance and adequate physical performance are the only issues we should observe. Movement-related issues that demonstrate risk could go undetected in these examples and others like them.

Another assumption is that poor performance is an issue only remedied by further conditioning. Poor movement and body mechanics can create increased energy expenditure and unnecessary physical load, often incorrectly measured as poor physical conditioning. In this scenario, movement dysfunction can go undetected and can result in a poor fitness or performance score. The athlete may aggressively pursue conditioning that will most likely perpetuate the problem and further ingrain movement dysfunction.

The pre-participation medical examination is usually recommended or even required before athletic activities, exercise and occupations involving above-average exertion. The pre-participation medical examination is intended to rule out serious medical problems that might interfere with physical exertion and physical tasks. The focus is general health and the absence of problems in the cardio-respiratory and other vital systems, doing little or no movement screening. The pre-participation medical examination is not a movement appraisal since movement is not the focus, and therefore it does not rule out movement dysfunction, limitation or asymmetry.

Performance testing assumes movement is adequate and moves directly into physical capacity testing. This testing looks at capacities for strength, endurance, coordination, agility and specific abilities. It's routinely performed in athletics to set conditioning and skill baselines, and in fitness for generalized physical capacity to gauge and grade status and to set goals. In occupational settings such as the military, testing gauges minimum levels of physical condition and performance standards. This insures safety and physical preparedness during training and on the job.

The medical exam assures freedom from serious, identifiable medical problems, disease and disability. That is the most important factor to consider when people are cleared for activity.

A movement screen fits *between* the pre-participation medical examination and the performance testing. Each stage represents a step in a physical hierarchy. This process assumes general health following the pre-participation medical exam, but it does not take for granted complete and acceptable functional movement patterns. It is possible to have good health and still move poorly. The movement screen will check for pain with movement by evaluating the person in a dynamic and functional capacity without considering performance, which is evaluated later.

Mobility and stability problems are detected, and basic movement limitations and asymmetries are observed by screening movement. This means any problems seen later in physical performance testing can be considered performance problems when the movement screen is clear. In contrast, problems on movement screening may greatly affect performance—yet these are not correctable by working on performance, because they are movement dysfunction. Without a movement screen, poor performance measures may not be correctly represented. Professionals must detect movement problems before performance testing to confirm they are checking *movement capacity* and not *movement ability*.

The responsible hierarchy of considering function—

Physical examination or medical screening for basic health—proper vital systems functioning

Functional movement screening for basic movement—fundamental movement capabilities

Performance testing for general fitness and athleticism—basic performance, physical capacity for power, endurance, coordination, strength, speed and more

Skill testing for specific performance—specific physical capacity for a specialized activity

We can develop these attributes in parallel. It is not necessary to have perfect health before working on basic movement, nor is it necessary to have perfect movement before developing some degree of physical capacity or performance. We don't have to maximize basic performance before developing specific-performance skills; each complements the other. The goal is to manage minimums at each level, not optimization and perfection at every level, and improvement sought at one level shouldn't create a deficiency in another.

MOVEMENT SCREEN OUTCOMES

Certain health concerns outweigh any basic movement or functional movement goal.

A broken bone needs to be set and cast—**this is basic health.**

Once we see healing and alignment, we look to movement—**this is basic movement.**

When basic mobility and stability are in place, we seek general endurance, strength and speed of movement—**this is basic performance.**

Finally, having satisfied general functioning of muscle and joint quality and quantity, we develop or redevelop specific movement skills—**this is specific performance.**

I use this simple analogy of the broken bone because a standardized movement appraisal does not exist, and those in the exercise and rehabilitation professions constantly confuse the hierarchy. We have set medical minimums and even performance and skill minimums, but we have yet to set and use movement minimums.

I designed the movement screen to capture pain provoked by functional movement patterns that we may not see in daily life or during basic medical screening. Often people learn to avoid problematic movement—the screen pushes the functional movement extremes through multiple positions to uncover underlying issues that may have gone undetected.

If you don't see pain in the movement screen, but you do see serious limitation or asymmetry, the limitations are a deficiency until rectified. All other progressive attempts at basic performance and specific skill will fall victim to the poor efficiency of a body working around a movement problem.

A movement screen can have three basic outcomes—

Clients with pain provoked by movement will require further medical assessment. *Move to the SFMA or get these people diagnosed by a healthcare professional.*

Give clients who demonstrate movement dysfunction corrective exercises to resolve dysfunction. These people should avoid much generalized and specific conditioning until the movement screen denotes movement patterns not associated with elevated injury risk. *Work them here until a change is noted.*

People demonstrating movement patterns not associated with elevated risk of injury should be cleared before pursuing conditioning past their current physical fitness. *Move them into conditioning, but periodically recheck with medical physicals and movement screening.*

This is a new movement-screening paradigm.

PRE-PARTICIPATION AND PERFORMANCE REVIEWS

Historically, the sports medicine model suggests pre-participation physicals followed by performance assessments. This systematic approach doesn't provide enough baseline information when assessing an individual's preparedness for activity. Commonly, the medical pre-participation fitness examination only includes information such as vital system checks and screens for disease that will exclude a person from participating in certain activities. In these, there is little consideration for functional movement patterns since these are not part of a conventional medical examination.

The perception of many researchers was that there were no set standards in determining who is prepared functionally and physically to participate in activities. Recently, numerous medical societies have collaborated to establish more uniformity in this area, however, this only provides baseline medical information. There should also be collaboration in determining the baseline for basic movement and deciding whether people should be allowed to participate if they are unable to perform movements at this fundamental level.

In the traditional system, performance tests follow the pre-participation physical. Common performance tests include situp and pushup endurance, strength measurements, endurance runs, sprints, agility activities and other quantitative measurements of productive physical capacity. In many athletic and occupational settings, these performance activities become more specific to the tasks of defined performance areas.

Performance tests customarily gather baseline quantitative information, and then attempt to establish goals and make recommendations. These are recommendations based on standardized normative information, which may not be relative to the individual's specific needs.

Likewise, in many cases, performance tests provide objective information that fails to evaluate the efficiency with which people perform certain movements. Little consideration is given to functional movement deficits that can limit performance or predispose the participant to micro-traumatic injury. Prescribed strength and conditioning programs often work to improve agility, speed and strength without consideration of basic movement-pattern quality or efficiency.

For example, a person has an above-average score on the situp test, but with poor quality and inefficiency, compensates by initiating the movement with the upper body and cervical spine instead of with the trunk. Compare this person to an individual who scores above average and is performing very efficiently and who doesn't use compensatory movements. These two would both be above average with no notation of movement inefficiencies.

If we see major deficiencies in functional movement patterns, should we judge these performances as equal? These two would likely have significant differences in functional mobility and stability, but without assessing, we can't assume these differences.

The goals in performing pre-participation or performance screenings are to decrease injuries, enhance job performance and ultimately improve quality of life. Today's research is inconsistent on whether the pre-participation or performance screenings and standardized fitness measures have the ability to achieve those goals. The standardized screenings do not provide individualized movement analysis, but if you incorporate the FMS into your pre-screenings, you will be able to determine who has the ability to perform certain essential movements.

Observing the responses of a symptomless client who has difficulty performing functional movement patterns is an important lesson for any exercise and rehabilitation professional. By providing movement screening for active populations such as laborers, firefighters, athletes and other highly fit people, a wealth of opportunity exists to gain much-needed knowledge of functional movement patterns.

The techniques developed to restore functional movement patterns in an asymptomatic population will benefit the therapeutic protocols used to restore patterns in a symptomatic client. Exercise professionals now implement the FMS with all levels of fit individuals and athletes, as well as in military personnel, firefighter and other industry groups. Because it examines a missing piece of the health and human performance puzzle, the feedback regarding the screening has been positive and valuable.

*Please see www.movementbook.com/chapter4
for more information, videos and updates.*

5

FUNCTIONAL MOVEMENT SYSTEMS AND MOVEMENT PATTERNS

When you review the perspectives, descriptions and analogies of movement, you'll discover the Functional Movement Screen (FMS®) creates a systematic way to observe movement patterns. Not measuring in the normal sense of the word, with this system we rate and rank the patterns on a numerical scale to focus on significant limitations or asymmetries.

We can further investigate deficiencies in these patterns, but many subtle factors underlying a poor movement will correct themselves if we just focus on the pattern and not the parts. That is not to say we'll be able to fix a significant undiagnosed injury or serious abnormality by this approach, but it's quite effective for many movement problems.

Unlike the Selective Functional Movement Assessments (SFMA®), the FMS does not have a formal breakout or built-in movement reduction for each pattern because it is not a diagnostic system. Its role is to impose minimum standards on movement patterns in active populations. Attempting isolated diagnosis would create an extra step without offering greater corrective solutions, and could even offer fewer options in some cases. In the FMS, there's no need to identify anything other than a limited or asymmetrical pattern.

Correct instruction and attention to detail will demonstrate asymmetry, as well as significant limitations in mobility and stability. You must attend to the limitations and asymmetries in your programming until they no longer limit the pattern or patterns. The goal of the FMS is to resolve asymmetry and serious limitations, which are identified by a score of one.

The SFMA offers breakouts for each movement pattern; people see this and assume it means more power and effectiveness. The experienced user of both systems sees the corrective sequence in the FMS has many parallels to the SFMA. The FMS philosophy is simple: *We don't spend time in movement breakouts when we know correction is needed. The corrective sequence will expose the level.*

The SFMA offers less-intensive movement patterns than the FMS. The lunge, pushup and rotary stability in the FMS offer a difficulty not needed for the purposes of the SFMA. The SFMA does not use an ordinal scale, and therefore requires increased observation skill.

It is no accident that both systems require the observer to make one of four assignments to a movement. In the FMS, we assign a score from zero to three to each movement pattern, thereby providing four options. In the SFMA, we use one of four categories based on the presence or absence of pain or dysfunction—again four options. Keeping the number of assignments low improves reliability and consistent interpretation.

The SFMA navigates the musculoskeletal assessment when pain is present. It is helpful during the initial patient examination, although some acute problems make it impractical at the outset. Outside of exposing dysfunctional regions that may complicate the examination process, the SFMA offers a unique perspective for corrective exercise in a clinical setting.

The SFMA does not implicate the painful site as dysfunctional unless it proves to display actual dysfunction. Likewise, dysfunction may be present in movement patterns and body segments uncomplicated by pain. The addition of the SFMA forces a global approach to movement and refines the corrective exercise choices to maneuvers directly related to documented dysfunction.

At or near the end of the rehabilitation process, we again use the FMS to demonstrate risk of future injury. The necessary criterion for the FMS is simply the absence of pain with movement. As long as movement-related pain is present, the SFMA is the preferred tool for decisions regarding corrective exercise and functional progressions. As we begin our discussion of the two systems, a visual overview of what's to follow will be beneficial.

FMS

DEEP SQUAT | HURDLE STEP | INLINE LUNGE
SHOULDER MOBILITY REACHING | ACTIVE STRAIGHT-LEG RAISE
TRUNK STABILITY PUSHUP | ROTARY STABILITY

Deep Squat

Hurdle Step

Inline Lunge

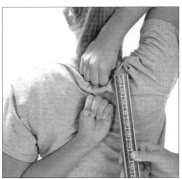
Shoulder Mobility

Active Straight-Leg Raise

Trunk Stability Pushup

Rotary Stability

SFMA TOP-TIER ASSESSMENTS

CERVICAL SPINE | UPPER EXTREMITY MOVEMENT PATTERN
MULTI-SEGMENTAL FLEXION | MULTI-SEGMENTAL EXTENSION
MULTI-SEGMENTAL ROTATION | SINGLE-LEG STANCE | OVERHEAD DEEP SQUAT

Cervical Spine

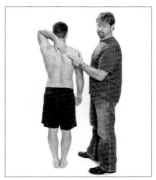

Upper Extremity

Multi-Segmental Flexion

Multi-Segmental Extension

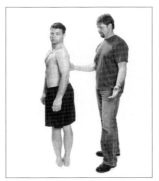

Multi-Segmental Rotation

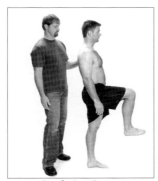

Single-Leg Stance

Overhead Deep Squat

SFMA BREAKOUTS

Cervical Spine

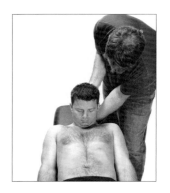 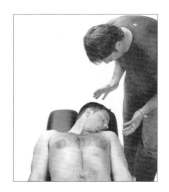 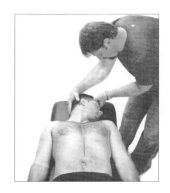 

Active Supine Cervical Flexion, page 138 | Passive Supine Cervical Flexion, page 139
Active Supine OA Cervical Flexion, page 139 | Active Supine Cervical Rotation, page 140
Passive Cervical Rotation, page 140 | C1-C2 Cervical Rotation, page 141 | Supine Cervical Extension, page 141

Upper Extremity

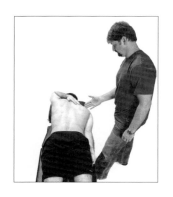 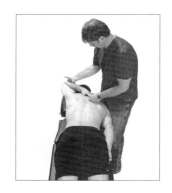 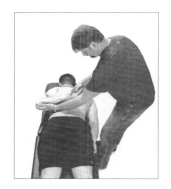 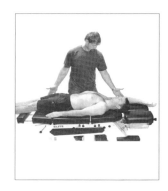

Active Prone Upper Extremity Patterns, page 142 | Passive Prone Upper Extremity Patterns, page 143
Supine Reciprocal Upper Extremity Patterns, page 144

Multi-Segmental Flexion

 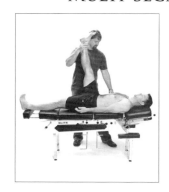 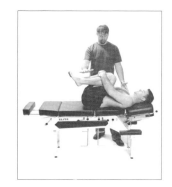 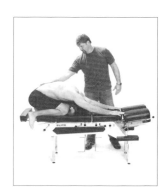

Single-Leg Forward Bend, page 145 | Long-Sitting Toe Touch, page 146 | Prone Rocking, page 146
Active Straight-Leg Raise, page 146 | Passive Straight-Leg Raise, page 147 | Supine Knees-to-Chest, page 148

Multi-Segmental Extension

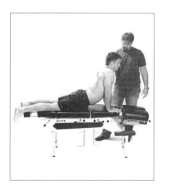  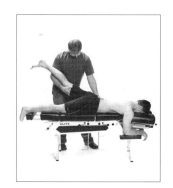 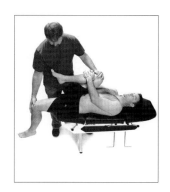

Spine Extension

Backward Bend without Upper Extremity, page 153 | Single-Leg Backward Bend, page 153 | Press-Up, page 154
Lumbar-Locked (IR) Active Rotation/Extension, page 154 | Lumbar-Locked (IR) Passive Rotation/Extension, page 155
Prone-on-Elbow Unilateral Rotation/Extension, page 156

Lower Body Extension

Standing Hip Extension, page 157 | Prone Active Hip Extension, page 157 | Prone Passive Hip Extension, page 158
FABER Test, page 158 | Modified Thomas Test, page 159

Upper Body Extension

Unilateral Shoulder Backward Bend, page 160 | Supine Lat Stretch, Hips Flexed, page 161
Supine Lat Stretch, Hips Extended, page 162 | Lumbar Locked (ER) Unilateral Extension/Rotation, page 162
Lumbar Locked (IR) Unilateral Extension/Rotation, page 163

Multi-Segmental Rotation

 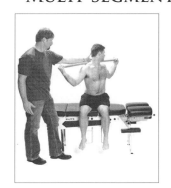 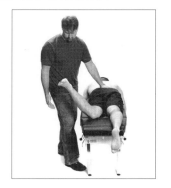 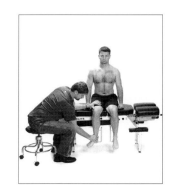

Limited Multi-Segmental Rotation

Seated Rotation, page 167 | Lumbar-Locked (IR) Active Rotation/Extension, page 154
Lumbar-Locked (IR) Passive Rotation/Extension, page 155 | Prone-on-Elbow Unilateral Rotation/Extension, page 156

Multi-Segmental Rotation, Continued

Hip Rotation

Tibial Rotation

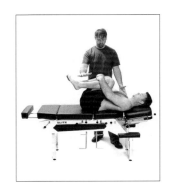
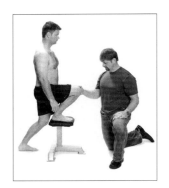

Single-Leg Stance

Vestibular and Core

Ankle

Overhead Deep Squat

Rolling Patterns

THE CONCEPT OF SCREENING

Functional Movement Systems can provide many benefits in your work, but the central objective is to remove unsupported judgments regarding movements fundamental to an active lifestyle.

Professionals with all levels of experience form unsupported assumptions regarding the effects of exercise and rehabilitation on movement. This subjectivity is only a problem if as professionals we do not acknowledge it. Studies have demonstrated how professional confidence can sometimes be unfounded.[36] Systems will help us all maintain our objectivity.

The FMS and SFMA are not complete evaluations; they complement other systems of movement appraisal. The FMS provides a basic perspective against performance and activity-specific information. The SFMA provides separation of painful movements and dysfunctional movements, and offers a full movement perspective against isolated impairment measurements and other testing procedures.

Most of the moves in the FMS, and some from the SFMA, are associated with exercises of the same name. However, this similarity should in no way indicate proficiency or poor performance with exercise. It only suggests that a problem may exist within the pattern, and if the exercise involves any part of the pattern, it potentially compromises the exercise performance.

The seven patterns in either system are interconnected and equally important. No single movement pattern is more or less important than the others. If the explanation of one pattern seems more involved or lengthy, do not let the exposition imply more importance or significance of that pattern over another.

Do all the screen tests, even if you think one might be less important for a client. During the short time you invest in the FMS to get a full perspective of the way a person moves, it's not worth skipping a few tests to save a few minutes. If the test you choose to remove is the one that identifies an individual's weakest link, you will later be inadvertently exercising around the problem and not working on it.

The first three tests of the FMS—the squat, the hurdle step and the lunge—are primarily important, because these demonstrate the representation of core stability in the three essential foot positions humans experience each day. However, the other four tests in the FMS will systematically help refine information, and it is the way in which all seven tests interact upon each other that helps identify the weakest link.

Those who use the FMS or SFMA in professional practice never question which patterns are most important. Only those with no experience in the screen will ask that question, and they ask to save time, without realizing how much they can save by using the system correctly. Remember the 80/20 rule—80 percent of your success is most likely related to 20 percent of your activity. We all need help finding the 20 percent, and that's exactly what the screen in its entirety will do for you.

One theme repeats itself regardless of the person, fitness or performance ability: The pattern with the greatest limitation or most significant asymmetry always seems to encapsulate the problem in a nice little package. This pattern is the weakest link.

Sometimes the pattern contains the reason for the weakest link, like a significant joint restriction, muscle tightness or poor segmental stability. More often, we'll see a tight prime mover and a poor joint stabilizer and a stiff joint, all in the same kinetic chain. Independently working on each problem will often not yield a more normal pattern. However, working on the entire pattern may simultaneously improve the flexibility of the prime mover, the timing and coordination of the stabilizer, and the mobility of the previously stiff joint.

Get used to saying *the pattern is the weakest link*. Within that pattern, you may find some supplementary elements that need individual attention, but never assume that attention will change the pattern, even though it might.

Always work the pattern, recheck the pattern and maintain the corrected patterns.

SCREENING CONSIDERATIONS

We refer to the squat, hurdle step and lunge as *the big three*. Colleagues originally referred to the first three tests in the screen as the big three and the name stuck. The other tests are not of less importance, but as mentioned, these represent the three primary foot positions.

We refer to the active straight-leg raise, shoulder mobility, trunk stability pushup and rotary stability as *the little four*. These represent more primitive function. Problems in these four tests may show or display compensation in the big three to some degree. All the tests in the FMS look at mobility and stability, but there is a bias in the little four toward one or the other. The straight-leg raise and the shoulder mobility tests both have a bias toward mobility, and the other two have a bias toward stability.

Always focus on the little four first, all scores being equal. This is important when a person has low scores across multiple tests. In this situation, a focus on the little four is the safest and most effective way to progress with your corrective strategy.

Another way to think of dividing the seven tests into two categories is by using the labels *functional and fundamental*. This will remind you not to attend to a functional pattern without first addressing a fundamental one. Think of the big three as functional, and the little four as fundamental.

The hurdle step, lunge, active straight-leg raise, shoulder mobility and rotary stability are split patterns because they are asymmetrical. These tests can be observed and appreciated in both a left or right pattern.

The squat and trunk stability pushup are straight-pattern symmetrical tests, since in these there is no opportunity to look at the left and right side of the body independently. When you observe low scores across multiple tests, the symmetrical tests should take a back seat to the asymmetrical, split-side tests.

After screening, you should plan the corrective strategy of a split pattern before a straight pattern. With straight patterns, there is a higher degree of neuromuscular control. The split pattern will allow the corrective exercise strategy to magnify a functional asymmetry with mobility or stability challenges. A faulty straight pattern may actually incorporate an asymmetry from a split pattern. Logically, all asymmetries should be managed before straight patterns are addressed.

Always assume an asymmetrical problem can cause poor mobility or stability in both a straight and a split pattern, but only a split-pattern corrective exercise strategy will capture the problem specifically. The corrective strategies targeted at straight patterns assume no lower-level asymmetries are present in the split patterns.

Let the weakest link guide you until it is no longer the weakest link.

FUNCTIONAL MOVEMENT SCREEN CRITERIA

The FMS grading system and apparatus are basic, effective and reliable.[37-38] The system ranks movements as complete, to be given a score of three; complete with compensation or deviation from the standard, or both, which is scored a two; and incomplete, which receives a score of one.

If you note pain on the FMS on any test, assign a score of zero, which nullifies all other scores. From here, either proceed to the SFMA if you're a licensed clinician, or make a referral for assessment by a healthcare professional.

Most people are aware of acute problems and should not to go through the FMS screening if they have pain. These clients deserve a clinical assessment by a healthcare provider. They are hurt, and they know it, but do not yet have a diagnosis. The next step is to assign a diagnosis.

Likewise, if a client talks of not having pain at rest, but has pain with movement, you should refer the person for assessment.

THE FMS DESIGN

The movement screening system is simple—the design is this screen's power. The screen uses four basic filters that create a system to capture pain and dysfunction. Since the basic patterns we test form the underlying properties of exercise and athletic movements, we suggest using the screen before conditioning and sports training.

Perform the movement screen when introducing clients to situations of increased activity, physical preparation or conditioning. As previously stated but because it bears repeating, the screen does not replace the pre-participation medical examination or physical examination, but instead adds a needed movement component.

PRIMITIVE MOVEMENT PATTERNS

The seven tests of the FMS are broken into two groups, primitive and higher level.

Primitive movement patterns

Basic mobility and stability movement patterns—

- Reciprocal reaching pattern

- Supine alternate leg-raising pattern

Transitional movement patterns in which a higher degree of stability, coordination and control are required—

- Trunk stability pushup pattern

- Quadruped rotational stability pattern

Higher-level movement patterns

- Squatting pattern

- Stepping pattern

- Lunging pattern

Primitive patterns take precedent over higher-level patterns—they support the higher patterns. It is possible to train the higher-level patterns without correcting a problem at a primitive level, but this is not advisable. Compensation and substitution usually occur with higher-level function in the presence of fundamental dysfunction.

The screen correction hierarchy reminds us to work on primitive patterns before higher-level patterns. Within the primitive patterns, we first correct basic mobility and stability patterns before the transitional.

The screen has built-in criteria to rate and rank sub-optimal patterns. You'll rate the movement pattern at three ability levels.

Score of three

The movement pattern is complete and consistent with the FMS test definition. This score changes to zero if pain is present.

Score of two

The completed movement pattern demonstrates compensation, faulty form or loss of alignment as consistent with the FMS test definition. The score again changes to zero if pain is present.

Score of one

The movement pattern is incomplete and was not performed consistent with the FMS test definition. As before, the score changes to zero if pain is present.

THE FOUR BASIC FILTERS

- Pain observed with movement screening

- Limited movement patterns observed with movement screening

- Asymmetrical movement patterns observed with movement screening

- Intentional redundancy, the duplication of inspection to reduce error and to demonstrate consistency or inconsistency within similar movements

PAIN OBSERVED

If pain presents with one or more of the tests within the screen, the screen has done its job—the screen is over. Limitation and asymmetry are not the primary issue. The first rule of movement is this: *Pain changes everything.*

The observation of movement complicated by pain is unreliable at best, and this we investigate with the SFMA. Pain is associated with behaviors that reduce the systematic gathering of objective information. It produces apprehension, inconsistency, magnification behavior, fear and denial.

Does pain cause movement problems, or do movement problems cause pain?

That is impossible to answer without an objective investigation tool or system.

That movement might be objective and reproducible in the presence of pain is not scientifically supported. The research is clear and the implications profound.

When you see pain consistently produced by movement, consider this an early warning sign. The body has the ability to work around pain and to override natural tendencies to give in to the discomfort. This is a great survival mechanism, but only as a temporary alternative when no other option is available. Pain is not the enemy, nor is it even the problem. Pain is simply a signal of invasion, infection, disharmony, misalignment, inappropriate muscular activity, inflammation and compromised structural integrity.

Don't kill the messenger—at least not until you get the message. The message is that pain is a signal to a problem. It's not the underlying problem.

Pain is a biological warning light, alerting you to a chemical or mechanical problem, or both. Covering up pain will not make the problem go away, but will instead reduce the ability to see what can actually affect the signal.

A *chemical problem* would be inflammation, infection, non-mechanical tissue irritation, swelling and effusion.

A *mechanical problem* could involve faulty alignment in joints, limited mobility in the joints and the surrounding tissue, and limited structural integrity or neuromuscular control. A mechanical problem could also include any physical, functional or structural limitation.

Many times the two problems occur together. Initially a sprained ankle will have chemical and mechanical problems. Significant inflammation or chemical pain will be present in the early stages of the injury and, over time, the muscle guarding and joint effusion will cause general stiffness in the tissues. The pain will diminish as time passes, but will still be present in specific movements. At this stage, the ankle is comfortable in mid ranges and painful at the end ranges. The swelling and muscle guarding cause an increase in mechanical tension earlier in the normal range of motion.

The movement screen detects mechanical problems. Placing a person who doesn't complain of pain into the seven tests can help expose most movement-based mechanical problems. If one or more of these tests produces pain, note the movement or movements that caused the pain and the pain location, and make an appropriate referral as needed for assessment of the problem.

Pain, even with one movement screen test, identifies a significant potential for an underlying injury or the increased risk of injury under increased activity. The body is already injured. The person just does not know it yet.

Pain is possible at any level of movement quality. The occurrence of pain has no bearing on the quality of movement observed, but for later reference, it's helpful to note the level of movement quality that produced the pain. Pain at any movement quality should receive appropriate consideration by a healthcare professional, meaning stop the FMS and begin the SFMA or refer out.

The screen's score will provide no significant insight when pain is involved, because it is unclear how pain and movement affect each other. The screen identified pain with movement that was previously undetected, unclear or unacknowledged. The first filter of the screen worked. This screen is over.

LIMITED MOVEMENT PATTERNS OBSERVED

The second filter is for significant limitation with movement. The seven tests in the FMS check and recheck the most common movements used in basic function. The joint motion requirements in each test fall at or slightly less than normal medical and rehabilitation goniometric measurement standards. The only difference is that movement screening requires the client to demonstrate multiple joints working simultaneously within normal ranges. This is how the screen catches compensation. If one segment is not contributing to the movement pattern, another segment must give up some degree of stability or demonstrate excessive mobility to complete the full pattern. This causes obvious deterioration in movement-pattern quality and is captured by the screen.

After you rate each pattern using the scoring criteria described on page 373, rank it within the group of seven patterns. The ranking system targets the most limited movement pattern. Score each movement pattern first, using the score sheet on page 89, and then rank the most limited pattern. The most limited pattern is the lowest score of the seven tests, and it contains the most significant mobility and stability compromises.

The less-efficient alternative is to measure the body's segments individually for normal or abnormal range of motion. Analyzing a single joint for normal measures does not effectively demonstrate that the joint functions normally within a whole movement pattern. This approach implies that the joint should function normally within a pattern, but fails to demonstrate that implication in a functional way.

By looking at movement patterns first, we see which one appears most limited. Once identified, this single pattern embodies the most problematic mobility or stability problem, or a combination of the two. You can then test the client's segments in isolation to show a gradient of mobility and stability at key areas if needed. If these test normal in isolation and poorly in a coordinated effort, isolated normalcy has little realistic implication.

The functional movement screen tests are not equal. Some are complex tasks and some are more primitive. The word *primitive* implies the movement's hierarchy in growth and development. This puts the axiom *crawl before you walk* in its literal interpretation.

ASYMMETRICAL MOVEMENT PATTERNS OBSERVED

Asymmetry suggests unevenness within functional movement patterns, and is the third filter built into the FMS, used to describe both structural and functional problems,. The FMS provides five opportunities to test and observe asymmetry. Of the seven tests, five require independent right- and left-side appraisal.

Historically, medical and rehabilitation professionals investigated structural asymmetry with more diligence and attention to detail than functional asymmetry or movement-based asymmetry. Structural asymmetry might be a leg-length discrepancy, abnormal spinal curvature, developmental abnormality, traumatic injury, arthritic changes or surgically altered anatomy. Functional asymmetry suggests measurably different function and movement ability between the left and right sides of the body.

It is initially more important to identify the asymmetry than to explain the dysfunction's complexities. We address asymmetries noted in primitive patterns before asymmetries within higher-level functional patterns.

Asymmetries can sometimes be a combination of a structural and a functional imbalance or irregularity. Our influence over the functional asymmetry is greater than our influence over structural asymmetry.

A good example is a slight structural scoliosis complicated by mobility and stability problems that intensify the curvature. Given time, these two problems can compound each other. Corrective exercise can have significant influence over the functional part of the asymmetry, and over time can even reduce structural decline. Therefore, corrective strategies can offer both improved function and a preventive measure against structural deterioration.

INTENTIONAL REDUNDANCY

The fourth filter is redundancy or repetition. We repeat the most common movements within different patterns to look for consistent limitations. Within each movement pattern, different segments perform different roles.

- Mobility—demonstrates unrestricted freedom of movement in a non-supportive situation

- Static Stability—demonstrates minimal to no movement, and maintains appropriate alignment in the presence of other segmental movement and mass displacement

- Dynamic Stability—demonstrates unrestricted freedom of movement in a supportive situation while also maintaining appropriate alignment

For example, five movement patterns in the movement screen use hip extension. The screen looks at hip extension in weight bearing and non–

weight-bearing positions, and with both a bent and a straight knee. This intentional redundancy helps create clarity.

If hip extension is consistently limited regardless of knee position or load, the information points toward a hip joint mobility problem. However, if hip extension is limited in some movement patterns and not in others, hip-joint mobility cannot be a large concern. The inconsistent limitation of hip mobility would point more toward a stability problem of the pelvic hip complex or to a specific flexibility problem.

In the case of the hip, the multi-joint rectus femoris can limit hip extension when the knee is flexed, because flexion creates significantly increased muscle tension and has a different length-to-tension influence on the hip when the knee is extended.

A FEW EXAMPLES OF
REDUNDANCY IN THE FMS

Hip Extension—five tests
Hip Flexion—four tests
Shoulder Extension—three tests
Shoulder Flexion—four tests
Knee Extension—three tests
Knee Flexion—four tests
Elbow Extension—three tests
Elbow Flexion—three tests
Ankle Dorsiflexion—three tests
Wrist Extension—two tests

Static spine stability with symmetrical and asymmetrical extremity activity is represented two or more times. The screen reviews both static and dynamic stability in the extremities at least three times. Stability of the upper and lower extremities is also demonstrated when each is loaded.

PAIN VERSUS DISCOMFORT
DURING SCREENING

Pain—a physical feeling that includes distress, tenderness, burning, aching, pinching, jamming, radiating, sharpness or soreness that is unqualified or unexplained

Give a zero score for a test in which the client encounters pain, and make a recommendation for assessment by a healthcare professional if you are not a licensed clinician. It is under your discretion to continue screening if the screen is incomplete, but either way, advise the person not to exercise or engage in activity involving any movement pattern that provokes pain.

PAIN CRITERIA CHECKLIST

- **Familiar**—occurring on a regular or on a consistent basis

- **Produced by common movements**—noticed in daily activities and exercises

- **Signs of concern or stress**—the person notes pain, seems focused on pain or is distracted or distressed by pain

- **Discomfort**—a non-distressing or alarming physical feeling that includes awkwardness, uneasiness, mild tightness or qualified soreness caused by exercise or massage

DISCOMFORT CRITERIA CHECKLIST

Unfamiliar—not occurring on a regular or consistent basis

Only produced with awkward movements—not noticed in daily activities and exercises

No sign of concern or stress—discomfort is noted, but not distressing or distracting in any way and will usually subside with repeated movements

Risks of continuing a screen when pain occurs before the end of the screen—

- The situation causing the pain could be aggravated

- The person might be fearful or apprehensive

- The painful episode may alter movement and not give a clear picture of the client's current functional status

Benefits of continuing a screen when pain occurs before the end of the screen—

- A complete screen can provide a thorough representation of movement

- More than one movement pattern can cause the same pain

- A different pain may be discovered if all movement pattern tests are performed

- A complete movement screen can serve as a baseline for future reference

Score discomfort the same as pain. No professional referral is necessary, but it is your responsibility to monitor the finding. Let the zero score signify the discomfort noted within a particular pattern, and recheck the discomfort after each exercise session.

Do not center on a correction strategy for this pattern. Instead, focus on the lowest non-zero score or the greatest asymmetry, and after corrective efforts, recheck the discomfort. If the feeling is persistent and remains unchanged, and the corrective strategy of the non-zero score has not positively affected the target movement pattern, move on to the SFMA if you are qualified, or provide a referral to a healthcare professional.

SCORING THE FMS— RESULTS HIERARCHY

Score of three—unquestioned ability to perform a functional movement pattern

Score of two—ability to perform a functional movement pattern, but some degree of compensation is noted

Score of one—inability to perform or complete a functional movement pattern

Score of zero—pain, a problem requiring SFMA breakouts or a referral to a healthcare professional

Medical or rehabilitation professionals with backgrounds in the SFMA system will be the most helpful and the most informative when pain is present. If pain is noted and you are not licensed and qualified in the system, a clinician who knows the SFMA represents the referral of choice.

PROCEEDING THROUGH UNCOVERED LIMITATIONS

The inability to perform a movement pattern is a more significant problem than a successful pattern that displays compensation.

Limitations or inabilities must be addressed before compensations.

Asymmetrical or unilateral limitation is more important than symmetrical or bilateral limitation.

In screens where the right and left sides can be reviewed separately, inability on only one side is a greater potential problem than equal inability to perform the movement pattern. Asymmetrical compensation is a greater problem than symmetrical compensation.

The FMS score matters. However, the interpretation of the score is more important for the exercise prescription. The FMS has a hierarchy designed to eliminate risk.

- First, eliminate pain by moving to the SFMA or by referring to a healthcare provider.

- After pain has been addressed, eliminate the greatest asymmetry in the most primitive movement pattern.

- Then eliminate the next asymmetry, and continue until there are no others.

- Only when there are no asymmetries should an exercise program target achieving a score of three.

CLEARING TESTS

Three movement patterns have additional clearing tests—the *shoulder mobility, trunk stability pushup* and *rotary stability*. The clearing tests are unlike the seven movement-pattern tests, as they are not graded on the *3—2—1—0* scale. These you will report as positive or painful, or as negative or non-painful.

These tests offer extra insight into dysfunction by looking at key areas where range-of-motion extremes are indicators of poor mobility or stability or both. The shoulder complex and lumbar-pelvic region routinely compensate by giving up some degree of stability when neighboring body segments have reduced mobility. These areas need an extra degree of screening scrutiny.

Shoulder Mobility Movement Pattern
Impingement Clearing Test

Trunk Stability Pushup Movement Pattern
Prone Press-Up Clearing Test

Rotary Stability Movement Pattern
Posterior Rocking Clearing Test

The clearing tests will be described following the screens with which they are associated.

RANGE OF MOTION CONSIDERATIONS

The joint positions of the FMS tests do not exceed normal movement range. The unique perspective of movement pattern screening looks at multiple joints working within normal ranges. Some movement screen positions require weight bearing and therefore offer observation of joint abilities in loaded and unloaded positions.

Screening will expose obvious and subtle limitations. Some limitations will be the result of habit, lifestyle and activity choices. Other limitations will come from injuries no longer presenting symptoms of pain or swelling, but display weakness and tightness from imperfect healing.

RANGE OF MOTION AS AN INDICATOR OF DECLINE IN MOBILITY AND STABILITY

The thoracic spine, ankles and hips display a general decline in normal mobility in the presence of excessive activity, significantly reduced activity and in a lack of variety in activity. Excessive activity can often cause stiffness, which is the body's default when stability does not appropriately match activity needs. In situations where activity is limited, the body conserves movement and energy and loses mobility, stability and coordinated function between the stabilizers and movers. When a lack of variety is present, such as in a person who participates in only one activity, it is also possible to have movement-pattern atrophy.

Past injuries to vulnerable body segments such as wrists, shoulders, low back and knees often retain the residual limitations of tightness and weakness.

FUNCTIONAL MOVEMENT SCREENING TEST DESCRIPTIONS

As you view the specific criteria for each movement screen test, note the similarity to the patterns and postures of growth and development. Review the redundancy of joint movements within differing patterns, providing extra opportunity to observe limitation and asymmetry. Remember, the criteria for each test are not so much to measure the movement pattern as to grade it.

Key point: If even one of the tested patterns produces pain, exercise and activity could further complicate the underlying problem.

The system is set up to rate and rank movement patterns to create clarity, communication, reliability and reproducibility in physical activity settings. The testing criteria provide quick set-up and efficient administration for individuals and groups. Once all scores are collected, the obvious training choice or deficit priority will present itself based on the rating and ranking system.

Please see www.movementbook.com/chapter5 for more information, videos and updates.

6

FUNCTIONAL MOVEMENT SCREEN DESCRIPTIONS

The Functional Movement Screen (FMS®) captures fundamental movements, motor control within movement patterns, and competence of basic movements uncomplicated by specific skills. It will determine the greatest areas of movement deficiency, demonstrate limitations or asymmetries, and eventually correlate these with an outcome. Once you find the greatest asymmetry or deficiency, you can use measurements that are more precise if needed.

The original idea of the screen was to portray movement-pattern quality with a simple grading system of movement appraisal; it's not intended to diagnose or measure isolated joint movement. Attempting to measure in isolation does a disservice to the pattern—the body is too complex to take isolated movements seriously in the initial stages of screening.

This system was developed to rate and rank movement patterns in high school athletes, but through a two-year refining process, we discovered uses beyond its original intended purpose. While we have not changed the screen since its official introduction in 1998, the information gathered from its use has broadened our scope of corrective exercise, training and rehabilitation. The screen has taught us how to use it, and helped us gain timely and valuable feedback from our attempts at movement correction.

Our collective expertise has come from working against the screen's standard, not from modifying the screen every time things got confusing or inconvenient. We have changed the way we look at the screen data many times, but we have not changed the way we collect the information. In a way, this work represents our evolution, not that of the screen. The screen patiently waited for us to see and understand all it was providing in return for about 10 minutes worth of time.

This chapter covers the FMS. It, along with the two Selective Functional Movement Assessment (SFMA®) chapters that follow, is the juicy part of this book. Take the time to read this section to gain a complete understanding of the screen before implementing it with your clients.

THE FMS TESTS

The FMS is comprised of seven movement tests that require a balance of mobility and stability. The patterns used provide observable performance of basic, manipulative and stabilizing movements by placing clients in positions where weaknesses, imbalances, asymmetries and limitations become noticeable by a trained health and fitness professional.

When the screen's movements mimic athletic moves, it is merely coincidence. The screen is not a training tool, nor is it a competition tool. It's purely an instrument for rating and ranking movements.

The screen's usefulness is its simplicity, practicality and ability to fill a void in the toolbox we use to judge performance and durability. It is not intended to determine why a dysfunctional or faulty movement pattern exists. Instead, it's a discovery of which patterns are problematic. The FMS exposes dysfunction or pain—or both—within basic movement patterns.

Many people are able to perform a wide range of activities, yet are unable to efficiently execute the movements in the screen.[39] Those who score poorly on the screens are using compensatory movement patterns during regular activities. If these compensations continue, sub-optimal movement patterns are reinforced, leading to poor biomechanics and possibly contributing to a future injury.

The public's knowledge of the intricacies of the FMS is minimal at best. To introduce your client to the process, suggest a visit the Functional Movement Systems website at *functionalmovement.com* to watch the introductory video. The website also has video demonstrations of the seven FMS moves and the three clearing tests.

You'll find scripted instructions for use with your client testing in the appendix beginning on page 381.

KEYS TO THE SCREEN

To administer the FMS correctly, you'll need to be familiar with the following bony structures or superficial landmarks.

- **Tibial tuberosity**
- **Anterior superior iliac spine (ASIS)**
- **Lateral and medial malleolus**
- **The most distal wrist crease**
- **Joint line of the knee**

FMS KIT EQUIPMENT AND ASSEMBLY

The optional test kit equipment is self-contained in a two-by-six box, however you are able to use your own testing tools. There is a cap on one end of the two-by-six that can be removed so the pieces used for the FMS can slide out. The pieces are—

- **A four-foot dowel rod**
- **Two smaller dowel rods**
- **A small-capped piece**
- **An elastic band**

Once removed, the two small dowel pieces are inserted in holes in the two-by-six. The dowel pieces must be forced into the two holes in the box in order to be snug. The small-capped piece is inserted into a small hole at the end of the two-by-six, which balances the hurdle once it is upright. The elastic band is then placed around the two upright pieces, making the hurdle.

Two-by-six box—used to carry equipment and to add compensation for the deep squat test. It is also used in the inline lunge and rotary stability tests for reliability and for reference during testing.

Four-foot dowel—used for the deep squat, inline lunge, hurdle step, shoulder mobility measurement and active straight-leg raise. The dowel is used in these tests for reliability and for more efficient scoring.

Hurdle—composed of the board serving as the base, two two-foot PVC dowels and an elastic band that goes around the dowels. It is used for the hurdle step, and allows for body-relative testing and improvement in scoring accuracy.

WHERE TO STAND DURING SCREENING

Where to stand during testing is a common question, because you might have three or four different criteria to review during each test, each putting you in a quandary of trying to be in two places at once. This is one of the reasons the client will perform three repetitions in each movement. If needed, this allows more than one opportunity to see the pattern.

Two things to consider when observing the movements of the screen are distance and movement. Considering these two things will take care of most of the issues involved in trying to see everything during the screen.

DISTANCE

Step back from the client to create enough distance, allowing you to see the whole picture at once. Most of the confusion over where to stand comes from being too close and too focused on one area of the test. Stand far enough away to allow a more global focus. View the entire movement and let the test criteria become evident.

MOVEMENT

The client has three attempts to perform each test, so don't be afraid to move around during the test. There are certain tests where standing to the side or facing the person provide the best vantage points. Take advantage of all three trials and move around if the score is not obvious from one point of view.

LIST OF FMS TESTS
Deep Squat Movement Pattern
Hurdle Step Movement Pattern
Inline Lunge Movement Pattern
Shoulder Mobility Movement Pattern
Active Straight-Leg Raise Movement Pattern
Trunk Stability Pushup Movement Pattern
Rotary Stability Movement Pattern

THE FUNCTIONAL MOVEMENT SCREEN

SCORING SHEET

NAME _____ DATE _____ DOB _____

ADDRESS _____

CITY, STATE, ZIP _____ PHONE _____

SCHOOL/AFFILIATION _____

SSN _____ HEIGHT _____ WEIGHT _____ AGE ____ GENDER ____

PRIMARY SPORT _____ PRIMARY POSITION _____

HAND/LEG DOMINANCE _____ PREVIOUS TEST SCORE _____

TEST		RAW SCORE	FINAL SCORE	COMMENTS
DEEP SQUAT				
HURDLE STEP	L			
	R			
INLINE LUNGE	L			
	R			
SHOULDER MOBILITY	L			
	R			
IMPINGEMENT CLEARING TEST	L			
	R			
ACTIVE STRAIGHT-LEG RAISE	L			
	R			
TRUNK STABILITY PUSHUP				
PRESS-UP CLEARING TEST				
ROTARY STABILITY	L			
	R			
POSTERIOR ROCKING CLEARING TEST				
TOTAL				

Raw Score: This score is used to denote right and left side scoring. The right and left sides are scored in five of the seven tests and both are documented in this space.

Final Score: This score is used to denote the overall score for the test. The lowest score for the raw score (each side) is carried over to give a final score for the test. A person who scores a three on the right and a two on the left would receive a final score of two. The final score is then summarized and used as a total score.

DEEP SQUAT
MOVEMENT PATTERN

PURPOSE

The *deep squat pattern* is part of many functional movements. It demonstrates fully coordinated extremity mobility and core stability, with the hips and shoulders functioning in symmetrical positions. While full deep squatting is not often required in modern daily life, general exercise and sport moves, active individuals still require the basic components for the deep squat.

Extremity mobility, postural control, pelvic and core stability are well represented in the deep squat movement pattern. The deep squat is a move that challenges total body mechanics and neuromuscular control when performed properly. We use it to test bilateral, symmetrical, functional mobility and stability of the hips, knees and ankles.

The dowel held overhead calls on bilateral, symmetrical mobility and stability of the shoulders, scapular region and the thoracic spine. The pelvis and core must establish stability and control throughout the entire movement to achieve the full pattern.

DESCRIPTION

The client assumes the starting position by placing the instep of the feet in vertical alignment with the outside of the shoulders. The feet should be in the sagittal plane with no lateral outturn of the toes. The client rests the dowel on top of the head to adjust the hand position resulting in the elbows at a 90-degree angle.

Next, the client presses the dowel overhead with the shoulders flexed and abducted and the elbows fully extended. Instruct the client to descend slowly into the deepest possible squat position, heels on the floor, head and chest facing forward and the dowel maximally pressed overhead. The knees should be aligned over the feet with no valgus collapse.

As many as three repetitions may be performed, but if the initial movement falls within the criteria for a score of three, there is no need to perform another test. If any of the criteria for a score of three are not achieved, ask the client to perform the test with the board from the earlier described FMS kit under the heels. If any of the criteria for the score of two are not achieved while using the FMS board, the client receives a score of one.

TIPS FOR TESTING

1. Observe the client from the front and side.

2. All positions including the foot position should remain unchanged when the heels are elevated, with either the FMS kit or a similar size board.

3. Do not judge the pattern or interpret the cause of the score while testing.

4. Do not coach the movement; simply repeat the instructions if needed.

5. Was there pain?

6. When in doubt, score low.

IMPLICATIONS OF THE
DEEP SQUAT
MOVEMENT PATTERN

•   Limited mobility in the upper torso can be attributed to poor glenohumeral or thoracic spine mobility, or both.

•   Limited mobility in the lower extremities, including poor closed kinetic chain dorsiflexion of the ankles or poor flexion of the knees and hips can cause poor test performance.

•   People might perform poorly because of poor stabilization and control.

Deep Squat 3 Front View

Deep Squat 3 Side View

Deep Squat 2 Front View

Deep Squat 2 Side View

Deep Squat 1 Front View

Deep Squat 1 Side View

HURDLE STEP MOVEMENT PATTERN

PURPOSE

The *hurdle step movement pattern* is an integral part of locomotion and acceleration. Although we do not step to this level in most activities, the hurdle step will expose compensation or asymmetry in stepping functions. The step test challenges the body's step and stride mechanics, while testing stability and control in a single-leg stance.

The movement requires proper coordination and stability between the hips, moving asymmetrically with one bearing the load of the body while the other moves freely. The pelvis and core must begin with and maintain stability and alignment throughout the movement pattern. The arms are still as they hold a dowel across the shoulders, giving the observer further representation of the static responsibility of the upper body and trunk in the stepping movement.

Excessive upper body movement in basic stepping is viewed as compensation; it is not seen when proper mobility, stability, posture and balance are available and functioning. The hurdle step challenges bilateral mobility and stability of the hips, knees and ankles. The test also challenges stability and control of the pelvis and core as it offers an opportunity to observe functional symmetry.

DESCRIPTION

Take a height measurement of the client's tibia to begin this test. Since it can be difficult to find the true joint line between the tibia and femur, the top center of the tibial tuberosity serves as a reliable landmark.

To adjust the previously described hurdle to the correct height, have the client stand with the outside of the right foot against the base of the hurdle, in line with one of the hurdle uprights. Slide the hurdle's marking cord to the center of the tibial tuberosity, and adjust the other side until the cord is level and displays accurate tibial tuberosity height on both indicators.

The other measurement option is to use the dowel to measure the distance from the floor to the tibial tuberosity, and raise the cord to that level.

Have the client stand directly behind the center of the hurdle base, feet touching at both the heels and toes, and with the toes aligned and touching the base of the hurdle.

Position the dowel across the shoulders, below the neck. Ask the client to step over the hurdle to touch the heel to the floor while maintaining a tall spine, and return the moving leg to the starting position. The hurdle step is performed slowly and under control.

If any of the criteria for a score of three are not achieved, the client receives a score of two. If any of the criteria for the score of two are not achieved, score this a one.

TIPS FOR TESTING

1. Ensure the cord is aligned properly.
2. Tell the client get as tall as possible at the beginning of the test.
3. Watch for a stable torso.
4. Observe from the front and side.
5. Score the hurdle-stepping leg.
6. Make sure the toes of the stance leg stay in contact with the hurdle during and after each repetition.
7. Do not judge the pattern or interpret the cause of the score while testing.
8. Do not coach the movement; simply repeat the instructions if needed.
9. Was there pain?
10. When in doubt, score low.

IMPLICATIONS OF THE HURDLE STEP MOVEMENT PATTERN

- Problems may be due to poor stability of the stance leg or poor mobility of the step leg.

- The main thing to consider is that no single part is being tested; a pattern is being tested. Imposing maximal hip flexion of one leg while maintaining apparent hip extension of the opposite leg requires relative bilateral, asymmetric hip mobility and dynamic stability.

Hurdle Step 3 Front View

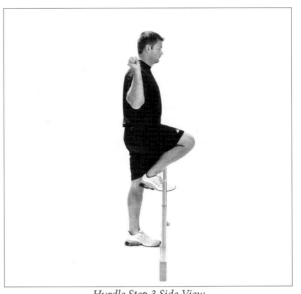

Hurdle Step 3 Side View

Hurdle Step 2 Front View

Hurdle Step 2 Side View

Hurdle Step 1 Front View

Hurdle Step 1 Side View

INLINE LUNGE
MOVEMENT PATTERN

PURPOSE

The *inline lunge pattern* is a component of deceleration movements and direction changes produced in exercise, activity and sport. Although the inline lunge explores more movement and control than many activities require, it provides a quick appraisal of left and right functions in the basic pattern. It is intended to place the body in a position to focus on the stresses as simulated during rotation, deceleration and lateral movements. The narrow base requires appropriate starting stability and continued dynamic control of the pelvis and core within an asymmetrical hip position equally sharing the load.

The inline lunge places the lower extremities in a split-stance position while the upper extremities are in an opposite or reciprocal pattern. This replicates the natural counterbalance the upper and lower extremities use to complement each other, as it uniquely demands spine stabilization. This test also challenges hip, knee, ankle and foot mobility and stability, at the same time simultaneously challenging the flexibility of multi-articular muscles such as the latissimus dorsi and the rectus femoris.

True lunging requires a step and descent. The inline lunge test only provides observation of the descent and return; the step would present too many variables and inconsistencies for a simple movement screen. The split-stance narrow base and opposite-shoulder position provide enough opportunities to discover the mobility and stability problems of the lunging pattern.

DESCRIPTION

Attain the client's tibia length by either measuring it from the floor to the top center of the tibial tuberosity, or acquiring it from the height of the cord during the hurdle step test. Tell the client to place the toe of the back foot at the start line on the kit. Using the tibia measurement, have the client put the heel of the front foot at the appropriate mark on the kit. In most cases, it's easier to establish proper foot position before introducing the dowel.

Place the dowel behind the back, touching the head, thoracic spine and sacrum. The client's hand opposite the front foot should be the hand grasping the dowel at the cervical spine. The other hand grasps the dowel at the lumbar spine. The dowel must maintain its vertical position throughout both the downward and upward movements of the lunge test.

To perform the inline lunge pattern, the client lowers the back knee to touch the board behind the heel of the front foot and returns to the starting position.

If any of the criteria for a score of three are not achieved, the client receives a score of two. If any of the criteria for the score of two are not achieved, the client receives a score of one.

TIPS FOR TESTING

1. The front leg identifies the side you're scoring—this simply represents the pattern and does not imply the functional ability of a body part or side.

2. Always remember you are screening patterns, not parts.

3. The dowel remains vertical and in contact with the head, thoracic spine and sacrum during the movement.

4. The front heel remains in contact with the board, and the back heel touches the board when returning to the starting position.

5. Watch for loss of balance.

6. Remain close to the client to prevent a complete loss of balance.

7. Do not judge the pattern or interpret the cause of the score while testing.

8. Do not coach the movement; simply repeat the instructions if needed.

9. Was there pain?

10. When in doubt, score low.

IMPLICATIONS OF THE INLINE LUNGE MOVEMENT PATTERN

- Ankle, knee and hip mobility may be inadequate for either the front or the rear leg.

- Dynamic stability may not be adequate to complete the pattern.

- There may also be limitations in the thoracic spine region, inhibiting the client from performing the test well.

Inline Lunge 3 Front View

Inline Lunge 3 Side View

Inline Lunge 2 Front View

Inline Lunge 2 Side View

Inline Lunge 1 Front View

Inline Lunge 1 Side View

SHOULDER MOBILITY REACHING MOVEMENT PATTERN

PURPOSE

The *shoulder mobility reaching pattern* demonstrates the natural complementary rhythm of the scapular-thoracic region, thoracic spine and rib cage during reciprocal upper-extremity shoulder movements. Although the full reciprocal reaching pattern is not seen in basic activities, it uses each segment to its range of active control, leaving little room for compensation. Removing compensation provides a clear view of movement ability.

The cervical spine and surrounding musculature should remain relaxed and neutral, and the thoracic region should have a natural extension before doing the alternate upper-extremity patterns.

This pattern observes bilateral shoulder range of motion, combining extension, internal rotation and adduction in one extremity, and flexion, external rotation and abduction of the other.

DESCRIPTION

First, determine the client's hand length by measuring the distance from the distal wrist crease to the tip of the longest digit. The client will stand with the feet together, and make a fist with each hand, thumbs inside the fingers. The client then simultaneously reaches one fist behind the neck and the other behind the back, assuming a maximally adducted, extended and internally rotated position with one shoulder, and a maximally abducted and externally rotated position with the other.

During the test, the hands should move in one smooth motion, and should remain fisted. Measure the distance between the two closest points of the hands to determine the client's symmetrical reach.

Have the client perform the shoulder mobility test a maximum of three times bilaterally. If any of the criteria for a score of three are not achieved, the client receives a score of two. If any of the criteria for the score of two are not achieved, score this a one.

TIPS FOR TESTING

1. The top shoulder identifies the side being scored. This simply represents the pattern and does not imply the functional ability of a body part or side.

2. If the hand measurement is the same as the distance between the two points, score low.

3. If pain is present in the clearing test, the client receives a zero.

4. Make sure the client does not try to walk the hands toward each other following the initial placement.

5. Do not judge the pattern or interpret the cause of the score while testing.

6. Do not coach the movement; simply repeat the instructions if needed.

7. Was there pain?

8. When in doubt, score low.

IMPLICATIONS OF THE SHOULDER MOBILITY REACHING MOVEMENT PATTERN

- The most obvious is the widely accepted explanation of increased external rotation gained at the expense of internal rotation in overhead throwing athletes. Although this is true to some extent, this is not the first thing to consider.

- Scapular stability depends on thoracic mobility. This should be the primary focus.

- Excessive development and shortening of the pectoralis minor, latissimus dorsi and rectus abdominus muscles can cause the postural alterations of forward or rounded shoulders. This postural problem leaves unrestricted mobility of the glenohumeral joint and scapula at a disadvantage.

- A scapulothoracic dysfunction may be present, resulting in decreased glenohumeral mobility secondary to poor scapulothoracic mobility or stability.

- The test requires an asymmetric movement because the arms travel in opposite directions. The test also requires both arms reaching simultaneously, coupled with postural control and core stability.

CLEARING EXAM

There is a clearing exam at the end of the shoulder mobility test. You're not scoring this, but instead are watching for a pain response. If pain is produced, a positive (+) is recorded on the score sheet, and a score of zero is given to the entire shoulder reach test.

The client places a palm on the opposite shoulder and lifts the elbow as high as possible while maintaining the palm-to-shoulder contact. This clearing exam is necessary because shoulder impingement will sometimes go undetected by shoulder mobility testing alone.

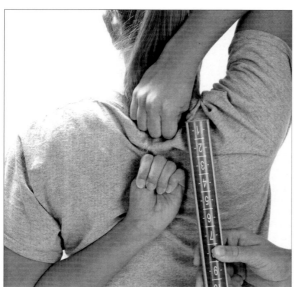

Shoulder Mobility 3 Right

Shoulder Mobility 1 Right

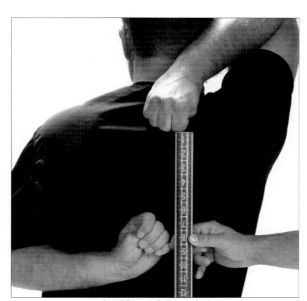

Shoulder Mobility 2 Right

Shoulder Pain Provocation Test

ACTIVE STRAIGHT-LEG RAISE MOVEMENT PATTERN

PURPOSE

The *active straight-leg raise* may appear to be the least functional screen, but don't be fooled by its simplicity. This pattern not only identifies the active mobility of the flexed hip, but includes the initial and continuous core stability within the pattern, as well as the available hip extension of the alternate hip. This is not so much a test of hip flexion on one side, as it is an appraisal of the ability to separate the lower extremities in an unloaded position. This movement is often lost when flexibility of multi-articular muscles is compromised.

The glute maximus/iliotibial band complex and the hamstrings are the structures most likely to result in flexion limitations. Extension limitations are often seen in the iliopsoas and other muscles of the anterior pelvis. This pattern challenges the ability to dissociate the lower extremities while maintaining stability in the pelvis and core. The movement also challenges active hamstring and gastroc-soleus flexibility, while maintaining a stable pelvis and active extension of the opposite leg.

DESCRIPTION

The client lies supine with the arms by the sides, palms up and the head flat on the floor. A board is placed under the knees; this can be either the FMS kit board, or a board of similar dimensions as described earlier. Both feet should be in a neutral position, the soles of the feet perpendicular to the floor.

Find the point between the anterior superior iliac spine (ASIS) and the joint line of the knee, and places a dowel at this position, perpendicular to the ground. Next, the client lifts the test limb while maintaining the original start position of the ankle and knee.

During the test, the opposite knee should remain in contact with the board; the toes should remain pointed upward in the neutral limb position, and the head remains flat on the floor.

Once reaching the end-range, note the position of the upward ankle relative to the non-moving limb. If the malleolus passes the dowel, record a score of three. If the malleolus does not pass the

dowel, move the dowel, much like a plumb line from the malleolus of the test leg, and again score per the criteria.

Perform the *active straight-leg mobility test* a maximum of three times bilaterally. If any of the criteria for a score of three are not achieved, the client receives a score of two. If any of the criteria for the score of two are not achieved, score this a one.

TIPS FOR TESTING

1. The moving limb identifies the side being scored.

2. If there is difficulty finding the joint line, identify the line by flexing and extending the knee.

3. Make sure the non-moving limb maintains a neutral position.

4. Do not judge the pattern or interpret the cause of the score while testing.

5. Do not coach; this is not exercise. This means if there's fault in the execution, simply repeat the instructions, not offering corrections.

6. Was there pain?

7. When in doubt, score low.

IMPLICATIONS OF THE ACTIVE STRAIGHT-LEG RAISE MOVEMENT PATTERN

- Pelvic control may not be sufficient for the execution of the pattern.

- The client may have inadequate mobility of the opposite hip, stemming from inflexibility associated with limited hip extension.

- The client may have poor functional hamstring flexibility in the moving limb.

- A combination of these factors will be exhibited if an client has relative bilateral, asymmetric hip mobility. The non-moving limb is at work during the optimal pattern; when the pattern is correct, the non-moving limb demonstrates stability, an automatic task, while the moving limb demonstrates mobility, a conscious task.

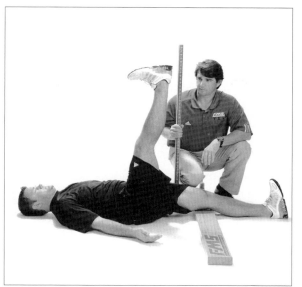

Active Straight-Leg Raise 3

Active Straight-Leg Raise 2

Active Straight-Leg Raise 1

TRUNK STABILITY PUSHUP MOVEMENT PATTERN

PURPOSE

The *trunk stability pushup* is a unique, single-repetition version of the common floor-based pushing exercise. It is used as a basic observation of reflex core stabilization, and is not a test or measure of upper-body strength. The goal is to initiate movement with the upper extremities in a pushup pattern without allowing movement in the spine or hips.

Extension and rotation are the two most common compensatory movements. These compensations indicate the prime movers within the pushup pattern incorrectly engage before the stabilizers.

The push-up movement pattern tests the ability to stabilize the spine in the sagittal plane during the closed kinetic chain, upper body symmetrical pushing movement.

DESCRIPTION

The client assumes a prone position with the arms extended overhead. During this test, men and women have different start positions. Men begin with their thumbs at the top of the forehead, while women begin with their thumbs at chin level. The thumbs are then lowered to the chin or shoulder level per the scoring criteria. The knees are fully extended, the ankles are neutral and the soles of feet are perpendicular to floor.

Ask the client to perform one pushup in this position. The body should be lifted as a unit; there should be no sway in the spine during this test. If the client cannot perform a pushup in the initial position, the hands are lowered to an easier position. Give a score of three if all criteria are met with the hands at the forehead, a score of two if done with the hands at the chin, and a one if the client can't complete the move.

Perform the *trunk stability pushup test* a maximum of three times. If any of the criteria for a score of three are not achieved, the client receives a score of two. If any of the criteria for the score of two are not achieved, score this a one.

TIPS FOR TESTING

1. The client should lift the body as a unit.
2. On each attempt, make sure the client maintains the hand position and the hands do not slide down as the client prepares to push.
3. Make sure the chest and stomach come off the floor simultaneously.
4. If pain is present in the clearing test, the client receives a zero.
5. Do not judge the pattern or interpret the cause of the score while testing.
6. Do not coach; this is not exercise.
7. Was there pain?
8. When in doubt, score low.

IMPLICATIONS OF THE TRUNK STABILITY PUSHUP MOVEMENT PATTERN

- Limited performance during this test can be attributed to poor reflex stabilization of the core.

- Compromised upper-body strength or scapular stability—or both—can also be a cause of poor performance during this test.

- Limited hip and thoracic spine mobility can affect an client's ability to achieve the optimal start position, also leading to poor performance during the test.

CLEARING EXAM

We use a clearing exam at the end of the *trunk stability press-up test*. This movement is not scored; it is performed to observe a pain response. If pain is produced, a positive (+) is recorded and a score of zero is given to the entire press-up test. Clear spinal extension with a press up from the pushup position. If the client receives a positive score, document both scores for future reference.

Trunk Stability Press-up Extension Test

Trunk Stability Pushup Male 3 Start

Trunk Stability Pushup Male 3 Finish

Trunk Stability Pushup Male 2 Start

Trunk Stability Pushup 2 Female Finish

Trunk Stability Pushup 2 and 1 Female Start

Trunk Stability Pushup 1 Female Finish

ROTARY STABILITY MOVEMENT PATTERN

PURPOSE

The *rotary stability pattern* observes multi-plane pelvis, core and shoulder girdle stability during a combined upper- and lower-extremity movement. This pattern is complex, requiring proper neuromuscular coordination and energy transfer through the torso. It has as its roots the creeping pattern that follows basic crawling in our developmental sequence.

The test has two important implications. It demonstrates reflex stabilization and weight shifting in the transverse plane, and it represents the coordinated efforts of mobility and stability observed in fundamental climbing patterns.

DESCRIPTION

The client gets into the quadruped position with a board, either the FMS kit board or one of similar size, on the floor between the hand and knees. The board should be parallel to the spine, and the shoulders and hips should be 90 degrees relative to the torso, with the ankles neutral and the soles of the feet perpendicular to the floor.

Before the movement begins, the hands should be open, with the thumbs, knees and feet all touching the board. The client should flex the shoulder while extending the same-side hip and knee, and then bring elbow to knee while remaining in line over the board. Spine flexion is allowed as the client brings the knee and elbow together.

This is performed bilaterally for a maximum of three attempts if needed. If one repetition is completed successfully, there is no reason to perform the test again.

If a score of three is not attained, have the person perform a diagonal pattern using the opposite shoulder and hip in the same manner described above. During this diagonal variation, the arm and leg need not be aligned over the board; however, the elbow and knee do need to touch over it.

TIPS FOR TESTING

1. The upper moving limb indicates the side being tested.

2. Make sure the unilateral limbs remain over the board to achieve a score of three.

3. The diagonal knee and elbow must meet over the board to achieve a score of two.

4. Make sure the spine is flat and the hips and shoulders are at right angles at the start.

5. Do not judge the pattern or interpret the cause of the score while testing.

6. Do not coach; this is not exercise.

7. Was there pain?

8. When in doubt, score low.

IMPLICATIONS OF THE ROTARY STABILITY MOVEMENT PATTERN

- Limited performance during this test can be attributed to poor reflex stabilization of the trunk and core.

- Compromised scapular and hip stability can also cause poor performance.

- Limited knee, hip, spine and shoulder mobility can reduce the ability to perform the complete pattern, leading to a poor test score.

CLEARING EXAM

A clearing exam is performed at the end of the rotary stability test. This movement is not scored; it is performed to observe a pain response. If pain is produced, a positive (+) is recorded on the sheet and a score of zero is given to the entire rotary stability test. We clear spinal flexion from the quadruped position, then rocking back and touching the buttocks to the heels and the chest to the thighs. The hands remain in front of the body, reaching out as far as possible. If there is pain associated with this motion, give a zero score. If the client receives a positive score, document both scores for future reference.

Rotary Stability Posterior Rocking Clearing Test

Rotary Stability 3 Extension

Rotary Stability 3 Flexion

Rotary Stability 2 Extension

Rotary Stability 2 Flexion

Rotary Stability 2 Extension

Rotary Stability 1 Flexion

FUNCTIONAL MOVEMENT SCREEN CONCLUSION

The exercise and rehabilitation professional should use these seven screens as a means to interact with normal and highly fit populations to clarify exercise program design. This will refine the training and develop a more complete approach by including the appreciation of movement patterns. Knowledge gained from the FMS can help professionals work with normal populations who want to improve efficiency and performance while increasing resistance to injury.

It's important to watch the distribution of scores and performances by a normal population throughout the movements during the initial stages of screening. This is all about movement; resist the temptation to create a screen for each population. In other words, sport does not need a screen different from a dance screen, different from one for elite-level athletes, or different from that used with fit grandmothers.

It is of greater statistical merit to have a common test battery presenting the researcher with biomarkers and predictors of function and risk. Ideally, subsequent researchers will understand this practical and clinical need in physical therapy, and focus on these biomarkers to develop more powerful assessment tools.

To get there, we need to start by capturing functional movement patterns, discussing their distribution—which are functional or dysfunctional, and distributed across which structures—and creating information that holds us to a higher standard.

INCREASED ACTIVITY RISK

It is important to understand the FMS provides both positive and negative information. Both are important and both must be incorporated in your training and rehabilitation programs to manage risk in exercise and activity.

- The positive information provides the corrective strategy needed to improve a deficient movement pattern. It also demonstrates which movement patterns can be performed effectively and can therefore be conditioned and trained.

- The negative aspect suggests the constructive, temporary cessation of activities that can increase risk or delay the progress of the corrective strategies.

Don't assume that a faulty movement pattern only requires the addition of corrective exercise. Movement habits, exercise programs, activities, occupational duties and athletics can all perpetuate faulty movement patterns. Consistently repeating these behaviors, even in the presence of corrective exercise, becomes a tug of war within the central nervous system.

Corrective exercise will attempt to reset mobility and stability, while the other behaviors undermine any positive gains due to the compensations they promote. The solution is to discontinue these activities temporarily until you have established a functional movement platform. After rescreening demonstrates functional movement competency, you can reintroduce these activities or can replace them with others that provide less risk and equal or greater reward.

The list of activities on the next page represents movements counterproductive to efficient and effective movement correction, and should be removed temporarily since they offer no corrective potential. These activities would require some degree of compensation if performed correctly—modification might reduce risk and compensation, but even then would not offer the most efficient path to correction.

These recommendations apply to all asymmetries and all scores of one on the FMS. A minimum score of two is necessary to move into these exercises.

FUNDAMENTAL MOBILITY

Active Straight-Leg Raise—Heavy closed-chain loaded activities, running and plyometrics

Shoulder Mobility—Heavy arm pushing and pulling, overhead pushing and pulling

SUB-MAXIMUM STABILIZATION

Rotary Stability—Conventional core training, training that would cause high threshold core control

HIGH-THRESHOLD STABILIZATION

Pushup—Heavy upper and lower body loads, vigorous plyometric activity

FUNCTIONAL MOVEMENT PATTERNS

Inline Lunge—Exercise and loads involving the lunge pattern

Hurdle Step—Exercise and loads involving the single-leg stance

Squat—Exercise and loads involving part or all of the squatting pattern

MOVEMENT SCREEN MODIFICATIONS

The movement screen provides a complete perspective of active people, and it is appropriate to use whenever you expect the population to have full functional movement ability and capacity within all movement planes. It is also appropriate for people lifting heavy loads and performing activities where you anticipate balance, coordination and normal flexibility. It is suitable for occupations requiring moderate to high physical demand, those participating in advanced exercise programs, athletics and all other activities of high physical capacity. And it is a useful way to predict risk in situations where musculoskeletal injury is prevalent.

There are people with limitations that are self-imposed or imposed by a physical incapacity, disability or medical restriction. Because of these limitations, a complete FMS cannot be performed, and you shouldn't have high expectations of it in those circumstances. Still, screening might provide valuable information and can offer a risk appraisal and the information necessary for effective exercise program design.

Complete screening may not be possible, but modified screening can still set baselines. Some clients may never achieve the physical ability to perform a full screen, but that is not the goal. Screening is a tool to observe functional movement for correction and maintenance purposes. If modifications to the full screen can support this perspective, we have achieved the goal—this modified tool will still have utility and purpose.

Some examples of these populations are—

- Those with medical restrictions, such as cardiac rehabilitation patients

- Wellness clients who have medical conditions and permanent medical restrictions

- Morbidly obese people with limitations as they attempt exercise

- A modified movement screen can be used until body composition and physical capacity allow more involved screening.

- Active seniors may not be able to complete a full FMS due to permanent degenerative changes, but modified screening can influence functional exercise choices and corrective strategies.

Some people think we should modify the FMS for children participating in exercise and athletics, but that is incorrect. Children are growing, developing and advancing their physical capacity, and each child reaches functional maturity at a different age. Full screening will present challenges to their growing bodies, but this is the best way to gain perspective. A full screen provides better risk appraisal and more accurately represents the demands of exercise and athletics.

If children are involved in organized athletics and formal supervised exercise, full movement screening is prudent and appropriate. It will expose deficiency associated with risk and provide for greater physical preparedness. They may not perform well on a full screen, but they are developing and their screen scores should progress as they grow, assuming they are not otherwise restricted.

Modification does not mean screen tests are changed or that we lower movement criteria. It means we administer fewer tests for the purposes of caution or safety. This is important because it provides movement perspective across the lifespan, the same way eye chart scales or blood pressure scales for testing remain the same. The very fact that the testing scale remains consistent allows us to observe progress and decline over the course of a life. We can consider different values acceptable across the lifespan, but still use the same test.

MODIFICATION HIERARCHY

Modifications on the FMS should only be done on an individual basis, necessitated by physical limitations or restrictions imposed for safety or professional caution. Individualized professional discretion is expected when screening movement. If the screen is modified, this will potentially alter the corrective approach because there is a corrective hierarchy.

Screening has a hierarchy for scoring consideration based on movement development and corrective principles. The hierarchy provides a platform for modification if it is needed. The most restrictive modification for the mobility screens are the shoulder mobility test and clearing exam and the active straight-leg raise.

The scoring criteria remain the same: Make these movement patterns a priority until asymmetry or a one score are replaced with symmetrical scores of two or three. Once you achieve the mobility goals, perform the rotary stability test and clearing exam if there are no medical restrictions or contraindications. As always, if there's a conflict, do not do the test.

You can also bypass the pushup test if you feel high-threshold core activity is not appropriate. This would include instances in which the client won't be performing heavy lifting or high-threshold core activities; otherwise, test it. We recommend the pushup clearing test unless prone passive extension is contraindicated. We don't use the clearing to assess mobility—its purpose is to identify pain in extension as a provocation sign for spinal dysfunction to be evaluated by a medical professional.

The next test to review in the modified screen is the hurdle step test. Once you've worked on the mobility issues, balance and motor control are the next steps to improve movement capacity.

You might skip lunging and squatting unless these represent movement patterns that comprise exercises and activities your client will be expected to perform.

SUMMARY OF MODIFICATIONS

If you modify the FMS, you must follow a systematic path. Randomly performing a test and then trying to correct movement based on that test alone would most likely undermine the developmental sequence and may also impose an inappropriate exercise disguised as a correction.

Modifications to the FMS should follow the path below. This will force corrective strategy toward the weakest link and reduce errors in making progressions.

- **Active Straight-Leg Raise Test**
- **Shoulder Mobility Reaching Test**
- **Pain Provocation Clearing Test**

Next, consider all that are applicable—

- **Rotary Stability**
- **Flexion Clearing Test**
- **Extension Clearing Test**

Then consider—

- **Pushup test, if appropriate**

Next, consider—

- **Hurdle Step Test**

Finally, consider—

- **Inline Lunge Test**
- **Deep Squat Test**

Please see www.movementbook.com/chapter6 for more information, videos and updates.

SFMA INTRODUCTION AND TOP-TIER TESTS

Expert rehabilitation professionals and researchers all have perspective regarding human posture, movement and function, as well as their complicating pain syndromes. These experts have been instrumental in describing and advancing the examination of structures and functions of movement, and have addressed structural alignment, instability and restrictions that contribute to pain and movement dysfunction.

Logical thinkers from earlier generations had primitive tools compared to those available today, yet they practiced with competence. Modern clinicians often rely on a broad assortment of modalities and diagnostic tools; our hands become soft and weak, as our brains lose the ability to quickly deduce and solve a myriad of movement problems. However, today's clinicians can enjoy the best of the past and the present if we can master the manual and deductive skills of our predecessors, and complement the practice with the modern methods and tools of testing and treating.

Early medical science uses a patho-anatomical approach to movement, where the limitations, degeneration and lesions of the anatomical structures were used to explain every movement-related pain or deficiency. Modern medical science is attempting to balance the scales of perspective and consider biomechanics, neuromuscular control and functional symmetry as elements for equal consideration. Applying both approaches creates balanced focus and is helpful in the complete explanation of painful or dysfunctional clinical conditions.

Anatomy encompasses structure, composition and framework, while physiology integrates function, processes and interrelationships. Weighted equally, these perspectives have helped rehabilitation professionals progress from the narrow-minded, basic examination of isolated muscles and joints, to an appreciation of the complex relationships that make movement possible.

This broad view considers painful and dysfunctional movement as both mechanical and behavioral. The appreciation of how movement systems are connected is the driving force in the shift toward an outcome-based practice of movement rehabilitation in modern physical medicine.

"There are numerous ways in which slight subtleties in movement patterns contribute to specific muscle weaknesses. The relationship between altered movement patterns and specific muscle weaknesses requires that remediation addresses the changes to the movement pattern; the performance of strengthening exercises alone will not likely affect the timing and manner of recruitment during functional performance."
~Dr. Shirley Sahrmann

CONSIDERING PATTERNS OF MOVEMENT CLINICALLY

Our bodies migrate toward predictable patterns of movement in response to injury, and also in the presence of weakness, tightness or structural abnormality. With a narrow-minded approach to either evaluation or treatment, we will not restore complete function, because pain-free movement restoration requires a working knowledge of acceptable movement patterns, and a map of dysfunctional patterns. We have to know what is bad in order to have any inkling of what is good and we need the structure of a standard operating procedure to build the map.

The goal of the Selective Functional Movement Assessment (SFMA®) is to capture the patterns of posture and function for comparison against a baseline. The SFMA is an organizational method to rank the quality of functional movements and, when sub-optimal, their provocation of symptoms.

"It has also been recognized that the dysfunctions of muscles and joints are so closely related, the two should be considered a single, inseparable functional unit."
~Dr. Vladimir Janda

The medical practice and published works of Dr. Vladimir Janda are clear. He acknowledged the need for a systematic breakdown of movement, but he also warned us not to lose perspective of clinically separate things that cannot be totally disconnected.

Influential contemporaries have guided and continue to guide the profession toward ever-improving models of human assessment and rehabilitation. This expanded knowledge can create confusion if not used in a systematic manner. A timeless direction in clinical systems comes to us from Dr. James Cyriax, considered by many as the father of non-operative orthopedic medicine.

"It is well to remember that the object of the physical examination is to find the movement that elicits the pain of which the patient complains, rather than some other nebulous symptom of which he was previously unaware.

"Only by sticking to a standard sequence will the physician be sure of leaving nothing out, and only by leaving nothing out are true findings feasible. The physician arrives at a diagnosis not from the evidence furnished by one painful movement, but by careful detection of a consistent pattern."

~Dr. James Cyriax

Dr. Cyriax uses his words carefully when he commands us to seek a consistent pattern. This demonstrates his appreciation of the crux and behavior of the problem, as well as its anatomical structural considerations.

Dr. Cyriax created a systematic method for classification of contractile tissue quality based on tension and irritability, and offered guidelines for the systematic diagnosis of soft tissue lesions.

Cyriax's Selective Muscle Tension Testing is often abbreviated as four muscle testing categories.

Strong and painless—*normal*

Strong and painful—*a minor lesion to the muscle or tendon*

Weak and painful—*a major lesion to the muscle or tendon*

Weak and painless—*a neurological problem*

This system was used to classify a contractile problem into categories. A more detailed examination could be performed, including specialized testing and impairment measures providing increased levels of objectivity and quantifiability after assigning a general classification. Cyriax's contribution gave us a system of soft tissue classification, but more notably, a clinician-friendly template for the development of clinical detective tools involving pain and function.

An efficient and reliable perspective of the problem is paramount to treatment, and the purity of the evaluation process should not be clouded by the treatment options. The treatment plan should be considered and initiated after appropriate attention is given to the diagnostic process. Cyriax demonstrated how a small handful of qualitative tests could rate and rank information in the examination in order for more involved quantitative tests to confirm and refine the problem.

His qualitative tests are still important and relevant today, but not because of technical refinement or measurement accuracy. The tests are valuable because they create perspective and categorization before narrowing the focus with precise measurements of suspicious structures and quantities.

General perspective before specialized measurement is a hallmark of clinical expertise.

A reconsideration of the Cyriax categories for tissue testing helped lay the framework for the SFMA. The systematic value of this type of classification is the presentation of only a few choices form which we are forced to pick. By using his first four categories and replacing the words strong and weak with function and dysfunction we create a system for movement pattern assessment. Now we have four terms that can be grouped to discuss movement.

By combining the words functional, dysfunctional, painful and non-painful, we get—

Functional and non-painful (FN)
Functional and painful (FP)
Dysfunctional and non-painful (DN)
Dysfunctional and painful (DP)

The word *functional* describes any unlimited or unrestricted movement. However, before a movement is deemed functional, the patient must complete a breath cycle at the end range for the movement pattern. If breathing is labored or breathing causes the patient to alter the movement pattern, it is deemed dysfunctional.

Dysfunctional describes limitations or restrictions that demonstrate a lack of mobility, stability or symmetry within a given movement. Whenever function is questioned consider it a dysfunction if it helps to use qualifiers like significant or minimal feel free if the clarity helps.

Painful denotes a condition in which the selective functional movement reproduces primary symptoms, increases primary symptoms, or brings about secondary symptoms.

Each time a functional movement is graded in this manner, side notes may be used to describe the origin and severity of the abnormality documented. This identification is the first step in the examination process, which focuses the remaining assessment to the tests and measures pertinent to the needs of the patient.

People visiting a clinician for treatment are primarily concerned with the resolution of pain. While the patient highlights the pain, the clinician must remain focused on clues, and the detection of consistent patterns to help explain the origin and behavior of the pain and dysfunction. It is important to understand that not every painful movement pattern is a dysfunctional pattern, and not every dysfunctional movement is painful. The clinician is obligated to identify both.

We know the pain is there. We can see it, or the patient easily demonstrates its location. What we need to find is the *cause* of the pain. By mapping movements that provoke pain and movements that are dysfunctional, but do not provoke pain we design a cleaner map of perception and behavior around pain and dysfunction.

CLINICAL SIGNS AND SYMPTOMS

The examination process should be a balance of the patient's signs and symptoms: the patient communicating symptoms while the clinician observes the signs. The patient's symptoms encompass descriptions of the offending complaint, and can include things that are annoying or hinder lifestyle and activities, usually associated with pain.

The signs are some of the subtle indicators that frequently accompany the symptoms, and these are often closely related. A bruise displays discoloration, a sign of trauma, and pain to the touch—a symptom of the inflammation resulting from the trauma.

In contrast, in conditions of chronic low back pain, the connection between signs and symptoms becomes more complicated. The patient might complain of low back pain with forward bending, but although forward bending produces symptoms, there may be no real signs of movement dysfunction. The movement hurts, but the patient can touch the toes and return to standing without movement dysfunction—uncomfortable perhaps, but with no signs of altered movement.

If the patient had not mentioned the pain, would the clinician note anything suspicious in the successful forward-bending movement? Since clinicians are human and prone to observational error and empathy, they often assume a patient is not forward bending correctly when pain was simultaneously reported. They may attempt to correct some part of the forward bending when nothing is wrong with the movement pattern—the movement is painful, but not dysfunctional from basic observation.

Although no other movement pattern may cause the symptoms of low back pain, there may be signs of movement dysfunction during the entire movement appraisal. The clinician may note that spinal extension does not cause pain, but it may be limited to less than 50 percent of normal spine extension for the patient's age and fitness level. This would suggest a significant restriction in a basic movement pattern. The restriction can produce perceptions and behaviors that will produce compensation and substitution throughout other body regions.

Another patient with the same forward-bending pain and no forward-bending dysfunction might have great mobility in all movements, but demonstrate noteworthy difficulty balancing on one leg for more than a few seconds. This issue is a sign of poor single-leg balance, body awareness and muscular control and stability, warranting further investigation but representing a completely different pattern of dysfunction associated with the exact same symptomatic complaint.

These patients have the same initial complaint of low back pain, but significantly different clinical signs with respect to movement patterns—*they have contrasting dysfunctions*. One displays reduced mobility or restriction, and one displays reduced stability or control.

Unfortunately, these two will often receive the same treatment, exercises and lifestyle instructions from the physician, therapist or chiropractor. Movement is not as easy to manage as a bruise, but many still treat it as if it was as rudimentary. They focus on the site of the pain and attempt to modulate the symptom by treating and exercising only that region.

Low back pain is the most common musculoskeletal complaint reported by adults in the US,[40] with more than one in four Americans reporting low back pain lasting at least one day in the previous three months.

This statistic begs the question: Do we have a low–back-pain epidemic or a low–back-pain management problem?

THE SFMA CHARACTERISTICS

We did not design the SFMA as a predictor of risk. The SFMA is instead used to gauge the status of movement–pattern-related pain and dysfunction. It uses movement to provoke symptoms and demonstrate limitations and dysfunctions; the information in a movement pattern deficiency relates to the patient's primary complaints. The SFMA is an opportunity to observe movement-pattern behavior before deconstructing these into impairments, measurements and other isolated testing. A complete functional profile emerges when the evaluation begins with the SFMA. The ability to quickly and easily revisit a movement test in the SFMA provides a systematic behavioral perspective that has not previously been used in sports and orthopedic outpatient rehabilitation.

The distinction between a screen and an assessment is this: The screen is done by health and fitness professionals on normal populations, whereas the assessment is performed by a medical or rehabilitation professional already aware of an abnormal condition. Pain is usually the primary complaint, which may or may not be complicated by movement dysfunction. Whether the patient has pain, is referred to a healthcare professional following a movement screen, or seeks medical advice directly, a systematic breakdown of the pain relative to movement is the first order of business.

MAPPING MOVEMENT PATTERNS

The movement assessments introduced next will use some of the same basic patterns found in the movement screen. This often creates surface confusion for the professional new to the complementary practices of screening and assessing movement. Don't get caught up in the differences of the two tools; they are designed to do two completely different things. Embrace the similarities and understand the differences. For the clinician, the tools provide movement perspective at intake (SFMA—diagnostic) and discharge (FMS—predictive).

The movement patterns look similar to the outsider because both screening and assessment review basic human movement. This assessment system filters pain, limitation and asymmetry and uses redundancy; it is selective because it creates opportunities to view movement patterns with varying degrees of bodyweight- and asymmetrical-load bias. It also works developmental posture in reverse. We are not often surprised when a right single-leg balance problem has its roots in a rolling-pattern motor-control problem we review 90 seconds later.

The information in the SFMA is rated and ranked in a completely different way than the FMS. We don't grade movement on an ordinal screening scale—*3, 2, 1, 0*. Since pain is involved, it must be considered to potentially exist in both functional and dysfunctional patterns.

Movement is not grouped based on quality alone, but on the way the two variables of pain and movement quality interact. This grouping creates two primary questions.

Does the movement pattern quality appear normal and functional, or limited and dysfunctional?

Does the movement produce pain or not?

If you over-think these responses or expand the answers, you are simply trying to make the system a comprehensive evolution tool instead of the movement-pattern component. We often see new individuals to the system investigate and discuss volumes of minutia and detail in the SFMA as if it is the only chance they will have to collect information. We tell them to force themselves to follow the steps and then proceed with the rest of the evaluation. Once the information is collected, the brain will start to connect the dots.

Ultimately, the SFMA will not diagnose a patient. It will only map movement patterns, but by mapping these patterns at a functional level through the top-tier testing or the developmental level through the breakout testing, the clinician can accurately identify successful interventions with treatment and exercise from those that are unsuccessful at changing movement behavior.

SFMA CRITERIA

Multiple movement patterns are reviewed, making it possible to observe each outcome within one assessment. The questions can produce four responses; study these because you'll use the responses throughout your scoring and recording. To help you become more familiar with these, I'll begin using them throughout the remainder of this text.

FN—*Functional or normal movement pattern, no pain*

FP—*Functional or normal movement pattern, with pain*

DP—*Dysfunctional or limited movement pattern, with pain*

DN—*Dysfunctional or limited movement pattern, no pain*

Seven basic movements are standardized for classification, and some patterns are broken down for clarity and perspective. A true functional diagnosis must start at this level, and further investigate at least one abnormal path.

Four potential scenarios emerge from the SFMA. Knowing each scenario is necessary, but two in particular will have the greatest clinical relevance for picking the path of further investigation.

Two patterns, FP (functional with pain) and DN (dysfunctional, no pain), are the clearest guides for the application of corrective exercise. These also serve as the most uncomplicated identifiers for successful manual therapy. Neglecting to use categorization could send the clinician down the wrong path; observation, testing and measurements taken in the wrong direction are time-consuming in the best circumstances and potentially dangerous in the worst.

The first step of the SFMA separates dysfunction and pain among the different movement patterns whenever possible. In the process, the clinician will discover that not all painful movement is dysfunctional, and not all functional movement is pain-free. Although this sounds like the most obvious fact in the world, you should know that your brain has a tendency to focus on painful movements and assume dysfunction and with non-painful movements neglect to see all but the most obvious dysfunction. Systems can protect us.

This simple breakdown sets off a second movement analysis designed to demonstrate mobility and stability problems within a particular pattern, and directs the user to specific review of pertinent movement-based impairments performed in a comprehensive evaluation that should always follow positive findings on the SFMA.

Chasing the pain and choosing the wrong path are common in clinical practice. This idea of chasing pain is seen by experienced clinicians as they witness the new graduate or intern get absorbed in the patient's symptoms, and ignore or value to a lesser degree other significant objective signs and clinical findings.

Let's consider the four answers above, because each represents a clinical path or journey into movement.

FUNCTIONAL AND NON-PAINFUL (FN)

Functional and non-painful signifies normal movement patterns without pain, *The Dead-End*.

Following this path proves everything is normal. Using the "find the flat tire" analogy, a slight abnormality observed here would represent low air pressure, not a flat tire. Remember the words of Cyriax:

"It is well to remember that the object of the physical examination is to find the movement that elicits the pain of which the patient complains, rather than some other nebulous symptom of which he was previously unaware."

The job here is to find the flat. Functional and non-painful (FN) is a dead-end.

We are not here to document movement perfection. We are here to find the weakest link in the movement pattern chain. FN does not mean perfect. It simply denotes where the weakest link is not. We have provided acceptable movement pattern standards based on biomechanical studies, goniometry, and lots of professional experience. The standards for each top tier test should provide you enough information to make the decision.

FUNCTIONAL AND PAINFUL (FP)

The functional and painful category represents normal movement patterns with pain, *The Marker*.

A patient's complaint may be out of the scope of a movement assessment if movement does not have any effect on pain. Constant pain, unaffected or modulated by movement or position, is not a good sign in an orthopedic rehabilitation setting. Outside the acute inflammation and muscle spasm associated with musculoskeletal injury or trauma, continuous unchanging pain can indicate a systemic problem, a non-orthopedic problem, or a serious psychological problem. Nothing further can be gained with a movement-based assessment regardless of origin, and other diagnostic services would be more appropriate.

However, the confirmation of pain *with* movement creates a pattern of provocation. The clinician then has a consistent marker for pain with movement and can proceed with the assessment, able to revisit the marker to establish change or variation. Establishing an indicator is helpful during treatment because it can be positively or negatively affected long before the patient will report a change in perceived symptoms.

These patterns are helpful because they demonstrate movement patterns that produce or provoke pain. To the observer there is no obvious problem with the movement, but the patient reports pain. This should confirm that the movement and the patient's pain are linked.

The painful, functional movement pattern can also be broken down to investigate sub-movements that may or may not provoke symptoms.

Corrective exercise intervention is not helpful or necessary in this case since the movement pattern is within functional parameters. No particular phase of movement needs to be rehearsed or exercised because it is already functional. However, it is painful and should not unnecessarily be repeated, as this can exacerbate symptoms and adversely affect treatment. This functional and painful pattern is your marker.

DYSFUNCTIONAL AND PAINFUL (DP)

With dysfunctional and painful, we have a limited movement pattern with pain, *The Logistical Beehive*.

These patterns are complicated because too much happens to create a reliable marker if not isolated. The findings in the breakdown may be different, and could show either an FP or a DN pattern.

Do not forget the question: *Is pain causing poor movement or is poor movement causing pain?* Answer this by marking and remembering the findings observed, but don't attempt corrective exercise in this pattern unless it is a last resort. Dysfunctional and painful is a logistical beehive when exercise in considered. An expert clinician could navigate these unpredictable paths, but never as the first option.

Common examples of DP can also occur as the result of post-surgical, and traumatic situations complicated by chemical pain. These patterns are best addressed with hands-on work instead of general exercise. Anti-inflammatory modalities, manual therapy and functional taping can shift these patterns into another category, but corrective exercise is not recommended until the pattern has changed.

If you are aware that chemical pain associated with trauma, surgery, or other cause is present, consider putting off the SFMA until the chemical component of the pain has been managed.

The limited approaches to exercise done in the presence of this pattern are mostly to improve or maintain circulation or joint and tissue mobility. These exercises are an attempt to maintain or improve a quantity of movement or metabolism, not to improve the quality of the pattern.

Lastly do not fail to consider the emotional component associated with pain. It can run the gamut from denial to extreme magnification.

DYSFUNCTIONAL AND
NON-PAINFUL
(DN)

Finally, we see limited movement patterns without pain, *The Corrective Exercise Path.*

These patterns are the key to understanding movement and movement dysfunction, and make it possible to consider the same filters used in the Functional Movement Screen. Without pain, we can look at movement uncomplicated by pain and the behaviors surrounding it. This means a skilled confident clinician will not have to always wait for symptom reduction to confirm the correct treatment path. Positive changes in movement prior to symptom reduction can be a powerful clinical confirmation.

Three remaining filters will help you refine the information: *limitation, asymmetry and redundancy.*

Choose the pattern with the greatest limitation, the pattern farthest from normal findings.

Pick the simplest pattern or the one with the lowest physical demand if more than one dysfunctional and non-painful pattern is present.

Asymmetrical dysfunction/limitation trumps symmetrical dysfunction/limitation.

Return to the next most-pronounced limitation or asymmetry once the first limitation is resolved.

Finally, redundancy is also built into the SFMA. Movements are checked and rechecked to observe consistency, which suggests potential **mobility** impairments, or inconsistency, which suggests potential **stability** impairments.

Working on corrective exercises for this pattern will require faith and confidence in the system. It can be hard to draw a correlation between the patient's symptoms and the limited non-painful movement identified, and there's even greater difficulty explaining this to the patient.

Be prepared for the following type of conversation.

Patient—*It's my neck, not my shoulder. Sure, my shoulder is a little stiff, but I came here for my neck. You are making me exercise my shoulder. Nothing is wrong with my shoulder. How can that be my problem?*

Calmly explain the neck movement patterns were painful, but not dysfunctional.

Clinician—*Yes, the movement of your neck was painful, but the neck was not limited in motion. There is nothing in the neck to work on from a mechanical or functional standpoint. Of course, we'll do something to help with the pain and inflammation, but there is nothing in your neck exercise will alter. The movement is normal; it just happens to be painful when you move. We will continue to monitor the neck for changes in pain.*

The shoulder was where the dysfunctional pattern was detected. That it doesn't hurt means you automatically compensate. This can be because of poor mobility and stability within the shoulder region, and means you do something unnecessary and unnatural with your neck and upper back to help the shoulder keep up with the demands. These little allowances create small amounts of stress and strain in the neck. We call these compensations.

Over time, the stress has caused irritation in the joints and muscles that support and move your neck. These structures are working overtime.

Working overtime is never a permanent solution to a labor problem. People and body parts can pull a little overtime occasionally, but when overtime becomes an everyday event, we see a breakdown.

The neck has had enough, but it is not the issue; the shoulder is. It's enjoying life under the radar, doing less than the job requires. The neck was picking up the extra work and started to break down. It was responding normally to an abnormal circumstance; when it had enough and started to complain, the complaints were justified.

The assessment I just performed called out the shoulder. Now we can investigate things with a little more clarity and detail. Your neck was the painful problem; the shoulder was probably the functional or mechanical problem that started this snowball.

ELEMENTS OF THE SFMA

The following is a description of suggested elements of the SFMA, the first in seven general categories of functional movement. Later we'll provide detail to describe the sequence and additional maneuvers contained within those categories. The system uses the four filters introduced in the FMS.

Pain—*provoked with movement patterns*
Limitation—*noted in movement patterns*
Asymmetry—*noted within movement patterns*
Intentional redundancy—*duplication of movements for inspection and consistency*

THE SFMA TOP-TIER TESTS

CERVICAL SPINE PATTERNS
Cervical spine movements

UPPER EXTREMITY PATTERNS
Two patterns and two pain-provocation signs

MULTI-SEGMENTAL FLEXION
Toe-touch maneuver

MULTI-SEGMENTAL EXTENSION
Overhead reach with spine extension

MULTI-SEGMENTAL ROTATION
Head, shoulder and pelvis rotation

SINGLE-LEG STANCE
Postural muscle response

OVERHEAD DEEP SQUATTING
Heels flat with shoulders flexed

Grade each with a notation of FN, FP, DN or DP. All responses other than FN can be further isolated to refine the movement information and direct the impairment testing to follow. The most obvious DN and FP are the best choices for initial breakdown.

Look for consistencies, inconsistencies and levels of dysfunction for each movement deduction compared with the original and other deductive movements of the same pattern.

FUNCTIONAL MOVEMENT ASSESSMENT RESULTS HIERARCHY

Separating painful movement and dysfunctional movement creates clarity and logistics when formulating a treatment strategy, and in most cases, we're able to observe the four movement categories described earlier. *The dysfunctional and non-painful movement pattern is the key.* This creates true synergy between the screening model and the assessment model.

Multiple DN findings also must be considered in a hierarchy represented by the order of the SFMA top-tier tests. This means a cervical DN must be addressed before a shoulder DN. A shoulder DN must be addressed before forward bending and backward bending DNs. The forward and backward DNs must be addressed before the rotation DNs. Rotation DNs must be addressed before DNs in single leg stance. Lastly all DNs must be addressed before squatting DNs. This hierarchy provides the best possible environment for re-acquisition of movement patterns. Each level of movement plays a role in the next patterns function and corrective exercise options. Therefore, whenever possible follow the hierarchy provided above.

Obviously DNs resulting from permanent restriction, long standing disability, extensive scaring, surgical fixation and joint replacements, as well as other factors may require the clinician to remove the DN from the scope of management. It should also be noted that improvements are possible in regions of chronic restriction. Clinical judgment is always the best guide.

As you will see later in the book, the FMS and SFMA corrective exercise strategies for dysfunctional movement are similar, even identical in some cases, although we do screening and assessment for different reasons and under different conditions.

We direct corrective strategies at movement dysfunction, not pain. We don't deal with pain in screening, and we perform assessments specifically so we do not exercise into pain. We map it and treat it but we don't try to exercise it away. We exercise dysfunction uncomplicated by pain whenever possible. We treat the chemical and mechanical causes of pain, but our corrective strategies and movement learning opportunities are directed at dysfunctional patterns that do not produce pain. This produces confusion for some clinicians who are taught manual techniques that do produce pain and discomfort, but we remind them that manual treatments do not require unassisted motor control by the patient. We must deal with pain during treatment, but we should not expect patients to have high levels of motor control and motor learning while experiencing pain. We should reserve most of the corrective exercise for dysfunctional non painful or patterns where the pain has been effectively managed.

Many well-educated people are confused by the simplicity of this movement-management model. Both functional screening and assessments are arranged to remove pain from the equation. Pain is obviously considered, but separated from dysfunction whenever possible to create consistency and clarity. Even though pain can alter or influence function, it is a *symptom* and not a functional entity.

If all function is restored and pain persists, it's obvious a problem still exists, but it's not a functional problem. This problem belongs to another profession; it will not follow functional logic or attempts at corrective exercise strategies. Cancer, systemic disease, physiological and psychological disorders can all produce pain outside the parameters of functional movement. This system can only suggest a referral to another medical specialty when the problem is not a movement-based problem.

The system is set up to drive corrective strategy to the greatest functional limitation, and to remove the inconsistencies that, when mixed with active movement, chasing pain will inevitably produce. At the same time, it will immediately highlight a non-functional problem, and it creates the logic for a referral when the pain cannot be linked to function.

The philosophy of screening and assessment is that functional movement patterns must be restored to a degree of normalcy for true movement homeostasis.

The ultimate goal is the restoration of normal movement-pattern minimums whether screening personal training clients, athletes or laborers, or when assessing patients referred with pain. Assessing and screening are the best ways to rehabilitate, reduce risk and minimize chances of recurrence. We have done nothing functional if movement patterns don't change beyond a minimum standard.

The truly objective professionals are those who know they are not—and therefore use systems and strategies to avoid the pitfalls of professional bias, partiality and subjectivity.

The system in a nutshell—

Set a movement-pattern baseline
Locate and observe the movement problem
Use corrective measures directed at the problem
Revisit the baseline

Maintain this general structure and the functional movement system will work. We are not naive and know that complete resolution of pain or normal function may not be possible for some. However we should still set goals against a current level of function and pain. Ultimately partial attainment of a goal is always better than no progress at all.

One consideration must be proposed. The clinicians who have helped me organize and refine the SFMA have all significantly improved the function and pain status of patients who were told there was nothing more that could be done. As we developed our expertise, our patient load became more complicated not easier. We dealt with patients with previous unsuccessful outcomes on a weekly basis. The point here is not that we were successful all the time because we were not. The point is that sometimes we were successful and the beneficiaries of that success appreciated our honest attempts. Our commitment to the system made this possible.

THE UTILITY OF THE SFMA

This chapter discusses in detail how to perform the SFMA. We will review the concepts and discus the implications of breaking down the information. The SFMA relies on a two-step, sequentially dependent process. The two steps create the moniker *selective* that precedes *functional movement assessment.*

Step one identifies the status of basic movement patterns in the categories of the absence or presence of pain, and the appearance of function or dysfunction. This is the four-step filter.

Step two introduces a system of reduction of the movement pattern or patterns in question. The reduction consists of systematic removal of load and regression though some developmental postures and patterns.

The requirement to rate and rank before measurement is ever-present as with any other screening, even though the SFMA is a clinical tool. The clinician can perform a breakdown process within each abnormal pattern by strategically selecting those that are dysfunctional and painful, dysfunctional and non-painful, and functional and painful.

The reduction process provides a clearer picture of the dysfunction and the characteristics of the movement-based pain. The selective process will examine movement and components of movement for symmetry, limitation and restriction, and variations between loaded and unloaded structures.

With the addition of the SFMA, the clinician benefits in the following ways:

The SFMA helps the clinician refine and rank the patient information to develop a more comprehensive functional diagnosis.

It adds perspective as it balances behavioral movement patterns with mechanical breakdowns necessary for structural isolation.

The SFMA will help the clinician use the most beneficial therapeutic and corrective exercise choices based on movement dysfunction.

It provides a method of capturing examples of regional interdependence, which demonstrate how structures and functions far from the site of the symptoms affect and influence pain and utility.

This is important because exercises are often wrongly used, based on isolated movements at or near the site of pain, or on generalized protocols related to a basic diagnostic classification of the patient.

It provides a systematic process that intentionally avoids the provocation of symptoms with exercise because of the negative effect on motor control and compensation.

THE SFMA VERSUS THE FMS

The power and utility of the SFMA can create confusion among those who work with patients and also advise non-patient clients in training in a non-clinical relationship. While some suggest the FMS and SFMA could be interchangeable, both should be used in a clinical setting, but they are not interchangeable: *The line remains at pain.*

Initially on patient intake, the SFMA will help the clinician navigate painful and dysfunctional movement patterns with simple classifications.

The FMS can provide an excellent discharge appraisal as the active person emerges from the problem and displays no remarkable pain associated with movement. The FMS is more appropriate on discharge because it is a prediction tool, not a diagnostic tool. A person can pass the SFMA and still have many issues on the FMS, but an excellent score on the FMS almost assures a good performance on the SFMA, outside of cervical assessment.

Most clinicians realize this after applying both systems and can easily see how they complement each other. One tool is not more complex, thorough or accurate; you must consider situational correctness. It's simply a case of diagnosis versus prediction.

The SFMA should not be routinely substituted for the FMS in non-clinical situations. The SFMA is not designed to provide predictive information. It is useful if the FMS demonstrates pain with movement, which immediately changes the relationship to that of referral or evaluation and treatment.

The sensitivity of the SFMA creates unnecessary breakdown of movements not related to the weakest link in the non-painful person, and leaves many moderate- to high-demand movements such as the inline lunge and the push-up unchecked. Single-leg stance will not be graded with the accuracy or reproducibility of the hurdle step, and symmetrical trunk control in quadruped will not be viewed in consideration with other movements. Consider the SFMA a low-demand movement filter when there is pain, and the FMS moderate to high demand for people who have no complaint of pain and plan to return to active lifestyles.

The act of movement screening is predictive for basic function and injury risk during training, activity and competition. The FMS is a tool for risk management and performance, the first step in responsible physical development, training and conditioning. It is introduced to people who have chosen to become active or to remain active; their intention is performance and they have no presence of pain.

The FMS can stand alone as a movement appraisal within normal populations intending to work on performance and conditioning, and it can be a prudent and conservative measure before performance or sports-specific testing prior to athletic seasons. It can also serve as a screen for potential risk prior to training for military, fire service, or other high demand occupation.

PAIN AND MOTOR CONTROL

Musculoskeletal pain is why most patients seek medical attention. The contemporary understanding of pain has moved beyond the traditional tissue-damage model to include the cognitive and behavioral facets. Most scientists accept that pain alters motor control, although the mechanism of these changes has not been clearly identified.

Pain causes changes in coordination during functional movements; the interaction between pain and motor control depends on the motor task. Researchers are now focusing on how pain alters the timing of muscle activation and movement patterns.

For example, Zedka et al[41] studied the lumbar paraspinal muscle response in subjects during natural trunk flexion movements before and after induced pain. The study demonstrated an altered level of paraspinal muscle activity and a 10- to 40-percent decrease in range of motion during the painful condition. Interestingly, when hypertonic saline was injected unilaterally, EMG changes were seen bilaterally, suggesting that pain alters the entire movement strategy.

The pain-adaptation model as described by Lund et al[42] predicts that pain will alter muscle activity depending on a given muscle's role as an agonist or antagonist to control movement. This model was the first attempt to explain how pain may either increase or decrease muscle activity. Further research has shown that this theory does not hold true under all conditions, and in some movements it's difficult to clearly identify the role of a given muscle. The pain-adaptation model has provided a mechanism to further study and explains changes beyond simple peripheral reflexes as previously believed.

The central nervous system (CNS) response to painful stimuli is complex, but motor changes have consistently been demonstrated and seem to be influenced by higher centers consistent with a change in the transmission of the motor command. Richardson et al[43] summarizes the evidence that pain alters motor control at higher levels of the CNS than previously thought.

"Consistent with the identification of changes in motor planning, there is compelling evidence that pain has strong effects at the supraspinal level. Both short- and long-term changes are thought to occur with pain in the activity of the supraspinal structures including the cortex. One area that has been consistently found to be affected is the anterior cingulated cortex which has long thought to be important in motor responses with its direct projections to motor and supplementary motor areas."
~Richardson, Hodges and Hides

Because it has been shown that pain alters motor control as high up as the motor planning level of the central nervous system, current research is focused on how muscles respond in painful conditions during different tasks. For example, Kiesel et al[44] demonstrated increased lumbar multifidus activation during a simple arm-lifting task and decreased multifidus activation during a weight-shifting task under an induced pain condition. These data suggest the CNS can instantaneously alter muscle activation during pain depending on the movement task at hand. Current thinking of how pain alters motor control specific to patients with low back pain is best captured by van Dieen et al,[45] who state that "motor control changes in back pain patients are functional in that they enhance spinal stability and are likely to be task dependent."

Because evidence suggests that pain alters motor control, likely at the planning level and dependent upon the motor task performed, our assessment of functional movement patterns must consider this. Pain-attenuated movement patterns can lead to protective movement and fear of movement, resulting in clinically observed impairments such as decreased range of motion, muscle length changes, declines in strength, and ultimately may contribute to the resultant disability.

Many components comprise the pain-free functional movement desired in occupation and lifestyle activities. Impairments of each element could potentially alter movement because of or resulting in pain. After assessment, we use traditional muscle length and strength tests, as well as other measurements to help identify the impairments associated with dysfunctional movement patterns.

The SFMA approach is not a substitute for existing exams and intervention, but a model to integrate the concepts of posture, muscle balance and the basic patterns of movement into the contemporary medical and rehabilitation practice. The approach assumes the clinician applying this model adheres to accepted indications for manual therapy and therapeutic exercise intervention, and has ruled out CNS lesion, progressing nerve root compression or a peripheral nerve problem when neurological signs are present.

A FUNCTIONAL PERSPECTIVE OF THE CLINICAL MUSCULOSKELETAL EVALUATION

The efficient and effective progression from qualitative tests and assessments to quantitative measurement is the bedrock of manual musculoskeletal evaluation. This sequence allows the qualitative tests and screens to control the direction of the evaluation, while quantitative measurements define and quantify specific information of anatomical structure, mechanical and physiological function, state of utility and the severity of the symptoms.

Much like a compass points, the traveler toward a destination, the SFMA gives direction to the problem-solving process used in musculoskeletal evaluation. Quantitative data such as time, speed and distance have relevance once the traveler is headed in the right direction, while these are of little use when moving off course. The novice clinician too often collects a large amount of quantifiable data without identifying the directional nature of the basic problem, much like the traveler making excellent time along the wrong path. Fixing this fundamental problem with logical systems is the purpose of this chapter.

LEVELS OF INFORMATION

We must collect information about movement at three levels.

The practical level shows disability; we gather the information by history and observation of the daily routine, as well as sports, leisure and work activities.

At the functional level, the information shows dysfunction within basic movement patterns, supports functional movements and will be identified by the SFMA.

On the clinical level, we identify impairments as the result of specific clinical observation and documentation that is quantifiable with testing and measurement.

Disabilities caused by the problem are the lifestyle limitations encountered, and are identified by taking a thorough history. Activities of daily living are included with other altered behaviors to constitute a person's functional status and movement ability. Disability can be measured by a variety of self-reported questionnaires ranging from generic measures of health such as the *SF-36* self-administered pain test, to more specific tools such as the *Modified Low Back Pain Disability Questionnaire* and the *Disabilities of the Arm, Shoulder & Hand (DASH)* form. There are also easy-to-use patient-specific tools for disability measurement, such as the *Patient Specific Functional Scale*. Samples of a few of these forms are provided in the forms appendix on page 391.

You should understand that measurement of disability with self-reported tools such as the tools recommended here is considered an objective measurement of function. Each tool has known measurement properties that have been established through scientific study. It is important the clinician utilize and understand the measurement properties of each tool.

Dysfunctions are identified by the clinician's ability to use functional and fundamental movements to demonstrate asymmetries, limitations or to provoke symptoms. Attempts to identify dysfunction involve an opportunity to relate functional and basic movement patterns to the previously gained knowledge of practical movements in daily life. The SFMA is one way to identify dysfunction.

Impairments are abnormalities or limitations noted at particular segments of the body. Limitations are measured with respect to strength, range of motion, irritability, size, shape and symmetry. We then compare this information with normative data and bilaterally when appropriate.

EXAMPLE

A 60-year-old female sedentary office worker who denies any remarkable or recent injury to the right knee complains of knee pain with certain activities.

Disability example—right knee pain specifically while descending stairs with a LEFS score of 62%; see sample LEFS form in the forms appendix

Dysfunctions examples—right knee pain while squatting with full range of motion, no limitation noted, poor static control with right single-leg stance (less than five seconds) compared with left (greater than 15 seconds)

Impairments examples—reduced medial rotation and abduction of the right hip with goniometric measurement, and right hip abduction strength deficit with isometric dynamometry

SUMMARY OF INFORMATION

- No joint effusion or tissue swelling is noted and ligament testing is within normal limits.

- Right hip mobility limitations for abduction and internal rotation with generalized abduction weakness or poor stability contribute to compromised femoral alignment under eccentric loading of the quadriceps.

- Repetitive stress on patellofemoral and tibio-femoral joints has resulted in inflammation and muscle dysfunction.

- The inflammation has not limited the squatting movement pattern, but has created pain in squatting due to poor dynamic hip stability.

PLAN OF CARE

Normalize hip mobility with muscle and joint mobility techniques, and perform progressive static-to-dynamic stabilization exercises to normalize abductor function.

Revisit single-leg stance and squatting to check functional status, and assess knee symptoms while descending stairs.

Initially consider manual techniques to increase hip mobility complemented by and followed with exercise to establish control.

Consider taping the knee to provide protection, support and symptom management in the early rehabilitation process. Each of the three levels of function should be reappraised to mark progress.

EVALUATION HIERARCHY

The evaluation hierarchy should start with a case review and patient history that specifically notes those activities that provoke the symptoms associated with the primary complaint. A clinician is obligated to reproduce some of the painful movements in various patterns to draw conclusions about functional mobility and stability if most of the symptoms happen with movement.

However, if the symptoms are provoked in static postures, such as extended standing or extended sitting, we look at the structures stressed in these postures. The history will provide the first indicator for clinicians, and will direct the next series of questions as well as the assessment maneuvers. The clinician must look at functional movement in as many postures as the patient's symptoms will allow, after establishing the primary complaint is associated with static or dynamic problems, or both. This creates a feedback system to confirm the functional diagnosis, as well as to validate treatment practices.

A quick assessment of the patient's available mobility in the upper and lower quarters and in the spine is an effective starting point for a functional assessment.

For example, within the limits of pain, a patient is asked to perform a forward bend toe touch, a backward bend, a squat and single-leg stance to note the provocation of symptoms and limitations or poor stability in movement. These movements can be replicated in an unloaded position by having the patient perform a long sitting reach, a prone press-up and a posterior rock in quadruped or bilateral knees-to-chest in supine.

The clinician can thus deduce the interplay between the patient's available mobility and stability in both loaded and unloaded positions. Functional movement clues are apparent when the first four movements in a loaded position are painful, restricted or limited in some way before the end range of motion.

When these unloaded movements do not provoke symptoms or have limitation, appropriate joint range of motion and muscle flexibility testing should be done to confirm mobility problems are not present. If the movements are performed easily in an unloaded position, a stability problem may be the cause of inability to perform the movements while in a loaded position. No specific diagnosis has been made, but a general behavior scenario or a consistent pattern is beginning to emerge. The patient has the requisite biomechanical ability to go through the necessary range of motion to perform the task, but the neuromuscular control and responses needed for stabilization to create dynamic alignment and postural support are not present.

A second scenario might be a patient who has limitation, restriction or pain, both loaded and unloaded; the patient demonstrates consistent abnormal biomechanical behavior of one or more moving segments. This situation also requires further specific clinical assessment of each joint and muscle and their related tissues to identify the barriers restricting movement and responsible for the pain.

These four movements may not be appropriate for all patients, but are used as an example because for almost any circumstance the interplay between loaded and unloaded conditions can be created and assessed in most musculoskeletal conditions.

Try maneuvers such as these before specific segmental clinical assessment and impairment measurements to direct the evaluation in an appropriate manner; start with general movement patterns before getting specific. Inexperienced clinicians are hindered in their diagnosis because too often they are focused on impairment measurements to confirm a medical diagnosis and fail to refine, qualify and quantify the functional parameters.

The therapeutic plan of care must start with a full representation of the symptoms and limitations shown in the initial examination. The functional diagnosis demonstrates postures and movement patterns that result from or cause the medical diagnosis and should expose whether a static or dynamic problem exists.

The diagnosis should also demonstrate whether a stability or mobility problem is present. From there, the clinician can use clinical tests, impairment tests and specific musculoskeletal examination techniques to deduce the structures and functions responsible for movement limitations.

Here are the specific keys the clinician should note during a musculoskeletal examination when function and posture are initially the primary concern.

Faulty alignment

Loss of spine stability

Indicators of tonic holding or the absence of co-contraction, or both

Provocation of symptoms

Functional asymmetries

Significant restriction and agreement or ambiguity between loaded and unloaded functional range of motion

The hallmark of the SFMA design is to use simple, basic movements to expose natural reactions and responses by the patient. These movements should be viewed loaded and unloaded whenever possible, and bilaterally to display functional symmetry.

FUNCTIONAL PATTERNS AND POSTURES

The hierarchy of the human balance strategy will often show a reduction in functional posture as well as in movement. Automatic responses are a much more objective indicator of a patient's function than the ability to perform organized tasks. For example, when a person is pushed gently from behind, the first and most primary balance strategy is a quick response to the closed-chain dorsiflexion of the ankles using a concentric contraction of the plantar flexors.

However, spine and hip stability and alignment must also be maintained so a hip hinge does not occur; the body remains upright, reactively rigid

FUNCTIONAL PATTERNS AND POSTURES

and taut for the plantar flexors to right the body with a perfectly timed and adjusted contraction.

If a greater perturbation is performed with more stress, as in a forward push, the individual will use a hip-hinge strategy, creating an angle between the upper and lower body. Once again, balance is managed; the spine maintains its stability while the extensors of the hip use concentric force to adjust. This response occurs in the sagittal plane to shift of the body mass behind the foot and some in front, thus creating a delicate balance of adjustment. A more exaggerated perturbation will bypass the first two strategies and create a step strategy that will significantly widen the base to check the momentum of falling forward, creating protection and balance.

These examples may seem like simple balance strategies, but when examined using the functional perspective, parts of basic movement patterns emerge.

Consider the partial patterns of a squat, a toe touch and a single-leg weight shift. The hardest of these to visualize is the squat. How does the dorsiflexion and plantar flexion balance response equate with squat mechanics? This is easily observed if the squat is already part of the evaluation process.

Those who cannot squat without significant limitation or difficulty do not use a coordinated effort between the hip and ankle. These people start the squat strategy with knee flexion, moving the mass of the body behind the feet; this is not a squat in which the tibia has freedom to move forward and adjust and where the body mass balances over the feet. This position requires significantly elevated activity of the quads, and altered activity of the plantar flexors and hip extensors as control mechanisms.

In contrast, using the optimal ankle strategy, the person who can squat fully starts the squatting process with dorsiflexion and core stabilization. This person progresses by adding knee and hip flexion, and can easily go below parallel maintaining alignment with trunk stabilization and closed-chain dorsiflexion—the two primary elements of the ankle-balance strategy.

Next, we look at forward bending and the hip hinge. When done correctly, forward bending starts with a hip hinge and spine stabilization. The spine should only flex in a segmental manner after using the available range of motion in hip flexion. People with significantly limited forward-bending ability or the lack of a toe touch will commonly start the forward-bending motion in the thoracic spine instead of at the hip. They abandon the requisite core stabilization and spine stability needed for the maneuver, and try to perform flexion by using segments that should be stable and rigid through the first part of the motion. This takes them halfway through the motion using spinal flexion, however once the spine flexes to its limits, they experience significant tension in either the spine or the hamstrings, both of which serve to protect against further flexion of these faulty mechanics. Without effective posterior weight shifting, further flexion would result in loss of balance.

The common assumption that a limited forward bend indicates tight hamstrings is based on symptom complaint alone, an assumption that should instead be based on a functional length assessment of the hamstring muscle group. The hamstrings will be the structures on greatest tension in a forward bend whether done correctly or incorrectly because those are the muscles in the most direct position to gauge tension and rate of change for feedback and perception.

Finally, the single-leg weight shift or step strategy for balance is a lunge pattern that involves a symmetrical stance followed by single-leg stance and then an asymmetrical lunge stance. Spine stability, core stability and balance between the hip adductors and abductors need to exist at the beginning of the step and single-leg stance patterns. The person must also have adequate preparatory muscle activation and stabilization before landing on the opposite leg in order to regain balance.

THREE EXAMPLES OF LINKS TO CORRECTIVE EXERCISE

When considering these three automatic balance reactions, it should be obvious the squat, deadlift and components of the lunge are not simply exercises reserved for sport or athletic competition. These are key rehabilitation techniques, assuming appropriate modification is made based on the age and activity level of each patient.

Clinicians routinely remove squatting, dead-lifting and lunging exercises from rehabilitation because these are not considered age-appropriate for the elderly or activity-appropriate for the non-athlete. However, without these three balance strategies, the elderly person is at greater risk for a fall and dysfunction or micro-trauma due to substitution and compensation. Without identification and retraining, a non-athletic person will not regain the full command of the lost movement pattern. Instead of appropriately adjusting these activities, these are often removed from rehabilitation programs at a significant cost to the treatment process.

Rehabilitation of the ambulatory patient who has an orthopedic musculoskeletal problem should follow the same continuum regardless of the activity level. The activity level only dictates how far along the continuum the patient is progressed.

PROVOCATION OF SYMPTOMS

Unfortunately, a functional orthopedic assessment must involve provocation of symptoms. Provocation of symptoms often occurs during the interplay of posture tests to observe movement in transition and in movement tests to observe the responding posture. Producing these symptoms creates the road map the clinician will follow to a more specific diagnosis. The patient usually accepts this when the provocation of pain is explained in a sensitive and logical manner.

- Once symptoms have been provoked, we work backward to more specific breakdowns of the component parts of the movement.

- Inconsistencies observed between symptom provocations that are not the result of symptom magnification—may suggest a stability problem.

- Consistent limitations and provocation of symptoms can be indicative of a mobility problem.

As an example, take the SFMA standing rotation test: A patient standing, feet planted side-by-side and stationary, makes a complete rotation using the segments of the entire body. Arms relaxed at the sides, the patient looks over the right shoulder and twists as far as possible. The patient then reverses direction.

When a consistent production of pain in the left thoracic spine is noted during standing left rotation, the same maneuver can be repeated in a seated posture. The stationary hip and leg position, although similar in spinal rotation, has many differences; with the hips and lower extremities removed from the movement, an entirely different level of postural control may result.

When noting nearly the same provocation of symptoms and limitations at the same degree of left rotation both standing and seated, the cause may be an underlying mobility problem somewhere in the spine. This mobility issue could be the result of a trigger point, increased or reduced muscle tone, joint restriction, faulty alignment or a combination thereof.

Alternatively, if the seated rotation does not produce a consistent limitation and provocation of symptoms in the same direction and at the same degree, this could be an indicator of a stability problem.

The position change results in a different degree of postural alignment, muscle tone, proprioception, muscle activation or inhibition and reflex stabilization. The clinician must now investigate the lower-body component of this problem. Once consistency or inconsistency is observed with respect to movement limitation or the provocation of symptoms, we continue to look for other instances that support this behavior.

Rule of thumb: Use provocation of symptoms in a functional musculoskeletal assessment within reason. When provoking symptoms, the clinician controls the degree and frequency, and has ample time to prepare and instruct the patient on the maneuvers to reach the desired outcome.

Likewise, the examiner must pay close attention during the exam to avoid over-provocation. Other than for reassessment and testing, functional activity and other active maneuvers should not provoke symptoms. Pain provocation is reserved for treatment when necessary, but once the patient starts an active rehabilitation program with corrective therapeutic exercise progressions, those exercises and progressions should not produce pain or symptoms.

THE SFMA TOP-TIER ASSESSMENTS

CERVICAL SPINE
UPPER EXTREMITY PATTERNS
MULTI-SEGMENTAL FLEXION
MULTI-SEGMENTAL EXTENSION
MULTI-SEGMENTAL ROTATION
SINGLE-LEG STANCE
OVERHEAD DEEP SQUAT

CERVICAL SPINE ASSESSMENT

OBJECTIVE

The first *cervical spine movement pattern assessment*—chin to chest—evaluates the amount of available cervical spine flexion, including occipital-atlas mobility.

The second *cervical spine movement pattern assessment*—face parallel to the ceiling—evaluates the amount of available cervical spine extension.

The third *cervical spine movement pattern assessment*—chin to left and right shoulders—evaluates the amount of available cervical spine rotation and lateral flexion. It is a combination pattern that incorporates side-bending and rotation.

DESCRIPTION

To perform pattern one: The patient assumes the starting position by standing erect with feet together and toes pointing forward. The patient then tries to touch the chin to the sternum, keeping the trunk erect during the movement.

To perform pattern two: The individual assumes the starting position by standing erect with feet together and toes pointing forward. The patient then looks up, aiming the face parallel with the ceiling.

To perform pattern three: The individual assumes the starting position by standing erect with feet together and toes pointing forward. The patient rotates the head as far as possible to the right, then flexes the neck, moving the chin toward the collarbone.

TIPS FOR TESTING

- Make sure the patient's mouth remains closed throughout the movement.

- Do not allow scapular elevation and protraction.

- Observe from the front and side.

- Do not coach the movement; simply repeat the instructions if needed.

- Was there pain?

- Could the movement be done? If not, proceed to the appropriate breakout.

ADDITIONAL INFORMATION

During pattern one, make sure the patient's mouth remains closed throughout the movement. The patient should be able to touch the sternum without pain.

During pattern two, the individual should be able to get within 10 degrees of parallel without pain.

During pattern three, normal range is mid-clavicle bilaterally without pain.

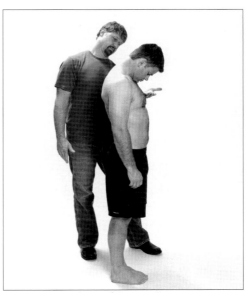

Cervical Pattern One

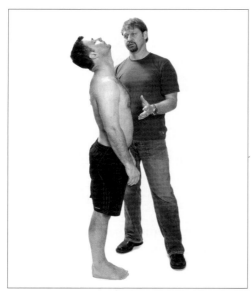

Cervical Pattern Two

Cervical Pattern Three

UPPER EXTREMITY MOVEMENT PATTERN ASSESSMENTS

Objective

The *upper-extremity movement pattern assessments* check for total range of motion in the shoulder.

Pattern one assesses internal rotation, extension and adduction of the shoulder.

Pattern two assesses external rotation, flexion and abduction of the shoulder.

Description

To perform pattern one: The patient assumes the starting position by standing erect with feet together and toes pointing forward. The patient then reaches back with the right arm trying to touch the inferior angle of the left scapula. Place a finger on the spot where the patient's fingers touch the back, and compare that spot to the left arm's test results. Note the distance from the scapula if the motion is reduced.

To perform pattern two: The individual assumes the starting position by standing erect with feet together and toes pointing forward.

The patient then reaches overhead with the right arm trying to touch the spine of the left scapula. Place a finger on the spot where the patient's fingers touch the back, and compare that spot to the left arm's test results. Note the distance from the spine of the scapula if the motion is reduced.

Tips for testing

- Observe from the rear and side.

- Do not coach the movement; simply repeat the instructions if needed.

- Was there pain?

- Could the movement be done? If not, proceed to the appropriate breakout.

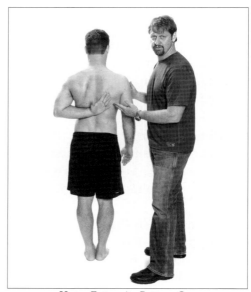

Upper Extremity Pattern One

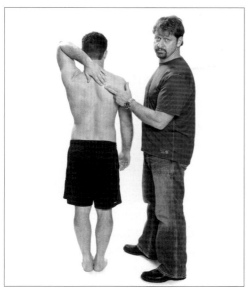

Upper Extremity Pattern Two

UPPER EXTREMITY PAIN PROVOCATION PATTERNS

OBJECTIVE

The first *upper-extremity pain provocation assessment* (pattern one, Yocum's impingement test) is for identifying rotator cuff impingement. The second *upper-extremity pain provocation assessment* (pattern two, the shoulder crossover maneuver) is for identifying AC joint pathologies.

DESCRIPTION

To perform pattern one: The individual assumes the starting position by standing erect with feet together and toes pointing forward. The patient then takes the right palm and places it on the left shoulder. Use your hand to stabilize the individual's hand against the shoulder. Have the patient slowly lift the elbow up to the sky. Repeat with left side.

To perform pattern two: The individual assumes the starting position by standing erect with feet together and toes pointing forward. The patient then takes an extended right arm and reaches across the chest. Have the patient use the left hand to help passively while horizontally adducting the right arm as far as possible. Repeat on the left side.

TIPS FOR TESTING

- Observe from the front and side.

- Do not coach the movement; simply repeat the instructions if needed.

- Was there pain?

- Could the movement be done? If not proceed to the appropriate breakout.

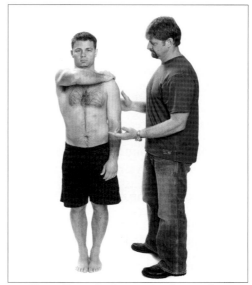

Upper Extremity Provocation Pattern One

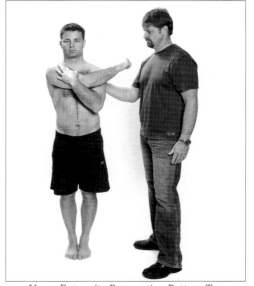

Upper Extremity Provocation Pattern Two

SFMA

MULTI-SEGMENTAL FLEXION ASSESSMENT

OBJECTIVE

The *multi-segmental flexion assessment* tests for normal flexion in the hips, spine.

DESCRIPTION

The patient assumes the starting position by standing erect with feet together and toes pointing forward. The patient then bends forward at the hips, trying to touch the ends of the fingers to the tips of the toes without bending the knees.

TIPS FOR TESTING

- Observe from the rear and side.

- Foot position should remain unchanged throughout the movement.

- Knees should remain straight.

- Do not coach the movement; simply repeat the instructions if needed.

- Was there pain?

- Could the movement be done? If not, proceed to the appropriate breakout.

ADDITIONAL INFORMATION

Look for the hips to move backward as the individual bends to touch the toes.

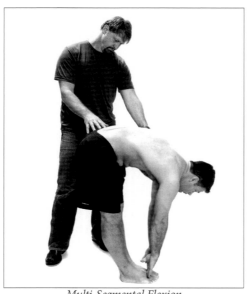

Multi-Segmental Flexion

MULTI-SEGMENTAL EXTENSION ASSESSMENT

OBJECTIVE

The *multi-segmental extension assessment* tests for normal extension in the shoulders, hips and spine.

DESCRIPTION

Have the patient assume the starting position by standing erect with feet together and toes pointing forward. The individual then raises the hands above the head with arms extended and with the elbows in line with the ears. Have the patient bend backward as far as possible, making sure the hips go forward and the arms go back simultaneously.

TIPS FOR TESTING

- Observe from the front and side.

- Foot position should remain unchanged throughout the movement.

- Do not coach the movement; simply repeat the instructions if needed.

- The spine of the scapula should clear the heels, shoulder blades behind the heels.

- The mid-hand line should clear the shoulder at the back of the extension with elbows remaining extended and in line with the ears.

- The pelvis stays in front of the toes.

- Was there pain?

- Could the movement be done? If not, proceed to the appropriate breakout.

ADDITIONAL INFORMATION

Mid-hand line should drop behind the shoulders at the top of the extension pattern.

Both ASISs should move past the toes, and the spine of the scapula on each side should go behind the heels.

Multi-Segmental Extension

MULTI-SEGMENTAL ROTATION ASSESSMENT

OBJECTIVE

The *multi-segmental rotation* assessment tests for normal rotational mobility in the neck, trunk, pelvis, hips, knees and feet.

DESCRIPTION

The patient assumes the starting position by standing erect with feet together, toes pointing forward and arms extended to the sides at about waist height. The patient then rotates the entire body—hips, shoulders and head—as far as possible to the right while the foot position remains unchanged. Have the patient return to the starting position, and rotate to the left.

TIPS FOR TESTING

- Observe from the rear and side.

- Foot position should remain unchanged throughout the movement.

- There should be at least 50 degrees of rotation in the lower quarter bilaterally.

- There should be at least 50 degrees of rotation from the thorax bilaterally.

ADDITIONAL INFORMATION

Since both sides are tested simultaneously with the feet together, the externally rotating hip is also extending, and this can limit rotation. Pay close attention to each segment of the body, the hips, trunk and head. One area may be hypermobile due to restrictions in an adjacent segment.

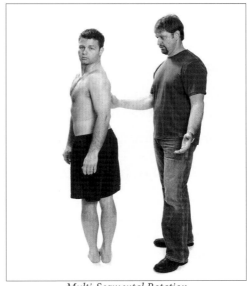

Multi-Segmental Rotation

SINGLE-LEG STANCE ASSESSMENT

OBJECTIVE

The *single-leg stance* assessment evaluates the ability to stabilize independently on each leg in a static and dynamic posture. Dynamic leg swings are also used in this assessment.

DESCRIPTION

The patient assumes the starting position by standing erect with feet together, toes pointing forward and arms extended toward the sides of the thighs. Have the patient lift the right leg so the hip and knee are both at 90-degree angles. The patient maintains this posture for at least 10 seconds. Repeat with eyes closed for 10 seconds, and repeat the test on the left leg.

To further assess the patient's single-leg stance in a dynamic posture, proceed to dynamic leg swings. For the dynamic leg swings, instruct the patient to stand with feet together, toes pointing forward and the arms extended by the sides but not touching the sides. Have the patient flex the right hip and begin to swing the right leg back and forth into flexion and extension of the hip while maintaining good posture and balance for at least 10 seconds. Repeat on the left leg.

TIPS FOR TESTING

- Observe from the front and side.

- The foot position should remain unchanged throughout the movement.

- Look for a loss of tall posture or height when moving from two legs to one.

- Look for the arms to flail.

ADDITIONAL INFORMATION

Tell the patient to *stand tall* before testing. The test is scored dysfunctional if the individual loses posture, has to move from the original foot position, falls or flails the arms. This requires good proprioception, muscular stability, and good hip and ankle strategies. Sometimes people can maintain balance statically, but not in motion. The back and forth movement in the sagittal plane of the dynamic leg swing can expose a dynamic stability problem.

In order to call this pattern FN, the patient being tested must be able to clear the assessments 10

seconds eyes open, 10 seconds eyes closed, and dynamic leg swings bilaterally.

SPECIAL NOTE ON VISION

Vision is never a handicap to balance—it always helps. Any vision is better than no vision; even a patient with severe cataracts will perform better with eyes open than eyes closed.

OVERHEAD DEEP SQUAT ASSESSMENT

OBJECTIVE

The *overhead deep squat* assessment tests for bilateral symmetrical mobility of the hips, knees and ankles. When combined with the hands held overhead, this test also assesses bilateral symmetrical mobility of the shoulders, as well as extension of the thoracic spine.

DESCRIPTION

The patient assumes the starting position by placing the instep of the feet in vertical alignment with the outside of the shoulders. The feet should be in the sagittal plane, with no external rotation of the feet. The patient then raises the arms overhead with the shoulders flexed and abducted and the elbows fully extended. Have the patient slowly descend as deeply as possible into a squat position. The squat position should be assumed with the heels on the floor, head and chest facing forward and the hands overhead. Knees should be aligned over the feet with no valgus collapse.

TIPS FOR TESTING

- Observe from the front and side.

- The hand width should not increase as the patient descends to the squat position. For testing repeatability, the patient should have a consistent hand width overhead. Two testing strategies are suggested—

    Have the patient place a dowel on the head and adjust the hands on the dowel so the elbows are at a 90-degree angle. Next, the dowel is pressed overhead with the shoulders flexed and abducted and the elbows fully extended.

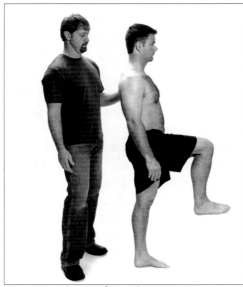

Single-Leg Stance

Have the patient start with elbows bent at 90-degree angle (90 degrees abduction, 90 degrees elbow flexion and 90 degrees external rotation), using that same hand width as the arms come overhead.

- Do not coach the movement; simply repeat the instructions if needed.
- Was there pain?
- Could the movement be done? If not proceed to the appropriate breakout.

ADDITIONAL INFORMATION

The ability to perform this test requires closed-chain dorsiflexion of the ankles, flexion of the knees and hips, extension of the thoracic spine, and flexion and abduction of the shoulders.

Overhead Deep Squat Start

Overhead Deep Squat

THE ART OF BALANCING MOVEMENT AND PAIN INFORMATION

The top-tier tests provide the clinician a practical and efficient snapshot of pain and movement behaviors. As you seek to understand and refine this information, follow the advice of Stephen Covey, author of the highly successful book *The Seven Habits of Highly Effective People*. Note his first three habits and realize you are applying them when you use the SFMA top-tier tests correctly.

The habits are—

Habit 1: Be Proactive
Habit 2: Begin with the End in Mind
Habit 3: Put First Things First

Here is how the habits apply to the correct use of the SFMA.

SFMA HABIT 1

You are being proactive since you are considering movement patterns for both dysfunction and pain provocation. The central purpose of the SFMA is to set a functional baseline for the course of rehabilitation and provide a systematic approach for the introduction of corrective exercise. Continuing the proactive theme: You have decided not to assign exercise based only on anatomical regions of impairment or generalized medical diagnosis. You will use movement patterns as a guide, and you have also considered the highly individual movement map that displays the unique perceptions and movement behaviors of each patient individually. This proactive professional act removes protocol-based corrective exercise and replaces it with an approach developed in direct complement to movement dysfunction while avoiding the complicating issues of pain provocation during corrective exercise.

SFMA HABIT 2

You have chosen to begin with the end in mind. You have not assumed all movements that are non-painful are functional. Likewise, you have not assumed all movements that are painful are dysfunctional. You have set a professional minimum standard for functional movement patterns in active populations, and you have decided to set goals based on functional movement patterns. You understand that addressing movement dysfunction is a valued service to patients entrusted to you, and it may be their only opportunity to have a complete functional movement-pattern assessment.

Movement-pattern dysfunction is emerging as a risk factor. Whether you are addressing dysfunctional patterns to improve a patient's current condition or simply reducing future risk, you are beginning with a compressive end in mind. Choosing to address the dysfunction is ultimately a clinical decision dictated by each individual case. As a movement specialist, your role is to inform, educate and offer treatment when appropriate. The SFMA can potentially form a link between a functional clinical perspective and a return to a more active lifestyle.

SFMA Habit 3

You have a new system and it actually complements the habit *first things first,* but sometimes you need two hierarchies. The first hierarchy removes the SFMA as an option altogether. The second approaches the hierarchy of the SFMA itself.

Hierarchy 1

Corrective exercise involving active movement is not indicated.

Not all rehabilitation situations warrant an SFMA. Acute trauma and post-surgical situations are complicated with chemical pain. This chemical pain is a result of inflammation, swelling, effusion, eccymosis and muscle guarding. Likewise, sub-acute and chronic conditions can also display levels of chemical pain that must be managed prior to corrective exercise involving active movements. The SFMA would also not be appropriate if an upper- or lower-quarter neurological scan or screen of dermatomes, myotomes, deep tendon reflexes or other testing reveals neurological compromise that was not part of the current diagnosis.

In situations of chemical pain and neurological compromise, appropriate treatment or further testing is more important than the map the SFMA can provide. The map would obviously be incomplete at best, and it would also incorporate movement dysfunction driven by clinical issues outside of non-irritable mobility and motor control limitations. Once these issues are effectively managed, the SFMA can be performed to better understand how movement patterns have been affected.

Some clinicians are often confused when the SFMA is not indicated on the normal intake examination, but sometimes it simply is not appropriate. The clinical application of corrective exercise should be the driver of the SFMA. It should be used prior to corrective exercise decisions and dosage. The SFMA assumes that serious medical issues like active chemical pain and neurological compromise have been effectively managed.

Hierarchy 2

Corrective exercise involving active movement is indicated.

The SFMA uses a hierarchy that introduces movement patterns that build on each other. The levels of movement-pattern involvement becomes more complex and progressively involves more anatomical regions and higher levels of motor control.

Obviously, it does not totally mirror the human developmental progression, but many of the principles are present. The SFMA begins with movement patterns of the cervical spine and moves to the shoulder movement patterns. This suggests that if both are found to be DN patterns, the cervical spine should be considered first. If possible, it should be managed to understand its influence on the shoulder movement patterns even if the shoulder patterns display more pronounced dysfunction. This in no way suggests the shoulder patterns should not be broken down or managed—it just implies that if both are treated and exercised simultaneously, it would be hard to attribute intervention to outcome.

Likewise, if a DN were noted in the shoulder movement pattern, it would probably influence the outcome and breakout information collected in the rotation patterns. Aggressively breaking down the rotation pattern and attempting correction before managing the shoulder patterns would be inappropriate. This does not suggest that the shoulder DN must be completely corrected; it indicates a segment involved in a larger movement pattern can be improved and therefore has influence over the larger pattern.

The hierarchy is part of the system to improve the clinical survey of regional interdependence. If multiple areas were initially managed simultaneously, this would reduce the observation of interdependence. Therefore, perform top-tier tests and note all DNs, as well as their degree of dysfunction. Address dysfunction within the hierarchy within time and treatment constraints. If the hierarchy is not followed, all information must be considered and qualified against an existing limitation.

EXAMPLE

The forward bending pattern might be the most obvious and dysfunctional DN limitation, but the cervical spine flexion pattern is also limited. The hierarchy suggests the C-spine should be addressed first if it can be done practically. Therefore, the C-spine pattern is broken down. Limitations are noted in the soft tissue and articular structures of the upper cervical spine. Following three minutes of mobilization and soft tissue work, the C-spine pattern is considered FN. When the forward bend top-tier test is repeated, three outcomes are possible.

FN—The C-spine tone was driving the limitation in the forward bending pattern. This is a common finding and demonstrates how the forward bending pattern incorporates the C-spine flexion pattern. Therefore, if it is also represented, it should be managed to remove its influence.

DN, partially improved—The C-spine tone was a partial influence and now the forward bending pattern can be broken down without complicating factors of the C-spine. This is also a common finding.

DN—The C-spine DN was an independent factor. The forward bending pattern can now be investigated without hierarchical consideration of the C-spine since the DN has been removed.

If the C-spine DN cannot be improved or managed, it must be considered a contributing factor since it cannot be ruled out. In this situation, it would be appropriate to break down the forward bending pattern, but the C-spine should be monitored and managed over the course of rehabilitation within the scope of the patient's condition and lifestyle.

This hierarchy assumes a clinical skill set is in place to efficiently and effectively manage DNs through manual techniques involving facilitation and inhibition. Otherwise, the suggestion to manage multiple DNs might seem overwhelming. If it can be agreed that the basic logic is correct, the limitation would be the skill set.

Please see www.movementbook.com/chapter7 for more information, videos and updates.

SFMA ASSESSMENT BREAKOUT DESCRIPTIONS AND FLOWCHARTS

The Selective Functional Movement Assessment (SFMA®) breakouts systematically dissect each of the major pattern dysfunctions described in the previous chapter. The hierarchy will dictate your investigation of all top-tier patterns scored as *Dysfunctional and Non-Painful* (DN), *Functional and Painful* (FP) or *Dysfunctional and Painful* (DP). It is most efficient to break out all DNs before testing the FPs and the DPs. You should test the DPs last since they can lead to further tissue inflammation and exacerbation of the symptoms. Breaking out the DPs can make further testing impossible or extremely uncomfortable for the patient.

The breakouts will either test all areas involved to isolate limitations or determine dysfunction by the process of elimination. The breakouts include active and passive movements, weight-bearing and non–weight-bearing positions, multiple-joint and single-joint functional movement assessments and unilateral and bilateral challenges.

The SFMA provides user-friendly testing to demonstrate large discrepancies between active and passive abilities whenever it can be efficiently performed. The breakouts are also performed to improve the efficiency of the SFMA decision tree. These assessments are mostly global, non-measured appraisals and are used to suggest the need for further clinical investigation. These tests prove a logical connection between functional movement and impairment measurements. The breakouts should never be considered a terminal point unless they provide negative information and no other risk factors are present.

In general, reduced movements seen in passive assessments suggest that mobility problems are likely. However, these must be confirmed with specific local testing. In contrast, normal motion in passive assessments suggests that potential mobility problems are unlikely. Once again, this can only be confirmed with specific local testing. A stability problem might be likely when active movement is limited in loaded or unloaded positions, or both, and when passive testing is normal.

In all cases, the SFMA will suggest that you perform local biomechanics testing to confirm normal range or that you clinically measure the level of mobility impairment. Biomechanical testing should also include appraisals of structural and neuromuscular integrity as well as motor control. The biomechanical tests should indicate if impairment is present or absent and help complete the functional diagnostic process. Local biomechanical testing is beyond the scope of the SFMA, used to specifically measure mobility with goniometric measurements, structural integrity with manual muscle testing, neuromuscular integrity and motor control. These tests should indicate if impairment is present or absent and should help complete the functional diagnostic process.

ADDITIONAL TERMINOLOGY

The SFMA breakout testing applies the same categorizations as its top-tier assessment, with isolated focus in each pattern demonstrating pain or dysfunction. This focus helps identify gross limitations in mobility and stability.

Unfortunately, the terms mobility and stability are not universally defined and can imply different things to clinicians of different backgrounds and training. For this reason, the SFMA will use subcategories when discussing breakout testing to help clarify the implications and intent of gross mobility and stability problems. The subcategories will improve communication and documentation, and reduce confusion by clearly defining or broadening the scope of each implication.

MOBILITY PROBLEMS

Mobility problems can be broken into two unique subcategories—

TED—*Tissue Extensibility Dysfunction*
JMD—*Joint Mobility Dysfunction*

TED
TISSUE EXTENSIBILITY DYSFUNCTION

Tissue extensibility dysfunction (TED) identifies tissues that are multi-articular. These tissues span more than one joint and therefore exert influence over more than one joint.

Examples of TED are—

- Active or Passive Muscle Insufficiency
- Neural Tension
- Facial Tension
- Muscle Shortening
- Hypertrophy
- Trigger Point Activity
- Scarring and Fibrosis

JMD
JOINT MOBILITY DYSFUNCTION

JMD identifies spinal articular segments having reduced mobility. The articular surfaces and the contractile and non-contractile tissues that connect them demonstrate reduced mobility with segmental testing and observation.

Examples of JMD are—

- Osteoarthritis
- Osteoarthrosis
- Uni-articular Muscle Spasm and Guarding
- Fusion
- Subluxation
- Adhesive Capsulitis
- Dislocation

STABILITY PROBLEMS

Stability problems have been renamed and reconsidered to reduce the assumption that we are referring only to an isolated strength problem. A stability problem in this sense may include an isolated weakness, but generally is more complex and refers to multiple systems motor control. To account for the complexity of a stability problem, we have coined the term *Stability or Motor Control Dysfunction (SMCD)*.

SMCD
STABILITY OR MOTOR CONTROL DYSFUNCTION

SMCD is a more correct description of *movement pattern stability problems*. Traditionally, stability dysfunction is often addressed by attempting to concentrically strengthen the muscle groups identified as stabilizers. This approach neglects the idea that true stabilization is reflex-driven and relies on proprioception and timing rather than isolated gross muscular strength.

By using the term SMCD to distinguish stability problems, we're forced to consider the central nervous system, peripheral nervous system, motor programs, movement organization, timing, coordination, proprioception, joint and postural alignment, structural instability and muscular inhibition, as well as the absolute strength of stabilizers.

This new moniker also reminds us of situations such as high-threshold strategy that because of pain, previous injury or chronic dysfunction, patients use global muscles to accomplish tasks more suited to local muscles. The broad concept of SMCD suggests that it may be necessary to break or manage a dysfunctional pattern before using exercises and techniques designed to improve stability or motor control.

It is also necessary to identify the level and involvement of SMCD. Subcategories of SMCD can be organized into two levels, static and dynamic. This demonstrates the hierarchy of motor control, which develops from static postural control to dynamic postural control.

To demonstrate functional motor control, it is necessary to establish static and dynamic stabilization. However, it is more important to identify compensatory behaviors that can provide pseudo-stabilization. Inefficient breathing, anxiety breathing and high-threshold strategies are examples of compensation behavior that can appear functional on the surface but be compensatory in nature. Excessive activity and tone in global musculature may also indicate local muscular dysfunction. The presence of neuromuscular trigger points in muscles that do not appear to be in a high tone or shortened state may contribute to poor stabilization, as appropriate timing and coordination may be compromised.

Examples of SMCD are—

- Motor control dysfunction
- Mechanical breathing dysfunction
- High threshold strategy
- Prime mover or global muscle compensation behavior or asymmetry
- Local muscle dysfunction or asymmetry
- Poor static stabilization, alignment, postural control, asymmetry and structural integrity
- Poor dynamic stabilization, alignment, postural control, asymmetry and structural integrity

FLOWCHARTS

The SFMA breakout flowcharts prepared by Dr. Greg Rose and found in the section beginning on page 337 are designed to separate pain and dysfunction when possible. They will help you identify movement patterns where exercise is indicated or contraindicated.

Unidentified mobility problems must be ruled out as underlying causes of SMCD. This forces us to always consider the regional interdependence model where limited mobility at one moving segment requires a distortion of motor control at nearby segments. Therefore, it is paramount in all SMCD scenarios to identify the compensatory behavior as well as the dysfunction.

Joint stiffness, high muscle tone, trigger points and all other forms of dysfunctional mobility may actually be compensations and natural responses in

the presence of poor motor control. This creates the *chicken or egg* scenario, which can be confusing, but the SFMA provides a logical management tool.

First, reduce or remove compensatory behavior and manage limited mobility. Demonstrate a measurable change in mobility if a limitation is initially present. Then, introduce activities and exercises that reestablish motor control at a level consistent with the SFMA functional status. This means exercises should be used at or below the functional level of the SFMA breakout test. Patterns involving pain should be treated with manual therapy techniques, but exercises in that pattern should not be used until the movement is pain-free.

Special circumstances may warrant exercise in some cases, but most models suggest that motor control will be distorted and outcomes will be inconsistent when exercise in performed in the presence of pain. The suggestions in each flowchart instruct you to look at the subcategories, and then confirm each with proper testing. These tests fall within the scope and responsibility of musculoskeletal evaluation and treatment, but are not part of the SFMA flowchart.

They can include, but are not limited to—

- Neurological testing for sensory motor integrity
- Muscle strength testing
- Joint stability testing
- Joint mobility testing
- Tissue tension testing of neurological structures, fascial structures and others
- Identification of neuromuscular trigger points
- Impairment measurements using goniometry, circumferential measurements and others

THE SFMA BREAKOUTS

We begin each breakout section with a few comments about the rationale behind the breakout, and follow that with a list of the assessments and a description of each. As you begin your study, refer regularly to the colored flowcharts at the back of the book.

SFMA BREAKOUTS

CERVICAL SPINE PATTERN BREAKOUT RATIONALE

See the *Cervical Spine Pattern Breakouts flowchart,* page 339.

CERVICAL SPINE FLEXION STABILITY OR MOTOR CONTROL DYSFUNCTION (SMCD)

Since the top-tier assessing involves motion in all three planes for the cervical spine, we use breakouts for each plane. This helps determine the primary dysfunction of the cervical spine. You'll only need to use the breakouts from the patient's dysfunctional top-tier assessment.

Reduce the postural stabilization requirements by having the patient lie supine with the head supported on a treatment table. In this position, the thoracic spine, shoulder girdle and cervical spine have minimal stability requirements. Instruct the patient to actively flex the neck. If standing cervical flexion is dysfunctional or painful, but active supine cervical flexion is normal, there is a postural and motor control dysfunction, or a stability and motor control dysfunction, or both, affecting cervical flexion.

If active supine cervical flexion is DN, DP or FP, check the same motion passively. If the passive supine cervical flexion test is now FN, the patient has either a stability or motor control dysfunction, or both, in active cervical flexion. If passive motion is still limited or painful, check for a cervical flexion mobility dysfunction.

OCCIPITOATLANTAL AND CERVICAL FLEXION JOINT MOBILITY DYSFUNCTION AND TISSUE EXTENSIBILITY DYSFUNCTIONS

We continue the breakdown of the cervical flexion mobility dysfunction by assessing the occipitoatlantal (OA) joint. Flexion dysfunctions often occur between the occiput and the atlas, and this can mimic pure cervical spine dysfunction. Test this range of motion using the active supine OA cervical flexion assessment. If OA flexion is FN, the patient has a cervical spine flexion joint mobility dysfunction or tissue extensibility dys-

function, or both. If OA flexion is dysfunctional, there is an OA joint mobility dysfunction or tissue extensibility dysfunction in conjunction with a possible cervical spine flexion mobility problem.

As with all the breakouts, if OA flexion is painful, stop and treat the problem.

CERVICAL ROTATION STABILITY AND MOTOR CONTROL DYSFUNCTIONS

Check active and passive cervical rotation to complete the breakout. Again, reduce the postural stabilization requirements by having the patient lie supine with the head supported on a treatment table. Instruct the individual to actively rotate the head. If the active supine cervical rotation assessment is normal, the patient has a postural motor control dysfunction or stability motor control dysfunction, or both, affecting cervical rotation.

If active supine cervical rotation is DN, DP or FP, check the same motion passively. If the passive supine cervical rotation assessment is now FN, the patient has an active cervical spine rotation stability dysfunction or a cervical spine motor control dysfunction, or both. If passive motion remains limited or painful, check for a cervical rotation mobility dysfunction.

ATLANTOAXIAL AND LOWER CERVICAL ROTATION JOINT MOBILITY DYSFUNCTION AND TISSUE EXTENSIBILITY DYSFUNCTIONS

Continue the cervical rotation mobility dysfunction by assessing the atlantoaxial joint. Half of cervical spine rotation comes from the atlantoaxial joint, so we assess this motion independently. Test this range of motion using the C1-C2 rotation test. If atlantoaxial rotation is FN, the patient has a lower cervical spine (C3-C7) rotation joint mobility dysfunction or a lower cervical (C3-C7) rotation tissue extensibility dysfunction, or both.

If atlantoaxial rotation is DN, the patient has an atlantoaxial joint mobility dysfunction or tissue extensibility dysfunction in conjunction with a possible lower C-spine rotation mobility problem.

If atlantoaxial rotation is painful, stop and treat the problem.

CERVICAL SPINE EXTENSION STABILITY OR MOTOR CONTROL DYSFUNCTION (SMCD)

Cervical extension is the last part of this breakout. This assessment allows two options to reduce the postural stabilization requirements. The patient can lie supine with the head just off the end of a treatment table, or you can have the patient side-lying as you support the patient's head. Either option allows the thoracic spine, shoulder girdle and cervical spine to have minimal stability requirements.

Instruct the patient to try to fully extend the neck. If the supine cervical extension assessment is normal and standing extension is poor, there is a postural and motor control dysfunction, or a stability and motor control dysfunction, or both, affecting cervical extension.

If supine cervical extension is DN, DP or FP, check for a cervical extension joint mobility or cervical extension tissue extensibility dysfunction, or both.

There is no need to perform passive versus active assessing for extension since the main assessment is virtually passive due to gravity.

CERVICAL SPINE MOVEMENT PATTERN BREAKOUTS

In the presence of a *limited cervical movement pattern assessment* in the top-tier testing, proceed with the following breakouts.

Active Supine Cervical Flexion
Passive Supine Cervical Flexion
Active Supine Occipitoatlantal Cervical Flexion
Active Supine Cervical Rotation
Passive Cervical Rotation
C1-C2 Cervical Rotation
Supine Cervical Extension

ACTIVE SUPINE CERVICAL FLEXION

OBJECTIVE

To assess cervical spine mobility or stability, or both, in a non-reduced postural position

DESCRIPTION

Instruct the patient to assume the supine position with the arms and hands by the thighs. Have the patient touch the chin to the sternum.

ADDITIONAL INFORMATION

During each movement, make sure the patient's mouth remains closed. Don't allow scapular elevation or protraction throughout the movements.

POSSIBLE FINDINGS

- An ability to perform the movement—chin to sternum (FN)

- An inability to perform the movements or only able to perform the movement in the presence of pain (DN, DP or FP)

If the finding is FN *and* the standing cervical flexion assessment is limited, there is a postural and motor control dysfunction, or a stability and motor control dysfunction, or both, affecting cervical flexion. This includes the cervical spine, thoracic spine and shoulder girdle postural dysfunction.

If the finding is DN, DP or FP, proceed to the *passive supine cervical flexion assessment.*

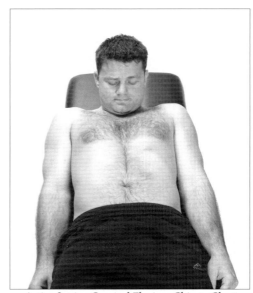

Active Supine Cervical Flexion, Chin to Chest

PASSIVE SUPINE CERVICAL FLEXION

OBJECTIVE

To assess cervical spine mobility or stability, or both, in an unloaded position

DESCRIPTION

Instruct the patient to assume the supine position with the arms and hands by the thighs. Move the patient's head such that the neck flexes, moving the chin toward the sternum.

ADDITIONAL INFORMATION

During the movement, make sure the patient's mouth remains closed. Don't allow scapular elevation or protraction throughout the movements.

POSSIBLE FINDINGS

- An ability to perform the movement—chin to sternum (FN)

- An inability to perform the movements or only able to perform the movements in the presence of pain (DP, DN or FP)

If the finding is FN, the patient has an active cervical spine flexion stability dysfunction or motor control dysfunction or both.

If the finding is DN, DP or FP, proceed to the *active supine occipitoatlantal cervical flexion assessment.*

ACTIVE SUPINE OCCIPITO-ATLANTAL CERVICAL FLEXION

OBJECTIVE

To assess the OA mobility or stability in an unloaded position

DESCRIPTION

Instruct the patient to assume the supine position with the arms and hands by the thighs. Have the patient rotate the head as far to the right as possible, then ask for a chin tuck. Have the patient repeat the movement on the left side.

ADDITIONAL INFORMATION

Look for at least 20 degrees of OA flexion bilaterally.

POSSIBLE FINDINGS

- Ability to achieve 20 degrees of flexion bilaterally (FN)

- Inability to reach 20 degrees of flexion (DN)

- Pain while trying to achieve 20 degrees of flexion (DP or FP)

If the finding is FN bilaterally, the patient has a postural motor control dysfunction or stability motor control dysfunction, or both, affecting cervical rotation.

If the finding is DN, there is an OA joint mobility dysfunction or tissue extensibility dysfunction in conjunction with a possible cervical spine flexion mobility problem.

If the finding is DP or FP, stop and treat the problem.

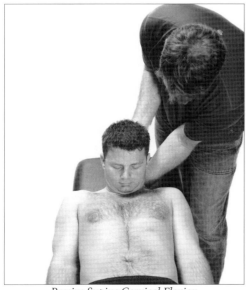

Passive Supine Cervical Flexion

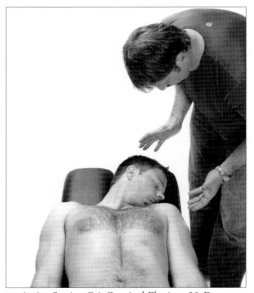

Active Supine OA Cervical Flexion, 20-Degree

ACTIVE SUPINE CERVICAL ROTATION

OBJECTIVE

To assess the ability to actively rotate the cervical spine

DESCRIPTION

Instruct the patient to assume the supine position with the arms and hands by the thighs. Have the patient rotate the head as far to the right as possible. Repeat the assessment on the left side.

ADDITIONAL INFORMATION

Look for at least 80 degrees of rotation bilaterally.

POSSIBLE FINDINGS

- An ability to achieve 80 degrees of rotation bilaterally (FN)

- An inability to reach 80 degrees of rotation bilaterally or pain while trying to reach 80 degrees (DN, DP, FP)

If the finding is FN, there is a postural and motor control dysfunction or stability and motor control dysfunction, or both, affecting cervical extension.

If the finding is DN, DP or FP, proceed to the *passive cervical rotation assessment.*

PASSIVE CERVICAL ROTATION

OBJECTIVE

To assess the ability to passively rotate the cervical spine

DESCRIPTION

Instruct the patient to assume the supine position with the arms and hands by the thighs. Rotate the individual's head to the right. Repeat the movement on the left side.

ADDITIONAL INFORMATION

Look for at least 80 degrees of rotation bilaterally.

POSSIBLE FINDINGS

- An ability to achieve 80 degrees of rotation bilaterally (FN)

- An inability to reach 80 degrees of rotation bilaterally or pain while trying to reach 80 degrees (DN, DP, FP)

If the finding is FN, the patient has an active cervical spine rotation stability dysfunction or a cervical spine motor control dysfunction or both.

If the finding is DN, DP or FP, proceed to the *C1-C2 cervical rotation assessment.*

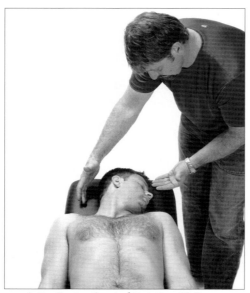

Active Supine Cervical Rotation, 80-Degree

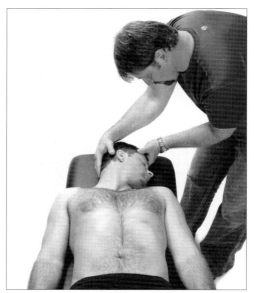

Passive Cervical Rotation

C1-C2 CERVICAL ROTATION

OBJECTIVE

To assess the rotation of the cervical spine at C1/C2.

DESCRIPTION

Instruct the patient to assume the supine position with the arms and hands by the thighs. Have the individual flex the neck, chin to the sternum. Then, while the chin is as close to the sternum as possible, instruct the patient to rotate the neck as far to the right as able. Repeat the assessment with a rotation to the left.

ADDITIONAL INFORMATION

Make sure the patient fully flexes before going into rotation. Look for at least 40 degrees of rotation *after* flexion.

POSSIBLE FINDINGS

- An ability to achieve at least 40 degrees of rotation bilaterally (FN)

- An inability to achieve at least 40 degrees of rotation bilaterally (DN)

- Pain during the movement (FP or DP)

If the finding is FN, the patient has a lower cervical spine (C3-C7) rotation joint mobility dysfunction or lower cervical spine (C3-C7) rotation tissue extensibility dysfunction, or both.

If the finding is DN, the patient has a C1-C2 joint mobility dysfunction or tissue extensibility dysfunction in conjunction with a possible lower cervical spine rotation mobility problem.

If the finding is FP or DP, stop and treat the problem.

SUPINE CERVICAL EXTENSION

OBJECTIVE

To assess the extension of the cervical spine in an unloaded position

DESCRIPTION

Instruct the patient to assume the supine position on a bench, head extending beyond the end of the bench. Have the individual extend the neck as far as possible.

ADDITIONAL INFORMATION

Look to see if the patient's face is perpendicular to the ground.

POSSIBLE FINDINGS

- An ability to extend the neck such that the face is perpendicular to the ground (FN)

- An inability to extend the neck and get the face perpendicular to the ground (DN)

- Pain while attempting to extend the neck (FP or DP)

If the finding is FN and standing extension is poor, there is a postural and motor control dysfunction or stability and motor control dysfunction, or both, affecting cervical extension.

If the finding is DN, the patient has a cervical extension joint mobility dysfunction or a cervical extension tissue extensibility dysfunction, or both.

If the finding is DP or FP, stop and treat the problem.

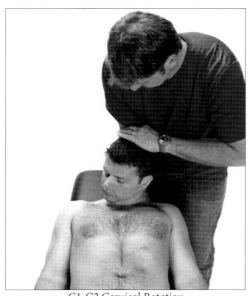

C1-C2 Cervical Rotation

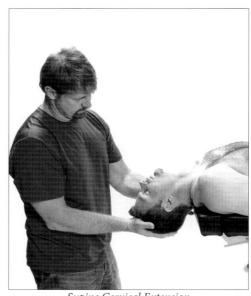

Supine Cervical Extension

SFMA

UPPER EXTREMITY PATTERN BREAKOUTS RATIONALE

See the *Upper Extremity Pattern Breakouts flowchart*, page 340.

UPPER EXTREMITY GIRDLE SMCD VERSUS JOINT MOBILITY AND TISSUE EXTENSIBILITY DYSFUNCTIONS

To better understand a patient's shoulder dysfunction, reduce the postural stabilization requirements by having the patient lie prone on a treatment table. In this position, the thoracic spine, shoulder girdle and cervical spine have minimal stability requirements. Next, instruct the patient to try to repeat the top-tier upper extremity patterns as described on page 124.

If the active prone upper extremity pattern test is now normal, the patient has a postural and motor control dysfunction or a shoulder-girdle stability and motor control dysfunction, or both, affecting the functional upper extremity pattern. If the active prone upper extremity pattern is DN, DP or FP, check the same motion passively.

If the passive prone upper extremity pattern test is now FN, the patient has some form of stability control or motor control dysfunction, or both. If passive motion is still limited, you should assume there is a local shoulder mobility dysfunction and proceed to local biomechanical testing of the shoulder girdle.

UPPER EXTREMITY PATTERN SMCD VERSUS ISOLATED GLENOHUMERAL OR SCAPULAR STABILITY SMCD

You can further break down a shoulder-girdle stability dysfunction by challenging the entire upper extremity pattern in the supine reciprocal upper extremity pattern test. This helps determine if a patient's stability dysfunction is isolated to movements of one shoulder versus the combined motion of both upper extremities working in a reciprocal pattern.

If the test is dysfunctional, assume the shoulder stability or motor control dysfunction, or both, is primary in end-range motion when the two shoulders are working in a reciprocal pattern.

Exercises for this finding should be focused on re-training the bilateral, simultaneous shoulder patterns rather than single-sided shoulder-based exercises. If the test is FN, assume the patient has an isolated glenohumeral or scapular stability issue or a motor control dysfunction, or both, in mid-range. For this finding, exercising isolated shoulder movements is appropriate.

In the presence of a limited *upper extremity movement pattern assessment*, proceed with the following breakouts.

Active Prone Upper Extremity Patterns
Passive Prone Upper Extremity Patterns
Supine Reciprocal Upper Extremity Patterns

ACTIVE PRONE UPPER EXTREMITY PATTERNS

OBJECTIVE

To assess shoulder mobility or stability, or both, in an unloaded position

DESCRIPTION

Instruct the patient to assume the prone position with the arms and hands by the thighs. Have the individual repeat the top-tier upper extremity patterns seen on page 124.

ADDITIONAL INFORMATION

In this position the thoracic spine, shoulder girdle and cervical spine have minimal stability requirements.

POSSIBLE FINDINGS

- An ability to perform each of the top-tier upper extremity tests (FN)

- An inability to perform each of the top-tier upper extremity tests or experiences pain while trying to perform them (DN, DP or FP)

If the finding is FN, the patient has a postural and motor control dysfunction or shoulder-girdle stability and motor control dysfunction, or both, affecting the functional upper extremity pattern in question.

If the finding is DN, DP or FP, proceed to the *passive prone upper extremity pattern assessments*.

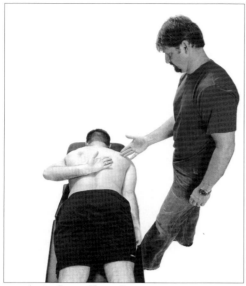

Active Prone Upper Extremity Pattern One

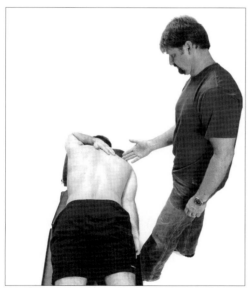

Active Prone Upper Extremity Pattern Two

PASSIVE PRONE UPPER EXTREMITY PATTERNS

OBJECTIVE

To assess shoulder mobility or stability, or both, in an unloaded position

DESCRIPTION

Have the patient assume the prone position with the arms and hands by the thighs. You'll then move the patient through the top-tier upper extremity movement tests as seen on page 124.

ADDITIONAL INFORMATION

In this position the thoracic spine, shoulder girdle and cervical spine have minimal stability requirements.

POSSIBLE FINDINGS

- An ability to perform each of the top-tier shoulder tests (FN)

- An inability to perform each of the top-tier shoulder tests (DN)

- Pain while performing or attempting the movement (DP or FP)

If the finding is FN, proceed to the *supine reciprocal upper extremity pattern test.*

If the finding is DN, there is a shoulder-girdle joint mobility dysfunction or a tissue extensibility dysfunction, or both. Proceed to local biomechanical testing.

If the finding is DP or FP, stop and treat the problem.

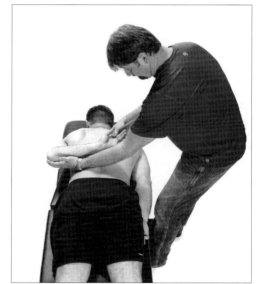

Passive Prone Upper Extremity Pattern One

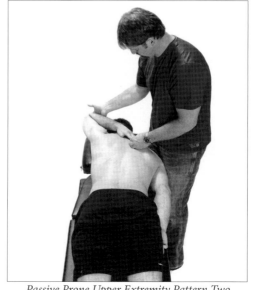

Passive Prone Upper Extremity Pattern Two

SFMA

SUPINE RECIPROCAL UPPER EXTREMITY PATTERNS

OBJECTIVE

To assess shoulder-girdle mobility or stability, or both, in an unloaded position

DESCRIPTION

Have the patient assume the supine position with one arm and hand by the side and the other arm and hand extended above the head. Put your hands on the upper forearms of the patient's, and instruct the patient to resist you trying to raise the arms. Apply gentle force, trying to lift both arms off the table simultaneously.

ADDITIONAL INFORMATION

In this position the thoracic spine, shoulder girdle and cervical spine have minimal stability requirements.

POSSIBLE FINDINGS

- Ability to resist movement in each arm (FN)

- Inability to resist movement in either arm (DN)

- Pain while trying to resist movement in either arm (DP or FP)

If the finding is FN, assume the patient has an isolated glenohumeral or scapular stability issue or motor control dysfunction, or both, in mid-range.

If the finding is DN, assume the shoulder stability or motor control dysfunction, or both, is primary in end-range motion when the two shoulders are working in a reciprocal pattern.

If the finding is DP or FP, stop and treat the problem.

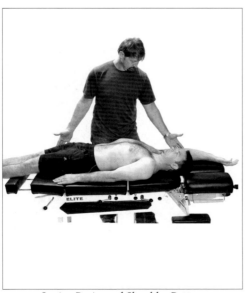

Supine Reciprocal Shoulder Pattern

MULTI-SEGMENTAL FLEXION BREAKOUTS RATIONALE

BILATERAL OR UNILATERAL FLEXION DYSFUNCTION

Begin by having the patient perform a single-leg forward bend to assess unilateral flexion. It is important to note that unilateral movement patterns can be consistent or inconsistent with whole patterns. Most people think an inability to touch the toes is a bilateral problem, but often, it is due to a unilateral dysfunction, and therefore must be investigated.

WEIGHT-BEARING POSTURAL STABILIZATION OR NON–WEIGHT-BEARING DYSFUNCTION

The next step is to reduce the postural stabilization requirements by having the patient perform a long-sitting toe touch. In this posture, the patient's non–weight-bearing flexion pattern can be compared to the hip weight-bearing flexion pattern. The limitations can be consistent or inconsistent with whole pattern findings. In this posture, it can be determined if the patient's dysfunction or pain is due to a hip flexion or a spinal flexion problem.

HIP FLEXION PROBLEM

Observing the patient's sacral angle in the long-sitting position can make a distinction between hip or spine dysfunction. If the sacral angle is limited or pain is noted, investigation of hip flexion is warranted. Assessing the patient's active straight-leg raise will help clarify findings.

If hip flexion is now functional, assume there is a weight-bearing spinal flexion dysfunction since long sitting still had the patient's spine in a weight-bearing position. Proceed to prone rocking to break down the spine flexion dysfunction.

Any dysfunction in the active straight-leg raise forces us to assume a hip flexion limitation, so you should next compare passive hip flexion to active in the passive straight-leg raise test. If hip flexion is now functional and non-painful, the patient has a core or active hip-flexion stability issue or a motor control dysfunction, or both. If passive hip flexion improves by more than 10 degrees compared to active but still does not reach a clinical norm of 80 degrees, it's a sign of potential core stability dysfunction or motor control dysfunction, or both.

If the patient still shows hip dysfunction in passive hip flexion, you can further dissect this by performing a supine knee-to-chest assessment. This will help differentiate if the individual has an active hip-flexion stability issue or a motor control dysfunction, or both, or solely a hip mobility limitation.

Spine Flexion Problem

If the sacral angle in long sitting is normal, but dysfunction or pain still occurs, assess non–weight-bearing spine flexion in the prone rocking assessment. This will help differentiate between a weight-bearing spinal-flexion stability dysfunction or a motor control dysfunction, or both, or solely a spinal mobility limitation. If the spine shows a uniform curve at the end range of the prone rocking maneuver, the patient has a weight-bearing spine stability or motor control dysfunction. If the spine lacks curvature at end range, there is a spinal joint mobility or tissue extensibility dysfunction.

Rolling

As with several of the SFMA breakouts, when mobility limitations are not identified, basic motor control stability dysfunction should be investigated. Rolling pattern assessments are a low-demand movement pattern we use to observe motor control and fundamental segmental stabilization.

It is only necessary to perform rolling assessments if the patient can perform an acceptable long-sitting toe touch or passive straight-leg raise without pain or dysfunction. Otherwise, limitations in mobility will compromise motor control and stabilization and create a false positive finding for motor control and stability.

MULTI-SEGMENTAL FLEXION BREAKOUTS

See the *Multi-Segmental Flexion Breakouts flowchart*, page 341.

In the presence of a limited *multi-segmental flexion assessment* in the top-tier testing, proceed with the following breakouts.

Single-Leg Forward Bend
Long-Sitting Toe Touch
Rolling Patterns (see page 187)
Active Straight-Leg Raise
Passive Straight-Leg Raise
Prone Rocking
Supine Knees-To-Chest Pattern

SINGLE-LEG FORWARD BEND

Use this assessment if the *multi-segmental flexion assessment* is limited.

Objective

Determines if the forward bend is a symmetrical or asymmetrical dysfunction or used as a pain provocation maneuver

Description

Instruct the patient to elevate the left leg on a step, straighten the right leg and place both hands on top of each other, palms together, elbows straight. The patient should try to touch both hands to the right toes without bending the right knee, and repeat on the other side.

Additional Information

Note the outcome, but continue on with breakout testing. This single-leg maneuver simply provides additional information with regard to asymmetry in the forward bending pattern. When forward bending is limited, it is often assumed to be a bilateral pattern. This test can demonstrate when the problem is only represented unilaterally.

SFMA

POSSIBLE FINDINGS

- Both functional and non-painful

- Bilateral dysfunctional or painful

- Unilateral dysfunctional or painful

Proceed to the *long-sitting toe touch breakout.*

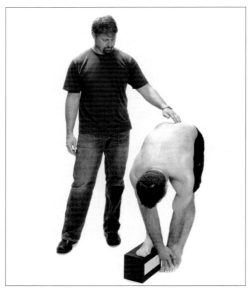

Single-Leg Forward Bend

LONG-SITTING TOE TOUCH

OBJECTIVE

Differentiates between true hamstring tightness and spinal flexion limitations, performed in a non–weight-bearing position

DESCRIPTION

The patient will be in a long-sitting position with the legs extended. Instruct the patient to bend forward to touch the toes.

ADDITIONAL INFORMATION

Take note of the sacral angle at the end of the reach. Many experts use 80 degrees as the norm for sacral base angle.

POSSIBLE FINDINGS

- Touches toes with at least 80 degrees of sacral angle (FN)—indicating a weight-bearing hip stability issue, poor coordination or poor sequencing of the toe touch

- Touches toes with less than 80 degrees of sacral angle (FP with limited sacral angle)—indicating limited hip flexion or hypermobile spinal flexion, or both.

- Can or cannot touch toes and has at least 80 degrees of sacral angle (FP, DP or DN with normal sacral angle)—indicating weight-bearing spinal stabilization dysfunction or just limited spinal mobility.

- Cannot touch toes and less than 80 degrees of sacral angle (FP, DP or DN with limited sacral angle)—indicating limited hip flexion or limited spinal flexion, or both

If the finding is FN, proceed to the *rolling breakouts.*

If the finding is DN, DP or FP with normal sacral angle, proceed to *prone rocking.*

If the finding is DN, DP or FP with a limited sacral angle, proceed to the *active straight-leg raise assessment.*

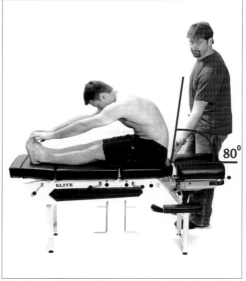

Long-Sitting Toe Touch

ACTIVE STRAIGHT-LEG RAISE

OBJECTIVE

Tests the ability to actively flex the hip with an extended knee.

DESCRIPTION

Have the patient lie supine with the palms up and by the sides and the head flat on the examination table. Both feet should be in a neutral position, the soles of the feet perpendicular to the table. Instruct the patient to lift the test limb while maintaining the initial ankle position.

During the test, the opposite knee should remain in contact with the surface, the toes should remain pointed upward in the neutral-limb position, and the head remains flat on the table.

Once the patient achieves the end range, note the position of the ankle relative to the non-moving limb. The angle of the raised leg relative to the other leg should be more than 70 degrees and within 10 degrees of the passive measurement if there's no dysfunction.

ADDITIONAL INFORMATION

This assessment combined with the passive straight-leg raise can differentiate between posterior chain TED or hip JMD versus the lack of stability or strength to actively move the hip. It also helps identify the restriction as a symmetrical or asymmetrical dysfunction in a non–weight-bearing position.

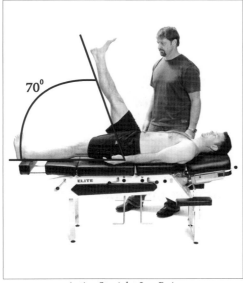

Active Straight-Leg Raise

POSSIBLE FINDINGS

- The ability to actively raise the leg to more than 70 degrees (FN)

- An inability to actively raise the leg to more than 70 degrees, or experiences pain while trying (DN, DP or FP)

If the finding is FN, proceed to the *prone rocking assessment*.

If the finding is DN, DP or FP, proceed to the *passive straight-leg raise assessment*.

PASSIVE STRAIGHT-LEG RAISE

OBJECTIVE

This helps differentiate between posterior chain TED or hip JMD versus the stability or strength to actively move the hip, and helps identify the restriction as a symmetrical or asymmetrical dysfunction in a non–weight-bearing position.

DESCRIPTION

Instruct the patient to lie supine with the arms down by the sides. With the patient's right leg extended and flat on the table, slowly lift the patient's left leg. Go as far as possible without letting the left knee bend or the pelvis shift, then measure the straight leg raise range of motion. The angle should be more than 80 degrees. Have the patient change legs and repeat the test.

ADDITIONAL INFORMATION

If the *passive straight-leg raise (PSLR) assessment* is less than 80 degrees, but is 10 degrees more than the *active straight-leg raise (ASLR)*, there can potentially be a core stability or a hip flexion strength problem, a hamstring high tone limitation, guarding or a hip mobility dysfunction.

If the *PSLR* test is greater than 80 degrees, there could be a core stabilization or a hip flexion strength problem.

If the *PSLR* test is less than or equal to the *ASLR*, proceed to the *supine knee-to-chest assessment*. This represents a possible hamstring high tone, guarding or hip mobility restriction.

POSSIBLE FINDINGS

- Ability to move the leg to greater than 80 degrees relative to the ground leg (FN)

- The leg can be raised more than 10 degrees higher than ASLR, but it is still less than 80 degrees total (FN). If this in the finding, check for a core stability and motor control dysfunction, then proceed to the *supine knee-to-chest assessment*.

- Has pain when performing the movement (FP or DP)

- Has an inability to perform the movement and has an angle less than or equal that of the ASLR (DN)

If the finding is FN, proceed to the *rolling breakouts.*

If the finding is FP, DP or DN, proceed to the *supine knee-to-chest assessment.*

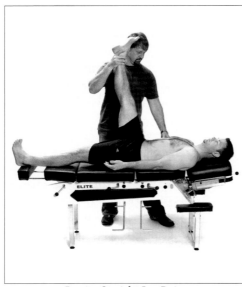

Passive Straight-Leg Raise

PRONE ROCKING

OBJECTIVE

This pattern identifies restrictions in spinal flexion in a non–weight-bearing or unloaded spine.

DESCRIPTION

Instruct the patient to start in the quadruped, all-fours position. Have the patient lower the buttocks onto the heels—rocking back toward full hip flexion. The lower rib cage should easily press into the thighs at the end of the rock-back.

ADDITIONAL INFORMATION

The knee can be a complicating factor here, so use the *supine knee-to-chest assessment* if the patient complains of knee discomfort. Look for the thighs to press against the lower rib cage at the end of the rocking maneuver.

POSSIBLE FINDINGS

- Ability to fully rock back (FN)

- Has limited spinal flexion while trying to rock back (DN)

- Able to, or unable to, rock back into the proper position, and experiences pain while trying (DP or FP)

If the finding is FN, there is a weight-bearing spine stability or motor control dysfunction.

If the finding is DN, there is a spinal joint mobility or tissue extensibility dysfunction.

If the finding is DP or FP, stop and treat the problem.

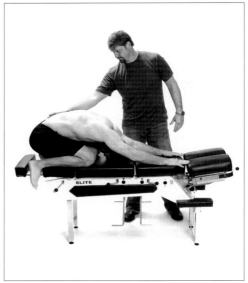

Prone Rocking

SUPINE KNEE-TO-CHEST PATTERN

OBJECTIVE

Allows the checking of the mobility of the hips in a non–weight-bearing position

DESCRIPTION

Have the patient lie supine and bring both knees up. The patient should then grab both thighs and pull the knees to the chest.

ADDITIONAL INFORMATION

This move will help differentiate between hip and hamstring mobility issues. The patient should be able to press both thighs against the chest.

POSSIBLE FINDINGS

- Ability to pull the thighs into position (FN)

- Unable to achieve the position, or able to with pain present (DP or FP)

- Unable to achieve the position (DN)

If the finding is FN, the patient has a posterior chain tissue extensibility dysfunction or an active hip flexion stability and motor control dysfunction, or both.

If the finding is DN, the patient has a hip joint mobility dysfunction or a posterior chain tissue extensibility dysfunction, or both.

If the finding is FP or DP, stop and treat the problem.

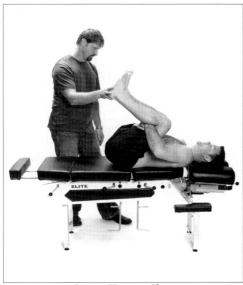

Supine Knee-to-Chest

MULTI-SEGMENTAL EXTENSION BREAKOUT RATIONALE

SPINE AND HIP EXTENSION PROBLEM

Begin by assessing spinal and hip extension. Have the patient perform the backward bend without upper extremities assessment to determine if the patient's extension dysfunction or pain is due to spinal or hip involvement. If the patient can now perform a functional and non-painful extension pattern, assume the problem is not spinal- or hip-based, and proceed to the upper-body extension flowchart. If extension is still dysfunctional or painful, there is a spinal or hip extension dysfunction, or both.

BILATERAL AND UNILATERAL SPINAL AND HIP EXTENSION PROBLEMS

Next, have the patient perform a single-leg backward bend to assess unilateral extension. If both sides are now functional and non-painful, the patient has a symmetrical-stance core stability or motor control problem, or both. In other words, the patient only has dysfunction with simultaneous, bilateral extension.

Since the possibility of an upper-body extension dysfunction has not been eliminated, proceed to the upper body extension breakout.

If the single-leg backward bend assessment is dysfunctional or painful, determine if the problem is from the hip or from the spine. Begin by looking at spinal extension independent of hip extension using the press-up.

WEIGHT-BEARING VERSUS NON–WEIGHT-BEARING PROBLEM

The next step should reduce the postural stabilization requirements by having the patient lie prone and performing a press-up. In this posture, we can assess how the patient's non–weight-bearing extension pattern compares to a full weight-bearing extension pattern.

Even though we begin the press-up in a non–weight-bearing spine position, the test is not considered a completely unloaded spine assessment. This is because the spine goes into a semi-weight-bearing position during the test. If the press-up is dysfunctional or painful, proceed to further spinal assessments.

If the press-up is functional and non-painful, the patient might have a weight-bearing spine extension stability or motor control dysfunction, but might also have a hip extension or shoulder flexion problem. Proceed to the lower- and upper-body extension flowchart.

THORACIC EXTENSION PROBLEM

The lumbar-locked (IR) active and passive rotation and extension assessments evaluate the tissue extensibility, joint mobility and stability or motor control of the thorax. They can also reduce the problem to a symmetrical or asymmetrical dysfunction. This includes the thoracic spine and soft tissues that limit thoracic spine extension.

The lumbar-locked position allows better isolation of the thoracic spine and reduces the contribution from the lumbar spine. Placing the arm behind the back in an internally rotated position allows a reduction in scapular involvement and minimizes anterior shoulder and chest dysfunction.

If thoracic mobility is functional and non-painful (greater than 50 degrees), the lumbar spine needs to be evaluated. Proceed to the prone-on-elbow assessment.

If there is dysfunction or pain, there is a thoracic extension mobility or stability problem. Retest the motion passively to determine the dysfunction. If the passive motion is still reduced, there is a unilateral or bilateral thoracic extension tissue extensibility or joint mobility dysfunction, or both.

If the passive motion is now functional and non-painful, there is a thoracic extension stability or motor control dysfunction, or both. There is no need to perform rolling patterns to rate the severity of this spinal stability dysfunction, since not all of the potential mobility limitations of extension in the upper and lower body have been cleared yet.

If there is pain with passive testing, stop and treat it.

Since hip extension and shoulder flexion dysfunction have not been ruled out yet, proceed to the upper- and lower-body extension flowcharts to further assess extension.

LUMBAR SPINE EXTENSION PROBLEM

If thoracic extension tests normal, you'll next assess the lumbar spine. Have the patient perform the prone-on-elbow unilateral extension assessment. In this position, the thoracic spine is already stretched, so the stress will now be placed on the lumbar spine. If extension from this position creates pain, stop and treat the lumbar spine.

If both directions are now functional and non-painful, there is a bilateral spine extension stability or motor control dysfunction. If there is dysfunction, the patient can have unilateral or bilateral tissue extensibility, joint mobility or stability and motor control problems, or both, in the lumbar spine. Use local biomechanical testing of the lumbar spine to further assess.

Proceed to the upper- and lower-body extension breakouts to assess the possibility of hip and shoulder dysfunction.

WEIGHT-BEARING HIP LOWER-QUARTER EXTENSION PROBLEM

Start with the single-leg hip extension assessment, with the leg being tested off the surface. This assessment allows evaluation of hip extension on a non–weight-bearing hip, while maintaining a fully loaded spine. If the hip now shows more than 10 degrees of extension, assume there is a weight-bearing lower-quarter stability dysfunction or motor control dysfunction, or both.

As a word of caution, this could be a sign of an ankle dorsiflexion limitation as well, so double-check the overhead deep squat and single-leg stance assessments.

If the standing hip extension still shows dysfunction, proceed to the prone active hip extension to assess hip function when the spine is unloaded.

SPINE WEIGHT-BEARING STABILITY OR MOTOR CONTROL DYSFUNCTION (SMCD)

This test reduces the postural stabilization requirements by having the patient lie prone and performing an active hip extension. In this posture, we can assess how the patient's non–weight-bearing spine affects the hip extension pattern. If the hip is now functional and non-painful, there is a stability problem.

Use rolling patterns to assess the severity of the patient's segmental stability and motor control problem. If rolling is normal, there is a spine weight-bearing stability and motor control dysfunction. If rolling is dysfunctional, there is a fundamental extension pattern dysfunction. If there is pain with rolling, stop and treat the problem.

CORE SMCDS AND ACTIVE HIP EXTENSION SMCDS

If the hip is still dysfunctional or painful in the prone active hip extension test, you should then assess hip extension passively. If hip extension is now functional and non-painful with passive testing, there is a stability problem.

Use rolling patterns to assess the severity of the segmental stability and motor control problem. If rolling is normal, there is a core stability dysfunction or active hip extension stability dysfunction, or both, or a motor control dysfunction. If rolling is dysfunctional, there is a fundamental extension pattern dysfunction. If there is pain with rolling, stop and treat the pain.

If hip extension is still dysfunctional or painful with passive testing, assume there's a hip mobility problem. Continue to the Modified Thomas test for further assessment.

HIP JOINT AND TISSUE EXTENSIBILITY DYSFUNCTIONS

First, examine the structural mobility of the hip and sacroiliac joint using the FABER test. If the FABER test is dysfunctional, the patient has a hip and sacroiliac joint mobility dysfunction or a possible core stability or motor control dysfunction, or a combination thereof.

Use local biomechanical testing of the hip and pelvis to further assess the dysfunction, and continue with the Modified Thomas test since muscular limitations may also be involved. If FABER is functional and non-painful, assume some form of tissue extensibility dysfunction and proceed to the Modified Thomas Test. If this test creates pain, stop and treat it.

To precisely determine which muscles or soft tissues are effecting hip extension mobility, use the Modified Thomas Test. In this test, you will systematically reduce the mobility requirements of each of the large muscles that can limit hip extension.

If the patient can only extend the hip with the knee straight, assume an anterior chain tissue extensibility dysfunction. The most common muscle involved in this situation is the rectus femoris of the quadriceps. If the patient can only extend the hip with the thigh abducted, assume a lateral-chain tissue extensibility dysfunction. The most common muscle involved here is the tensor fasciae latae (TFL). If the patient can only extend the hip with the knee straight and thigh abducted, assume an anterior- and lateral-chain tissue extensibility dysfunction.

If the patient is never able to extend the hip in the Modified Thomas Test, there is still a possibil-ity of a hip joint mobility problem, anterior-chain tissue extensibility limitation or a core stability or motor control dysfunction, or a combination thereof. Use local biomechanical testing of the hip and pelvis to further assess the dysfunction.

If the Modified Thomas test is functional and non-painful bilaterally, assume an underlying core stability or motor learning dysfunction mimicking a mobility problem. Since the patient is able to flex the one hip in this test, it can artificially stabilize the core and give a clean Thomas test.

As always, if there is pain with the Modified Thomas test, stop and treat it.

BILATERAL OR UNILATERAL UPPER-BODY EXTENSION PROBLEMS

Begin this flowchart by having the patient perform a unilateral shoulder backward bend. It is important to note that unilateral movement patterns can be consistent or inconsistent with whole patterns. If both sides are now functional and non-painful and were not when tested bilaterally, double-check the spine extension flowchart for limited bilateral thoracic spine extension.

The cervical spine can also cause this finding, so verify there is no cervical spine dysfunction.

If there is still dysfunction or pain, progress to the supine lat stretch with hips flexed.

WEIGHT-BEARING UPPER-QUARTER STABILITY OR MOTOR CONTROL PROBLEM

The supine lat stretch, hips flexed test reduces the postural stabilization requirements by having the patient perform full shoulder flexion in a non–weight-bearing position. In this posture, you can compare the patient's non–weight-bearing upper-quarter extension to the full weight-bearing extension pattern.

If the test is functional and non-painful, there is a stability problem. Use rolling patterns to assess the severity of the segmental stability and motor control problem. If rolling is normal, there is a weight-bearing upper-quarter stability or motor control dysfunction. If rolling is dysfunctional, there is a fundamental extension pattern dysfunction. If there is pain with rolling, stop and treat the pain.

If there is still dysfunction or pain with the supine lat stretch, continue with the flowchart.

LATISSIMUS AND POSTERIOR-CHAIN PROBLEMS

Reduce the tension applied to the lat and thoracolumbar fascia by extending the patient's hips. If the patient's arms now touch the surface below and show functional upper-quarter extension (shoulder flexion), assume there is a lat or posterior-chain tissue extensibility dysfunction.

It is important to note one special exception: If the patient has limited hip extension and goes into excessive lumbar lordosis as the hips extend, this can effectively shorten the lat and posterior chain. This would give a false positive, so work through the lower body extension breakouts flowchart found on page 343.

If the patient's arm position improves, but they don't completely touch the table, you can assume there is a lat or posterior-chain tissue extensibility dysfunction as part of the problem. Unfortunately, there remains another dysfunction limiting the full upper-quarter extension.

If in the supine lat stretch with hips extended assessment the arms never lower or only slightly improve, assess the thorax and shoulder girdle.

THORAX AND SHOULDER GIRDLE PROBLEMS

Place the patient in the lumbar-locked, unilateral extension position with the hand behind the head, in the external rotation position, to take advantage of two things. First, this identifies a symmetrical or asymmetrical dysfunction in the thorax. Second, it reduces the scapular stability requirements needed to extend the upper quarter.

If the patient can now demonstrate a functional and non-painful extension and rotation in the lumbar-locked, externally rotated position, assume there is a scapular dysfunction or glenohumeral stability dysfunction or motor control dysfunction, or all three.

If the patient still has dysfunction or pain, place the hand behind the back in the internal rotation position to maintain minimal scapular stability requirements, but instead reduce anterior shoulder and chest mobility requirements for thorax extension.

If the patient can now demonstrate a functional and non-painful extension and rotation in the lumbar-locked, internally rotated position, assume there is a shoulder girdle joint mobility or tissue extensibility dysfunction. If extension is still dysfunctional or painful, retest the motion passively to determine the dysfunction.

If the passive motion is still reduced, there is a unilateral or bilateral thorax tissue extensibility dysfunction or a joint mobility dysfunction, or both. If the passive motion is now functional and non-painful, there is a bilateral thoracic-extension stability dysfunction or motor control dysfunction, or both.

MULTI-SEGMENTAL EXTENSION BREAKOUTS

In the presence of a limited *multi-segmental extension assessment* in the top-tier testing, proceed with the following breakouts.

SPINE EXTENSION

Backward Bend Without Upper Extremity
Single-Leg Backward Bend
Prone Press-Up
Lumbar-Locked (IR) Active Rotation/Extension
Lumbar-Locked (IR) Passive Rotation/Extension
Prone-On-Elbow Rotation/Extension

LOWER BODY EXTENSION

Standing Hip Extension
Prone Active Hip Extension
Prone Passive Hip Extension
Rolling Patterns (see page 187)
FABER Test
Modified Thomas Test

UPPER BODY EXTENSION

Single-Shoulder Backward Bend
Supine Lat Stretch Test, Hips Flexed
Supine Lat Stretch Test, Hips Extended
Lumbar-Locked (ER) Rotation/Extension
Lumbar-Locked (IR) Active Rotation/Extension
Lumbar-Locked (IR) Passive Rotation/Extension

SPINE EXTENSION BREAKOUTS

See the *Spine Extension Breakouts flowchart,* page 342.

BACKWARD BEND WITHOUT UPPER EXTREMITY

OBJECTIVE

To remove the shoulder joints and musculature from the backward bend maneuver

DESCRIPTION

Instruct the patient to stand tall with the hands on the hips. Tell the patient to bend backward as far as possible. The patient should easily be able to get the shoulders past the heels, the ASIS past the toes, and return to a standing position without pain.

ADDITIONAL INFORMATION

Remember to limit the amount of knee flexion for accurate results. This testing modification will help rule out upper-extremity dysfunction or pain provocation from the backward bend.

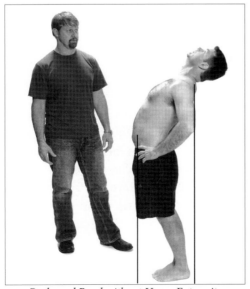

Backward Bend without Upper Extremity

POSSIBLE FINDINGS

- The ability to do the movement (FN)

- Ability or inability to perform the movement, and pain is present with both (FP or DP)

- An inability to perform the movement, and there is no pain (DN)

If the finding is FN, proceed to the *upper extension assessment.*

If the finding is DN, DP or FP, proceed to the *single-leg backward bend assessment.*

SINGLE-LEG BACKWARD BEND

OBJECTIVE

To reduce the problem to a symmetrical or asymmetrical dysfunction or used as a pain provocation maneuver

DESCRIPTION

Instruct the patient to elevate one leg on a step and place the hands on the hips. Have the patient bend backward as far as possible. The patient should easily be able to get the shoulders past the heel, the ASIS past the toes and return to a standing position without pain. Repeat with the opposite leg up on the step and compare the results.

ADDITIONAL INFORMATION

The hip is the focus of attention in this test, but unilateral spinal facet limitations can be a contributing factor.

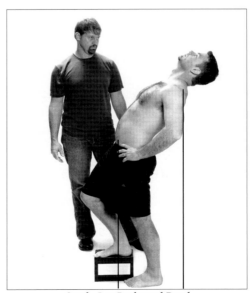

Single-Leg Backward Bend

POSSIBLE FINDINGS

- An ability to perform the movement (FN)

- An inability to perform the movement with or without pain (FP, DN or DP)

If the finding is FN, check for symmetry with the other leg. If both sides are now functional and non-painful, the patient has a symmetrical-stance core stability and motor control problem. Proceed to the *upper body extension breakout.*

If the finding is DN, DP or FP, proceed to the *prone press-up assessment.*

PRONE PRESS-UP

OBJECTIVE

To look at the backward bend in a non–weight-bearing position

DESCRIPTION

Instruct the patient to lie prone with the head face down, arms by the sides with elbows bent, and palms facing down underneath the armpits. Have the patient then try to extend the trunk up as high as possible using the hands and upper body for assistance and support. Elbows should fully extend at end of movement with the ASIS remaining in contact with the table.

ADDITIONAL INFORMATION

If the patient cannot fully extend the arms from the prone position and keep the ASIS on the table, place a two-and-a-half-inch dense foam pad under the pelvis and repeat the movement. If the patient can now fully extend the arms and keep the ASIS in contact with the pad, this is still considered normal movement.

When the hand placement is correct, the arms should be perpendicular to the table. This can also differentiate between extension limitations verses weight-bearing stability dysfunction.

POSSIBLE FINDINGS

- An ability to complete the movement, with or without pad (FN)

- Pain during the movement (FP or DP)

- An inability to perform the movement and no pain is involved (DN)

If the finding is FN—that is, the patient can perform a full extension in the prone position but not while standing—the patient might have a weight-bearing spine extension stability or motor control dysfunction. The patient might also have a hip extension or shoulder flexion problem. Proceed to the *lower- and upper-body extension flowchart.*

If the finding is DN, DP or FP, proceed to the *lumbar-locked (IR) unilateral extension assessment.*

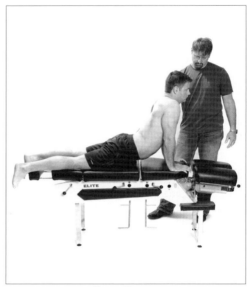

Prone Press-up

LUMBAR-LOCKED (IR) ACTIVE ROTATION/EXTENSION

OBJECTIVE

To look at combined extension and rotation in the thoracic spine combined with shoulder internal rotation in a non–weight-bearing position

DESCRIPTION

Instruct the patient to get into a full prone rocking position—butt touching heels—with the right hand and forearm behind the back, while the left forearm and hand are on the table centered in front of the knees. Have the patient rotate the right shoulder up and back as far as possible while maintaining the prone rocking position. Compare both sides.

ADDITIONAL INFORMATION

The angle of the raised shoulder should be at least 50 degrees relative to the table. The prone rocking position helps eliminate the lumbar spine from contributing to the extension. This can also differentiate between right and left upper thoracic restrictions.

POSSIBLE FINDINGS

- An ability to complete the movement (FN)

- An ability or inability to perform the move with pain present (FP or DP)

- An inability to perform the move and pain is not present (DN)

If the finding is FN, proceed to the *prone-on-elbow unilateral extension assessment.*

If the finding is DP, DN or FP, proceed to the *lumbar-locked (IR) passive rotation/extension assessment.*

DESCRIPTION

Instruct the patient to get into a full prone rocking position—butt touching heels—with the right hand and forearm behind the back, while the left forearm and hand are on the table centered in front of the knees. Rotate the patient's right shoulder up and back as far as possible while the patient maintains the prone rocking position with the left hand and forearm remaining on the table. Compare both sides.

ADDITIONAL INFORMATION

The angle of the raised shoulder should be greater than 50 degrees relative to the table. The prone rocking position helps eliminate the lumbar spine from contributing to the extension. This can also differentiate between right and left upper-thoracic restrictions.

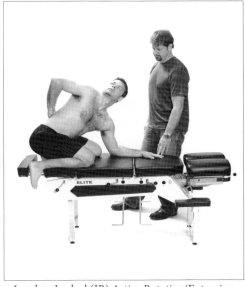

Lumbar-Locked (IR) Active Rotation/Extension

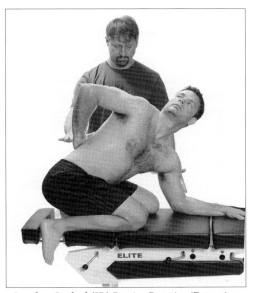

Lumbar-Locked (IR) Passive Rotation/Extension

LUMBAR-LOCKED (IR) PASSIVE ROTATION/EXTENSION

OBJECTIVE

To look at combined extension and rotation in the thoracic spine combined with shoulder internal rotation in a non–weight-bearing position

POSSIBLE FINDINGS

- An ability to complete the movement (FN)

- An inability to perform the movement without pain (FP or DP)

- An inability to perform the movement either on one side without pain (unilateral DN), or an inability to perform the movement on both sides without pain (bilateral DN)

If the finding is FP or DP, stop and treat the pain.

If the finding is FN, there is a thoracic bilateral extension stability or motor control dysfunction, or both.

If the finding is a unilateral DN, there is a unilateral thoracic extension tissue extensibility or joint mobility dysfunction, or both.

If the finding is a bilateral DN, there is a bilateral thoracic extension tissue extensibility or joint mobility dysfunction, or both.

PRONE-ON-ELBOW ROTATION/EXTENSION

OBJECTIVE

Performed as a clearing test and provocation maneuver for the lumbar spine

DESCRIPTION

Have the patient prone with the right hand behind the back and the left forearm on the table for support. The patient should rotate the right shoulder complex up and back as far as possible. Repeat and compare with the other side.

ADDITIONAL INFORMATION

If this motion is limited or induces pain or exacerbates the pain, you must consider the lumbar spine as a source of dysfunction. This can also reduce the problem to a symmetrical or asymmetrical dysfunction. Normal range is at least 30 degrees.

POSSIBLE FINDINGS

- An ability to complete the movement on both sides (FN)

- An inability to perform the move without pain on both sides (FP or DP)

- An inability to perform the move either on one side (unilateral DN), or an inability to perform the move on both sides (bilateral DN)

If the finding is FN on both sides, there is still potential for a bilateral spine extension stability and motor control dysfunction. Continue to the *upper- and lower-body extension flowchart.*

If the finding is DN on both sides (bilateral DN), there is a bilateral lumbar extension joint-mobility dysfunction or tissue extensibility dysfunction or stability and motor control dysfunction, or all three. Proceed to the *upper- and lower-body extension flowchart.*

If the finding is DN on one side (unilateral DN), there is a unilateral lumbar extension joint mobility dysfunction or tissue extensibility dysfunction or stability and motor control dysfunction, or all three. Proceed to the *upper- and lower-body extension flowchart.*

If the finding is a DP or FP, stop and treat problem.

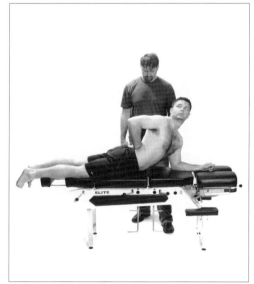

Prone on Elbow Rotation/Extension

LOWER BODY EXTENSION BREAKOUTS

See *Lower Body Extension Breakouts flowchart,* page 343.

STANDING HIP EXTENSION

OBJECTIVE

To reduce the problem to a symmetrical or asymmetrical dysfunction or used as a pain provocation maneuver, and to look at extension from the bottom up

DESCRIPTION

Instruct the patient to place both hands at the sides and slowly extend the right leg back as far as possible. Observe to make sure the leg stays extended and the extension comes from the hip and not from knee flexion. Have the patient return to the starting position, and repeat with the left leg.

ADDITIONAL INFORMATION

Tell the patient to hold the head steady, not allowing it to go down or forward. The hip is the focus of attention here. Free extension from the hip on this test helps differentiate a spinal extension problem from a hip dysfunction.

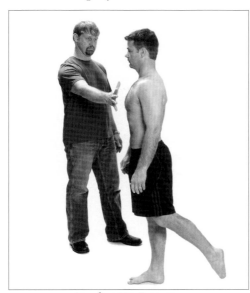

Standing Hip Extension

POSSIBLE FINDINGS

- An ability to extend both legs greater than 10 degrees relative to the stationary leg (FN)

- An ability to extend either or both legs greater than 10 degrees relative to the stationary leg, but with pain involved (FP)

- An inability to extend either or both legs greater than 10 degrees relative to the stationary leg (DN or DP)

If the finding is FN, assume that there is a weight-bearing lower-quarter stability and motor control dysfunction or limited ankle dorsiflexion, or both. Recheck the *overhead deep squat* and the *single-leg stance assessments.*

If the finding is DN, DP or FP, proceed to the *prone active hip extension assessment.*

PRONE ACTIVE HIP EXTENSION

OBJECTIVE

To reduce the problem to a symmetrical or asymmetrical dysfunction, or used as a pain provocation maneuver in the hip in an active, non–weight-bearing maneuver

DESCRIPTION

Instruct the patient to lie prone with the hands down by the sides or under the head. The patient will then actively extend the right hip as far as possible. After returning to starting position, have the patient repeat the action with the left leg.

ADDITIONAL INFORMATION

The patient is trying to get equal to or greater than 10 degrees of hip extension. Watch for anterior tilt of the pelvis, or external rotation and abduction of the foot.

POSSIBLE FINDINGS

- An ability to extend both legs greater than 10 degrees relative to the surface (FN)

- An ability to extend either or both legs greater than 10 degrees relative to the stationary leg with pain involved (FP)

- An inability to extend either or both legs greater than 10 degrees relative to the stationary leg (DN or DP)

If the finding is FN, proceed to the *rolling patterns breakout.* If the *rolling patterns breakout* is FP or DP, stop and treat the problem.

SFMA

If the rolling patterns breakout is FN, there is a spine weight-bearing hip extension stability and motor control dysfunction. If the rolling patterns breakout is DN, there is a fundamental extension pattern dysfunction.

If the finding is DN, DP or FP, proceed to the *prone passive hip extension assessment.*

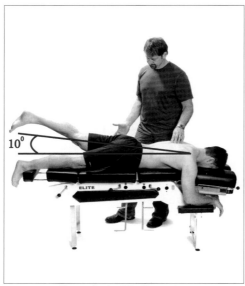

Prone Active Hip Extension

PRONE PASSIVE HIP EXTENSION

OBJECTIVE

To compare the patient's active hip extension to the passive hip extension

DESCRIPTION

Instruct the patient to lie prone with the hands down by the sides or under the head. Passively extend the right hip as far as possible. Compare this range to the patient's active hip extension measurement noted earlier. Return to the starting position and do the same with the left leg.

ADDITIONAL INFORMATION

These two measurements should be within 10 degrees of each other; if not, there is a dysfunction. If the patient cannot actively extend the hip within 10 degrees of passive hip extension, suspect a lumbo-pelvic-hip stability dysfunction or muscular weakness in hip extension.

POSSIBLE FINDINGS

- The angle will not be reached or there will be pain involved in trying to reach it (DN, DP or FP). Proceed to the FABER test.

- The angle could possibly be greater than 25 percent of the active hip extension. In this case, proceed to the rolling patterns breakout.

If the rolling patterns outcome is painful, stop and treat the problem.

If the rolling patterns outcome in FN, there is a core stability and motor control dysfunction or active hip extension stability and motor control dysfunction, or both.

If the rolling patterns outcome is DN, there is a fundamental extension pattern stability or motor control dysfunction, or both.

If the finding is FN, proceed to the *Modified Thomas test.*

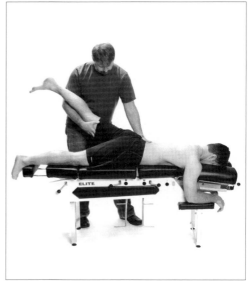

Prone Passive Hip Extension

FABER TEST

OBJECTIVE

To evaluate the affect of hip flexion, abduction and external rotation overpressure on the hip and lumbar spine

DESCRIPTION

Instruct the patient to lie supine and then flex, abduct and externally rotate the hip by placing the left foot over the right thigh or knee. Have the patient slowly lower the knee toward the table. Look for restrictions or signs of pain.

Repeat on the other side.

ADDITIONAL INFORMATION

The name of this test tells the story—Flexion, Abduction, External Rotation of the hip. If there's pain in the front of the hip, or joint discomfort or limited motion, suspect hip joint pathology such as degenerative joint disease (DJD), or anterior hip capsular tightness or labrum involvement. If pain develops in the back over the lumbar spine or SI joint, suspect limited hip-joint mobility creating lumbo-pelvic instability.

POSSIBLE FINDINGS

- Ability to perform the movement (FN)

- Inability to perform the movement (DN)

- Pain while performing the movement or pain while trying to perform the move (FP or DP)

If the finding is FN, go to the *Modified Thomas test* to further investigate.

If the finding is DN, there is a hip-joint mobility dysfunction or a tissue extensibility dysfunction or a core stability and motor control dysfunction, or all three. Use local biomechanical testing on the hip, then proceed to the *Modified Thomas test* for further assessments.

If the finding demonstrates pain, stop and treat the problem.

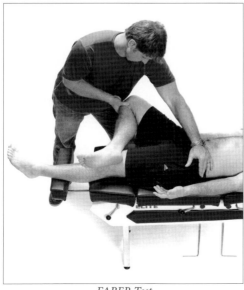

FABER Test

MODIFIED THOMAS TEST

OBJECTIVE

Determines the overall mobility of the hip flexor muscles and the anterior hip capsule

DESCRIPTION

Have the patient sit on the edge of a treatment table and roll backward, grabbing both knees while rolling. With the lower thoracic spine and sacrum flat, instruct the patient to grab one knee with both hands and pull back until the lumbar spine is flat on the table. Make sure the sacrum stays in contact with the table, not rolling back so far that the glutes come off the table and the lumbar spine flattens.

Take the non-supported leg and place it in 90 degrees of hip flexion and 90 degrees of knee flexion, and adduct it next to the supported leg.

Passively lower the leg. Ask the patient to place zero tension in the moving leg. Make sure the patient is not arching the back during the lowering phase, keeping a firm grasp of the opposite leg.

If the thigh does not go all the way down to the table without the knee straightening or thigh abducting, perform the following additional tests.

- Bring the leg back up to the starting position and extend the knee. Try to lower the thigh to the table, and note any difference in the distance from the table.

- Bring the leg back up to the starting position and abduct the thigh. Lower the thigh toward the table, and note any difference in the distance from the table.

- Bring the leg back up to the starting position, extend the knee and abduct the thigh. Again lower the thigh toward the table and note any difference in the distance from the table.

ADDITIONAL INFORMATION

This test will help differentiate between high tone or mobility dysfunction in the iliacus, rectus femoris and the tensor fasciae latae (TFL).

POSSIBLE FINDINGS

- An inability to completely lower the thigh to the table without pain (DP or FP)

- An ability to touch the table with a straight knee (DN)

- An ability to touch the table with an abducted hip (DN)

- An ability to touch the table with an abducted hip and a straight knee (DN)

- A FN *Modified Thomas test*

If with the knee straight, the thigh touches the table, there is a lower anterior-chain tissue extensibility dysfunction.

If with the hip abducted, the thigh touches the table, there is a lateral-chain tissue extensibility dysfunction.

If with the hip abducted and knee straight, the thigh touches the table, there is an anterior- and lateral-chain tissue extensibility dysfunction.

If the *Modified Thomas test* is good, there is a core stability and motor control dysfunction.

If the thigh never touches the table and there is no pain, there is a hip joint mobility dysfunction or tissue extensibility dysfunction or a core stability or motor control dysfunction. Perform local biomechanical testing of the hip.

If there is pain (DP or FP), stop and treat it.

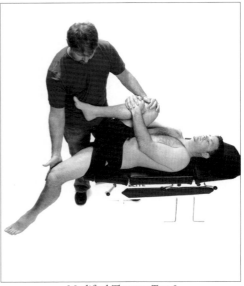

Modified Thomas Test 2

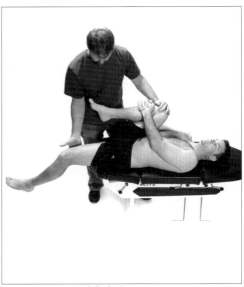

Modified Thomas Test 3

UPPER BODY EXTENSION BREAKOUTS

See the *Upper Body Extension Breakouts flowchart*, page 344.

UNILATERAL-SHOULDER BACKWARD BEND

OBJECTIVE

To remove one shoulder at a time from the backward bend maneuver

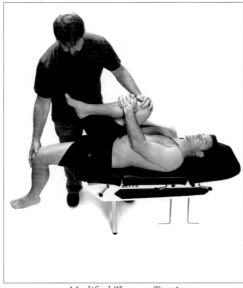

Modified Thomas Test 1

DESCRIPTION

Instruct the patient to stand tall with the right arm extended straight above the head and the left hand on the hip. The patient should be able to get the arm next to the ear. While in this position, have the patient bend backward as far as possible.

ADDITIONAL INFORMATION

This will help identify symmetrical or asymmetrical dysfunction, or is used as a pain provocation maneuver for the upper extremities. During the backward bend, the patient should be able to get the shoulders past the heels, the humerus in line with the ear, the ASIS past the toes, and return to a standing position without pain.

POSSIBLE FINDINGS

- An inability to perform the move, or pain while trying (DN or FP or DP)

- The patient is able to perform the move on both sides (FN)

If the finding is DN, DP or FP, proceed to the *supine lat stretch, hips flexed assessment.*

If the finding is FN, double-check the *press-up* in the *spine extension breakout* for possible thoracic spine involvement, and double-check cervical spine patterns to rule out cervical spine involvement.

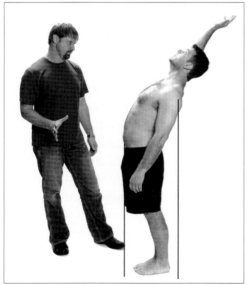
Unilateral Shoulder Backward Bend

SUPINE LAT STRETCH HIPS FLEXED

OBJECTIVE

To evaluate the length of the latissimus dorsi musculature

DESCRIPTION

Instruct the patient to lie supine with the arms held straight, vertical in front of the chest, palms facing the feet. Next, have the patient bring the knees to the chest and flatten the lower back against the surface below. See if the patient can lower the arms flat on the surface above the head, while keeping the arms extended.

ADDITIONAL INFORMATION

By flexing the hips, we put traction on the thoraco-lumbar fascia, which helps isolate the lat and posterior chain. The patient should easily be able to rest the arms flat on the surface overhead with the arms extended.

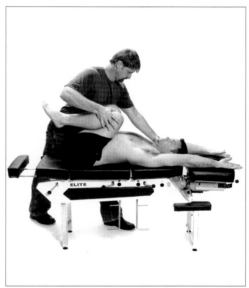
Supine Lat Stretch Hips Flexed

POSSIBLE FINDINGS

- Ability to lower the arms to the table (FN)

- An inability to lower the arms to the table, or pain while trying (DN, DP or FP)

SFMA

If the finding is FN, proceed to the *rolling patterns assessment*. If the rolling patterns outcome is FN, there is a weight-bearing upper-quarter extension stability and motor control dysfunction. If rolling pattern outcome is DP or FP, stop and treat the problem. If the rolling patterns outcome is DN, there is a fundamental extension pattern dysfunction.

If the finding is DN, DP or FP, proceed to the *supine lat stretch hips extended assessment*.

SUPINE LAT STRETCH HIPS EXTENDED

OBJECTIVE

To determine if a problem is solely a lat problem or another shoulder flexion limitation

DESCRIPTION

Instruct the patient to lie supine with the arms held straight out, vertical, in front of the chest and palms facing the feet. With the hips extended, have the patient lower the arms towards the surface above the head; note the distance between each arm and the surface below.

ADDITIONAL INFORMATION

Only use this test if the hips-flexed version demonstrated dysfunction or pain. If the problem is only in the lat or posterior chain, the patient should now be able to rest the arms flat on the surface overhead with the arms extended.

POSSIBLE FINDINGS

- An ability to rest the arms flat on the table (FN)

- No change or pain while trying to perform the move (DN, DP or FP)

- Shoulder flexion only slightly improves

If the finding is slight improvement of shoulder flexion, a lat or posterior-chain tissue extensibility dysfunction is part of the problem. Check the *lower-body extension flowchart* and proceed to the *lumbar-locked (ER) unilateral extension assessment*.

If the finding is DN, DP or FP, proceed to the *lumbar-locked (ER) unilateral extension assessment*.

If the finding is FN, there is a lat or posterior-chain tissue extensibility dysfunction. If the patient has limited hip extension and goes into excessive lumbar lordosis as the hips extend, this can effectively shorten the lat and posterior chain, giving a false positive. Work through the *lower body extension flowchart*.

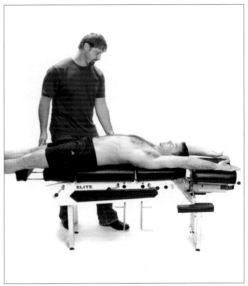

Supine Lat Stretch Hips Extended

LUMBAR-LOCKED (ER) EXTENSION/ROTATION

OBJECTIVE

To look at extension and rotation in the thoracic spine combined with shoulder external rotation and scapular retraction, all in a non–weight-bearing position

DESCRIPTION

Instruct the patient to get into a full prone rocking position—butt to heels—and place the right hand behind the head, while the left forearm and hand are on the table centered in front of the knees. Have the patient rotate the right elbow up and back far as possible while maintaining the prone rocking position; there should be no weight shifting and no leaning. The patient returns to the start position, and repeats on the other side.

ADDITIONAL INFORMATION

The prone rocking position helps eliminate the lumbar spine from contributing to the extension.

This can also differentiate between right and left upper-thoracic restrictions. The angle of the up-turned shoulder should be at least 50 degrees, and the elbow should clear the chest wall.

POSSIBLE FINDINGS

- Ability to perform the movement on both sides without pain (FN)

- An inability or pain while trying to perform the movement (DN, DP or FP)

If the finding is DN, DP or FP, proceed to the *lumbar-locked (IR) active rotation and extension assessment.*

If the finding is FN, there is a scapular dysfunction or glenohumeral stability or motor control dysfunction, or both.

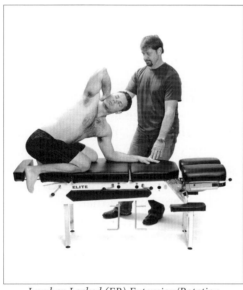

Lumbar-Locked (ER) Extension/Rotation

LUMBAR-LOCKED (IR) ACTIVE ROTATION AND EXTENSION

See description and instructions, page 154.

LUMBAR-LOCKED (IR) PASSIVE ROTATION AND EXTENSION

See description and instructions, page 155.

MULTI-SEGMENTAL ROTATION BREAKOUT RATIONALE

THORACIC ROTATION MOBILITY VERSUS STABILITY PROBLEM

Multi-segmental rotation dysfunctions can be due to spine, hip or below the knee problems, or all three. Start the breakout at the spine and progress to the lower quarter.

The first test is the seated rotation assessment. If the patient has a normal range of motion, assume the spine is clear and proceed to the hip rotation and tibia flowcharts. If the seated rotation assessment is dysfunctional or painful, proceed to the lumbar-locked (IR) active rotation assessment.

WEIGHT-BEARING VERSUS NON–WEIGHT-BEARING PROBLEM

The lumber-locked internal (IR) and external (ER) active rotation assessments evaluate the tissue extensibility, joint mobility and stability and motor control of the thorax and shoulder. The assessments can also reduce the problem to a symmetrical or asymmetrical dysfunction, and can differentiate between a shoulder girdle versus thoracic spine dysfunction.

The lumbar-locked position allows better isolation of the thoracic spine and reduces the contribution of the lumbar spine. Placing the arm behind the head (ER) challenges the shoulder girdle, while the arm behind the back (IR) allows a reduction of scapular involvement and minimizes anterior shoulder and chest involvement.

If the lumbar-locked (ER) active rotation assessment causes the thoracic rotational mobility dysfunction to switch directions as compared to the seated rotation, assume there is a stability problem and use rolling patterns to grade the severity.

If rolling is normal, there is a weight-bearing thoracic rotational stability dysfunction or a motor control dysfunction, or both. If rolling is dysfunctional, there is a fundamental spine rotational stability dysfunction or a motor control dysfunction, or both. If there is pain with rolling, stop and treat the pain

If lumbar-locked (ER) active rotation is functional and non-painful, evaluate the lumbar spine with prone on elbow unilateral assessment.

If lumbar-locked (ER) active rotation is dysfunctional or painful, proceed to lumbar-locked (IR) active rotation.

If the lumbar-locked (IR) active rotation assessment is functional and non-painful (greater than 50 degrees), assume there is a shoulder girdle tissue extensibility or joint mobility dysfunction.

If the lumbar-locked (IR) active rotation assessment is dysfunction or painful, the patient has a thorax or spine rotational mobility or stability problem. Retest the motion passively to determine the dysfunction. If the passive motion is still reduced, the patient has a unilateral or bilateral thorax rotational tissue extensibility or joint mobility dysfunction, or both. If the passive motion is now functional and non-painful, the patient has a thorax rotational stability or motor control dysfunction, or both.

Use rolling patterns to assess the severity of the thorax rotational stability or motor control problems. If rolling is normal, the patient has a thorax rotational stability or motor control dysfunction, or both. If rolling is dysfunctional, the patient has a fundamental spine rotational stability or motor control dysfunction, or both. If there is pain with rolling, stop and treat the it.

LUMBAR SPINE
EXTENSION PROBLEM

Even though the lumbar spine only possesses 10 degrees of rotation normally, if thoracic rotational testing is normal, the lumbar spine must be assessed using the prone-on-elbow unilateral extension assessment. In this position, the thoracic spine is already extended, so all the stress will be placed on the lumbar spine. If rotation and extension from this position creates pain, stop and treat the lumbar spine.

If both directions are now functional and non-painful, there is a spine rotational stability or motor control dysfunction. Use rolling patterns to grade the severity. If rolling is normal, there is a weight-bearing spine rotational stability dysfunction or a motor control dysfunction, or both. If rolling is dysfunctional, there is a fundamental spine rotational stability or a motor control dysfunction, or both. If there is pain with rolling, stop and treat the pain.

If there is dysfunction in the prone-on-elbow unilateral rotation assessment, the patient can have unilateral or bilateral tissue extensibility problems, joint mobility problems, or both, or stability and motor control problems in the lumbar spine. Use local biomechanical testing of the lumbar spine to further assess.

Note: Limited hip extension can cause a false positive in the prone-on-elbow position. Double-check the lower body extension flowchart for possible complicating problems.

Proceed to the hip rotation breakout.

HIP INTERNAL VERSUS
EXTERNAL ROTATION PROBLEM

The hip rotation breakout is divided into two parts. The first part breaks down external hip rotation dysfunctions. The second part breaks down internal hip rotation dysfunctions. Both hip breakouts follow the same logic. We'll start our review with hip rotation, part one of the breakout flowchart.

It is important to run both seated and prone testing before we can make any definitive stability diagnosis. The hip can function very differently from a flexed versus extended position. This is due to the way the soft tissues of the hip and pelvis cross the joint. Throughout the SFMA, when mobility problems appear and then disappear in different positions, it is a sign of stability problems. This is not the case here. Many times mobility dysfunctions will be picked up with the hip extended, but not flexed. When the hip dysfunction disappears flexed but reappears prone, it is still a mobility dysfunction. Those mobility dysfunctions need to be addressed before any stability exercises are performed. We can only make the diagnosis of hip stability dysfunction if both seated and extended hip mobility are functional and non-painful.

ACTIVE VERSUS
PASSIVE HIP EXTERNAL PROBLEMS

The first test we perform is the seated active external hip rotation assessment. This will allow us to evaluate external hip rotation in a non–weight-bearing position with the hip flexed, and it determines if the problem is unilateral or bilateral.

If this is functional and non-painful, proceed to the prone active external hip rotation to see if any limitations occur when the hip is extended. If this is dysfunctional or painful, we immediately compare passive hip external rotation to active in the seated passive external hip rotation assessment. If passive testing is now functional and non-painful, we assume there is a stability problem with the hip flexed, but must first check the hip extended before we label this a stability dysfunction. If passive mobility is still dysfunctional, there is a hip joint mobility or tissue extensibility dysfunction, or both, with external hip rotation when the hip is flexed. No matter what the finding is, proceed to the prone active external hip rotation to see how the hip performs in an extended position.

Next, perform the same tests with the hip extended. First, check active motion with the prone active external hip rotation assessment. If this is functional and non-painful and the seated active or passive external hip rotation assessments were also functional and non-painful, we can now say there is a stability or motor control problem, or both. Use the rolling pattern outcomes to determine the severity of the stability problem. If rolling is functional and non-painful, the patient has a weight-bearing external hip rotation stability or motor control dysfunction, or both. If rolling is dysfunctional and non-painful, the patient has a fundamental hip rotation stability and motor control dysfunction.

If the prone active external hip rotation assessment is functional and non-painful, but the seated passive external hip rotation assessment is dysfunctional, we need to address the mobility problem with the hip flexed first. This is still considered a mobility problem.

If the prone active external hip rotation assessment is dysfunctional or painful, we immediately compare passive to active in the prone passive external hip rotation assessment. If passive mobility is still dysfunctional, there is a hip joint mobility or tissue extensibility dysfunction, or both, with external hip rotation when the hip is extended. From here, we proceed to the tibial rotation flowchart to check below the knee for other possible lower-quarter rotation restrictions. Since the hip is extended in this assessment, there is also a possibility that poor hip extension is influencing the test results. For this reason, proceed to the lower body extension flowchart to double-check hip extension.

If the prone passive external hip rotation assessment is functional and non-painful and the seated active or passive external hip rotation assessments are also functional and non-painful, we can now say there is a stability or motor control problem, or both. Use the rolling pattern outcomes to determine the severity of the stability problem. If rolling is functional and non-painful, the patient has a weight-bearing external hip rotation stability or motor control dysfunction, or both. If rolling is dysfunctional and non-painful, the patient has a fundamental hip rotation stability and motor control dysfunction.

If the prone passive external hip rotation assessment is functional and non-painful, but the seated passive external hip rotation assessment is dysfunctional, we need to address the mobility problem with the hip flexed first. This is still considered a mobility problem.

ACTIVE VERSUS PASSIVE HIP INTERNAL PROBLEMS

This flowchart follows the same logic as stated earlier for external rotation and must be checked regardless of the results in hip rotation flowchart, part one.

The first test we perform is the seated active internal hip rotation assessment. This will allow us to evaluate internal hip rotation in a non–weight-bearing position with the hip flexed. If this is functional and non-painful, proceed to prone active internal hip rotation to see if any limitations occur when the hip is extended. If this is dysfunctional or painful, immediately compare passive hip internal rotation to active in the seated passive internal hip rotation assessment. If passive testing is now functional and non-painful, we assume there is a stability problem with the hip flexed, but must first check the hip extended before we label this a stability dysfunction. If passive mobility is still dysfunctional, there is a hip joint mobility or tissue extensibility dysfunction, or both, with internal hip rotation when the hip is flexed. No matter what the finding is, proceed to prone active internal hip rotation to see how the hip performs in an extended position.

Next, perform the same tests with the hip extended. First we check active motion with the prone active internal hip rotation assessment. If this is functional and non-painful and the seated active and/or passive internal hip rotation assessments were also functional and non-painful, we can now say there is a stability or motor control problem, or both. Use the rolling pattern outcomes to determine the severity of the stability problem. If rolling is functional and non-painful, the patient has a weight-bearing internal hip rotation stability or motor control dysfunction, or both. If rolling is dysfunctional and non-painful, the patient has a fundamental hip rotation stability and motor control dysfunction.

If the prone active internal hip rotation assessment is functional and non-painful, but the seated passive internal hip rotation assessment is dysfunctional, we need to address the mobility problem with the hip flexed first. This is still considered a mobility problem.

If the prone active internal hip rotation assessment is dysfunctional or painful, we immediately compare passive to active in the prone passive internal hip rotation assessment. If passive mobility is still dysfunctional, there is a hip joint mobility or tissue extensibility dysfunction, or both, with internal hip rotation when the hip is extended. From here, we proceed to the tibial rotation flowchart to check below the knee for other possible lower-quarter rotation restrictions. Since the hip is extended in this assessment, there is also a possibility poor hip extension is influencing the test results. For this reason, proceed to the lower body extension flowchart to double check hip extension.

If the prone passive internal hip rotation assessment is functional and non-painful and the seated active or passive internal hip rotation assessments are also functional and non-painful, we can now say there is a stability or motor control problem, or both. Use the rolling pattern outcomes to determine the severity of the stability problem. If rolling is functional and non-painful, the patient has a weight-bearing internal hip rotation stability or motor control dysfunction, or both. If rolling is dysfunctional and non-painful, the patient has a fundamental hip rotation stability and motor control dysfunction.

If the prone passive internal hip rotation assessment is functional and non-painful, but the seated passive internal hip rotation assessment is dysfunctional, we need to address the mobility problem with the hip flexed first. This is still considered a mobility problem.

TIBIAL ROTATION PROBLEM

The last step in the multi-segmental rotation breakout is to evaluate motion below the knee. Most people don't realize that tibial rotation can account for up to 20 degrees of rotation. We begin by performing the seated active internal and external rotation assessments. If the motion is functional and non-painful, tibial rotation is normal.

As stated earlier, it is possible that as the patient turns to the right in the multi-segmental rotation test, the dysfunction lies in the left hip not being able to extend. For this reason, when the seated active internal and external rotation assessments are functional and non-painful, check the lower body extension flowchart just to be sure.

If the seated active internal and external rotation assessments are dysfunctional or painful, compare passive to active motion using the seated passive internal and external tibial rotation assessments. If passive tibial rotation is functional and non-painful, there is a tibial rotation stability or motor control dysfunction, or both. If passive tibial rotation is dysfunctional, there is a tibial rotation joint mobility or tissue extensibility dysfunction, or both. If there is pain, stop and treat it.

Since poor foot mechanics can also contribute to multi-segmental rotation limitations, double-check single-leg stance findings to see if that is a contributing factor.

MULTI-SEGMENTAL ROTATION BREAKOUTS

In the presence of a limited *multi-segmental rotation assessment* in the top-tier testing, proceed with the following breakouts.

LIMITED MULTI-SEGMENTAL ROTATION

Seated Rotation
Lumbar-Locked (ER) Active Rotation/Extension
Lumbar-Locked (IR) Active Rotation/Extension
Rolling Patterns (See page 187)
Lumbar-Locked (IR) Passive Rotation/Extension
Prone-on-Elbow Rotation/Extension

HIP ROTATION

Seated Active External Hip Rotation
Seated Passive External Hip Rotation
Prone Active External Hip Rotation
Prone Passive External Hip Rotation
Seated Active Internal Hip Rotation
Seated Passive Internal Hip Rotation
Prone Active Internal Hip Rotation
Prone Passive Internal Hip Rotation

TIBIAL ROTATION

Seated Active Internal Tibial Rotation
Seated Passive Internal Tibial Rotation
Seated Active External Tibial Rotation
Seated Passive External Tibial Rotation

LIMITED MULTI-SEGMENTAL ROTATION BREAKOUTS

See the *Limited Multi-Segmental Rotation Breakouts flowchart,* page 344.

SEATED ROTATION

OBJECTIVE

Determines if the patient has good spinal rotation bilaterally

DESCRIPTION

Instruct the patient to get into a seated position with the thighs and knees shoulder width apart, body in an upright and erect posture. Place a dowel across the shoulders behind the neck and have the patient place the hands under the dowel at both ends. Have the patient rotate the trunk both to the right and to the left as far as possible. Once the patient reaches the farthest point, take a measurement with a goniometer.

ADDITIONAL INFORMATION

Do this assessment to remove the lower body joints from the rotation test. Look for 50 degrees bilaterally without pain. The dowel is used to limit scapular recruitment.

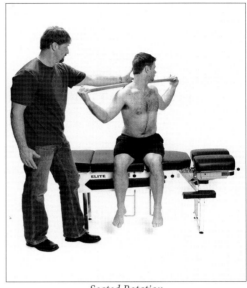

Seated Rotation

POSSIBLE FINDINGS

- An ability to turn more than 50 degrees bilaterally (FN)
- Inability to turn more than 50 degrees bilaterally with or without pain (DN, DP, FP)

If the finding is FN, proceed to the *hip rotation flowchart.*

If the finding is DN, DP or FP, proceed to the *lumbar-locked (IR) active rotation and extension assessment.*

LUMBAR-LOCKED (ER) ACTIVE ROTATION/EXTENSION

See page 162 for description and instructions.

LUMBAR-LOCKED (IR) ACTIVE ROTATION/EXTENSION

See page 154 for description and instructions.

LUMBAR-LOCKED (IR) PASSIVE ROTATION/EXTENSION

See page 155 for description and instructions.

PRONE-ON-ELBOW ROTATION/EXTENSION

See page 156 for description and instructions.

HIP ROTATION BREAKOUTS

See the *Hip Rotation Breakouts flowchart*, pages 346-347.

SEATED ACTIVE EXTERNAL HIP ROTATION

OBJECTIVE

To assess the dysfunction or pain with hip external rotation actively in a non–weight-bearing position with the hip flexed

DESCRIPTION

Put the patient in a seated position with knees and feet together, body in an upright and erect posture, hands on top of each ilium—this allows you to see any hiking of the pelvis during testing. Have the patient rotate the hip externally while keeping the knee flexed and the pelvis level. Once the patient reaches the outermost point of the movement, take a measurement with a goniometer. Repeat the test on the other side.

ADDITIONAL INFORMATION

Look for 40 degrees bilaterally without pain during the active external rotation.

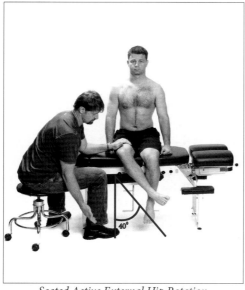

Seated Active External Hip Rotation

POSSIBLE FINDINGS

- An ability to externally rotate more than 40 degrees (FN)

- An inability to externally rotate more than 40 degrees (DN, DP or FP)

If the finding is FN, proceed to the *prone active external hip rotation assessment.*

If the finding is DN, DP or FP, proceed to the *seated passive external hip rotation assessment.*

SEATED PASSIVE EXTERNAL HIP ROTATION

OBJECTIVE

To assess the dysfunction or pain with hip external rotation passively in a non–weight-bearing position

DESCRIPTION

Put the patient in a seated position with knees and feet together, body in an upright and erect posture, hands on top of each ilium—this allows you to see any hiking of the pelvis during testing. Rotate the patient's hip externally, with the knee still flexed and the leg passive. Once the outermost point of the rotation is reached, take a measurement with a goniometer. Repeat on the other side and compare with the active external rotation.

ADDITIONAL INFORMATION

We are looking for 40 degrees bilaterally without pain during the rotation. The difference between active and passive external rotation should be within 10 degrees.

POSSIBLE FINDINGS

- An ability to perform the movement (FN)

- Pain during the attempt (FP or DP)

- Inability to perform the move, but no pain present (DN)

If the finding is FN, proceed to the *prone active external hip rotation assessment.*

If the finding is DN, there is a hip joint mobility dysfunction or tissue extensibility dysfunction with external rotation with the hip flexed. Proceed to the *prone active external hip rotation assessment*.

If the finding is DP or FP, stop and treat the problem.

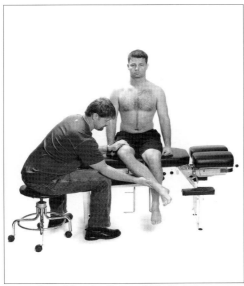

Seated Passive External Hip Rotation

PRONE ACTIVE EXTERNAL HIP ROTATION

OBJECTIVE

To assess the dysfunction or pain with hip external rotation actively in a non–weight-bearing position with the hip extended

DESCRIPTION

The patient should lie face down with the knees flexed. While stabilizing the pelvis, have the patient lower the leg or rotate the hip externally, keeping the knee flexed. Once the patient reaches the farthest point, take a measurement with a goniometer. Note the measurement and repeat on the other side.

ADDITIONAL INFORMATION

With the hip extended, there is a bigger stretch placed on the soft tissue and joint capsule. This gives a good picture of the hip musculature and supporting soft tissue mobility. Look for 40 degrees bilaterally without pain during the rotation.

POSSIBLE FINDINGS

- An ability to perform the movement to greater than 40 degrees (FN)

- An inability to perform the move or performing the move in the presence of pain (DN, DP or FP)

If the finding is FN *and the seated passive rotation* is DN, stop and treat the DN.

If the finding is FN and *seated active external hip rotation* or seated *passive external rotation* are FN, continue to the *rolling patterns*. If the rolling patterns outcome is DP or FP, stop and treat the problem.

If the rolling patterns outcome is FN, there is a weight-bearing external hip rotation stability and motor control dysfunction. Proceed to the *tibial rotation flowchart* and the *lower-body extension breakout*.

If the rolling patterns outcome is DN, there is a fundamental hip rotation stability and motor control dysfunction. Proceed to the *tibial rotation flowchart* and the *lower-body extension breakout*.

If the finding is DN, DP or FP, proceed to the *prone passive external hip rotation assessment*.

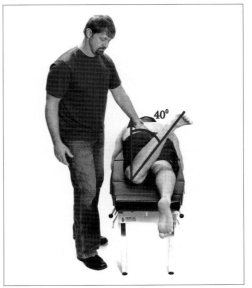

Prone Active External Hip Rotation

SFMA

PRONE PASSIVE EXTERNAL HIP ROTATION

OBJECTIVE

To assess the dysfunction or pain with hip external rotation passively in a non–weight-bearing position with the hip extended.

DESCRIPTION

Have the patient lie face down on the table with the knees flexed. While stabilizing the pelvis, lower or rotate the hip externally while the patient keeps the knee flexed. Once the farthest point is reached, take a measurement with a goniometer. Note the measurement, and repeat on the other side.

ADDITIONAL INFORMATION

With the hip extended, a bigger stretch is placed on the soft tissue and joint capsule. This provides a good picture of the hip musculature and supporting soft tissue mobility.

POSSIBLE FINDINGS

- Ability to perform the movement (FN)

- Inability to perform the movement (DN)

- Pain present while performing or attempting the move (FP or DP)

If the finding is FN *and* the *seated passive rotation* is a DN, stop and treat the DN.

If the finding is FN and *seated active* or *passive rotation* are FN, continue to the *rolling patterns*. If the rolling patterns outcome is DP or FP, stop and treat the problem.

If the rolling patterns outcome is FN, there is a weight-bearing external hip rotation stability and motor control dysfunction. Proceed to the *tibial rotation flowchart* and the *lower-body extension breakout*.

If the rolling patterns outcome is DN, there is a fundamental hip rotation stability and motor control dysfunction. Proceed to the *tibial rotation flowchart* and the *lower-body extension breakout*.

If the finding has any pain present (FP or DP), stop and treat the problem.

If the finding is DN, there is a hip joint mobility dysfunction or a tissue extensibility dysfunction, or both. Proceed to the *tibial rotation breakout* and the *lower body extension breakout.*

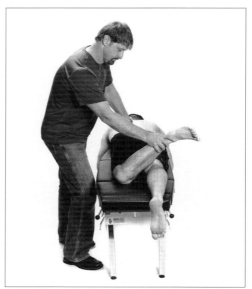

Prone Passive External Hip Rotation

SEATED ACTIVE INTERNAL HIP ROTATION

OBJECTIVE

To assess the dysfunction or pain with hip internal rotation actively in a non–weight-bearing position with the hip flexed

DESCRIPTION

Put the patient in a seated position with knees and feet together, body in an upright and erect posture, hands on top of each ilium—this allows you to see any hiking of the pelvis during testing. Have the patient actively rotate the hip internally while keeping the knee flexed and the pelvis level. Once the patient reaches the farthest point, take a measurement with a goniometer, and repeat the test on the other side.

ADDITIONAL INFORMATION

Look for 30 degrees of rotation bilaterally without pain during the movement. When the hip is flexed to 90 degrees, you're getting a good picture of total hip joint mobility.

POSSIBLE FINDINGS

- An ability to internally rotate the hip greater than 30 degrees (FN)
- An inability to internally rotate the hip or the presence of pain while attempting the movement (DN, DP or FP)

If the finding is FN, proceed to the *prone active internal hip rotation assessment.*

If the finding is DN, DP or FP, proceed to the *seated passive internal hip rotation assessment.*

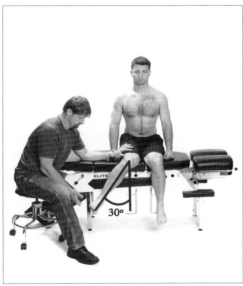

Seated Active Internal Hip Rotation

SEATED PASSIVE INTERNAL HIP ROTATION

OBJECTIVE

To assess the dysfunction or pain with hip internal rotation passively in a non–weight-bearing position

DESCRIPTION

Put the patient in a seated position with knees and feet together, body in an upright and erect posture, hands on top of each ilium—this allows you to see any hiking of the pelvis during testing. You'll then rotate the hip internally while keeping the knee flexed. Once the farthest point is reached, take a measurement with a goniometer and repeat the test on the other side.

ADDITIONAL INFORMATION

Flexing the hip to 90 degrees presents a good picture of total hip joint mobility. Look for 30 degrees of rotation bilaterally without pain. Compare passive rotation to the active rotation; these should be within 10 degrees.

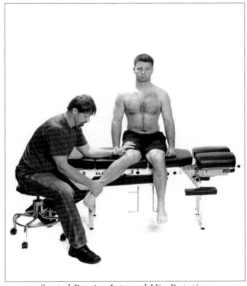

Seated Passive Internal Hip Rotation

POSSIBLE FINDINGS

- An ability to internally rotate the hip without pain (FN)

- An inability to internally rotate the hip or pain is present when trying to perform the movement (DN, DP or FP)

If the finding is FN, proceed to the *prone active internal hip rotation assessment.*

If the finding is DP or FP, stop and treat the problem.

If the finding is DN, there is limited hip joint mobility dysfunction or tissue extensibility dysfunction when the hip is flexed. Proceed to the *prone active internal hip rotation assessment.*

PRONE ACTIVE INTERNAL HIP ROTATION

OBJECTIVE

To assess the dysfunction or pain with hip internal rotation actively in a non–weight-bearing position with the hip extended

SFMA

DESCRIPTION

Have the patient lie face down with the knees flexed. While stabilizing the pelvis, instruct the patient to actively lower the leg or rotate the hip internally while keeping the knee flexed. Once the patient reaches the farthest point, take a measurement with a goniometer and repeat the test the other side.

ADDITIONAL INFORMATION

When the hip is extended, a bigger stretch is placed on the soft tissue and joint capsule, presenting a good picture of the hip musculature and supporting soft tissue mobility. Look for 30 degrees of rotation bilaterally without pain.

POSSIBLE FINDINGS

- An ability to internally rotate more than 30 degrees bilaterally (FN)

- The presence of pain or an inability to internally rotate (DN, DP or FP)

If finding is FN *and* the *seated passive internal rotation* was a DN, stop and treat the DN.

If *seated active* or *passive rotation* are FN, proceed to the *rolling patterns*. If the rolling patterns outcome is DP or FP, stop and treat the problem.

If the rolling patterns outcome is FN, there is a weight-bearing internal hip rotation stability and motor control dysfunction. Proceed to the *tibial rotation flowchart* and *lower-body extension breakout*.

If the rolling patterns outcome is DN, there is a fundamental hip rotation stability and motor control dysfunction. Proceed to the *tibial flowchart* and *lower-body extension breakout*.

If the finding is DN, DP or FP, proceed to the *prone passive internal hip rotation assessment*.

Proceed to *tibial rotation flowchart* and *lower-body extension breakout*.

PRONE PASSIVE INTERNAL HIP ROTATION

OBJECTIVE

To assess the dysfunction or pain with hip internal rotation passively in a non–weight-bearing position with the hip extended

DESCRIPTION

Have the patient lie face down on the table with the knees flexed. While stabilizing the pelvis, lower the leg or rotate the hip internally while keeping the knee flexed. Once the farthest point is reached, take a measurement with a goniometer. Repeat on the other side.

ADDITIONAL INFORMATION

When the hip is extended, the soft tissue and joint capsule are in a bigger stretch presenting a good picture of the hip musculature and supporting soft tissue mobility. Look for 30 degrees of rotation bilaterally without pain.

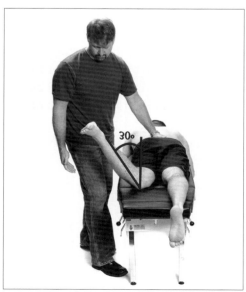

Prone Active Internal Hip Rotation

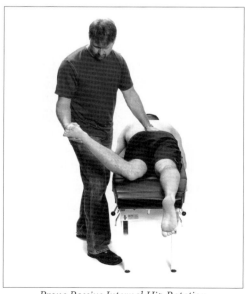

Prone Passive Internal Hip Rotation

POSSIBLE FINDINGS

- An ability to perform the movement (FN)

- Inability to perform the movement (DN)

- Pain present while performing the move or while trying to perform the move (DP or FP)

If *seated active* or *passive rotation* are FN, proceed to the *rolling patterns breakout.* If the rolling patterns outcome is DP or FP, stop and treat the problem.

If the rolling patterns outcome is FN, there is a weight-bearing internal hip rotation stability or motor control dysfunction. Proceed to the *tibial rotation flowchart* and *lower-body extension breakout.*

If the rolling patterns outcome is DN, there is a fundamental hip rotation stability and motor control dysfunction. Proceed to the *tibial flowchart* and *lower-body extension breakout.*

If the finding is DN, there is a hip joint mobility dysfunction or tissue extensibility dysfunction with internal rotation when the hip is extended. Proceed to *tibial rotation flowchart* and *lower-body extension breakout.*

If the finding is DP or FP, stop and treat the problem.

TIBIAL ROTATION BREAKOUTS

See the *Tibial Rotation Breakouts flowchart,* page 348.

SEATED ACTIVE INTERNAL TIBIAL ROTATION

OBJECTIVE

To assess the dysfunction or pain with tibial internal rotation actively in a non–weight-bearing position

DESCRIPTION

Instruct the patient to get into a seated position with knees flexed to 90 degrees, body in an upright and erect posture, and arms down by the sides. Have the patient rotate the foot internally keeping the knee flexed. Once the patient reaches the farthest point, take a measurement with a goniometer. Repeat on the other side.

ADDITIONAL INFORMATION

When the knee is flexed to 90 degrees, a good picture of the total tibial rotation is presented. Look for 20 degrees bilaterally without pain.

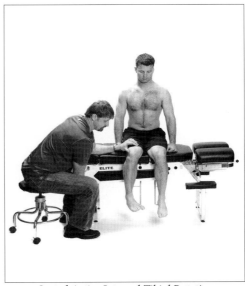

Seated Active Internal Tibial Rotation

POSSIBLE FINDINGS

- The ability to internally rotate at least 20 degrees bilaterally (FN)

- An inability to perform the internal rotation or presents with pain while trying to perform the movement (DN, DP or FP)

If the finding is FN, tibial internal rotation is normal. Recheck the *lower-body extension breakouts.*

If the finding is DN, DP or FP, proceed to the *seated passive internal tibial rotation assessment.*

SEATED PASSIVE INTERNAL TIBIAL ROTATION

OBJECTIVE

To assess the dysfunction or pain with tibial internal rotation passively in a non–weight-bearing position

DESCRIPTION

Instruct the patient to get into a seated position with knees flexed to 90 degrees, body in an upright and erect posture and arms down by the sides. Rotate the foot internally, keeping the knee flexed. Once the farthest point is reached, take a measurement with a goniometer and repeat the test on the other side.

ADDITIONAL INFORMATION

When the knee is flexed to 90 degrees, a good picture of the total tibial rotation is demonstrated. Look for 20 degrees bilaterally without pain.

Seated Passive Internal Tibial Rotation

POSSIBLE FINDINGS

- Ability to complete the internal rotation without pain (FN)

- An inability to perform the internal rotation or pain with trying to perform the movement (DN, DP or FP)

If the finding is FN, there is a tibial rotation stability and motor control dysfunction.

If the finding is DN, there is a tibial internal rotation tissue extensibility dysfunction or joint mobility dysfunction, or both.

If the finding is DP or FP, stop and treat the problem.

SEATED ACTIVE EXTERNAL TIBIAL ROTATION

OBJECTIVE

To assess the dysfunction or pain with tibial external rotation actively in a non–weight-bearing position

DESCRIPTION

Instruct the patient to get into a seated position with knees flexed to 90 degrees, body in an upright and erect posture, and arms down by the sides. Have the patient actively rotate one foot externally with the knee flexed. Once the patient reaches the farthest point, take a measurement with a goniometer and repeat the test on the other side.

ADDITIONAL INFORMATION

When the knee is flexed to 90 degrees, a good picture of the total tibial rotation is presented. Look for 20 degrees bilaterally without pain.

POSSIBLE FINDINGS

- The ability to externally rotate at least 20 degrees bilaterally (FN)

- An inability to perform the external rotation or pain when trying to perform the movement (DN, DP or FP)

If the finding is FN, tibial external rotation is normal. Recheck the *lower-body extension breakouts*.

If the finding is DN, DP or FP, proceed to the *seated passive external tibial rotation assessment*.

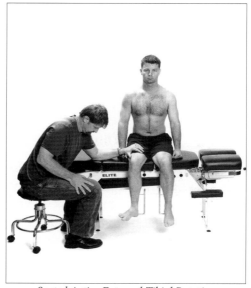

Seated Active External Tibial Rotation

SEATED PASSIVE EXTERNAL TIBIAL ROTATION

OBJECTIVE

To assess the dysfunction or pain with tibial external rotation passively in a non–weight-bearing position

DESCRIPTION

Instruct the patient to get into a seated position with knees flexed to 90 degrees, body in upright and erect posture and arms down by the sides. Rotate the foot externally, keeping the knee flexed. Once the farthest point is reached, take a measurement with a goniometer and repeat the test on the other side.

ADDITIONAL INFORMATION

When the knee is flexed to 90 degrees, a good picture of the total tibial rotation is seen. Look for 20 degrees bilaterally without pain.

POSSIBLE FINDINGS

- Ability complete the external rotation without pain (FN)

- An inability complete the external rotation or there is pain while trying to perform the movement (DN, DP or FP)

If the finding is FN, there is a tibial rotation stability and motor control dysfunction.

If the finding is DN, there is a tibial external rotation tissue extensibility dysfunction or joint mobility dysfunction, or both.

If the finding is DP or FP, stop and treat the problem.

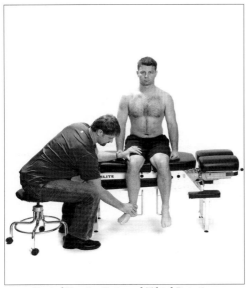

Seated Passive External Tibial Rotation

SINGLE-LEG STANCE BREAKOUT RATIONALE

VESTIBULAR PROBLEM

If the patient is bilaterally dysfunctional with eyes closed in the top-tier single-leg stance assessment as described on page 127, there is a potential for a vestibular problem. Unilateral and bilateral vestibular problems will always present themselves as a **bilateral** single-leg stance dysfunction. Any other findings could not represent a vestibular problem and therefore this portion of the breakout can be skipped—proceed directly to the half-kneeling narrow-base assessment.

Otherwise, check the vestibular system using the Clinical Test for Sensory Interaction on Balance (CTSIB). The CTSIB assesses the influence of vestibular, somatosensory and visual inputs on postural control. If the CTSIB is dysfunctional, a vestibular problem is affecting the single-leg stance. Stop and address the problem with referral and appropriate treatment.

If the CTSIB is functional or painful, the vestibular system is not the problem, and you should proceed to the half-kneeling narrow-base assessment.

SPINE, HIP AND CORE STABILITY PROBLEMS

After assessing the vestibular system, reduce the mobility and stability requirements needed to support single-leg stance. Take the patient to a half-kneeling narrow-based posture and recheck the ability to stabilize. If the patient is functional and non-painful in half-kneeling, and the dynamic leg swings assessment is normal, assume there is an ankle or proprioceptive problem and proceed to the single-leg stance ankle flowchart.

If half-kneeling is functional and non-painful, and dynamic leg swings is dysfunctional or painful, evaluate the hip for possible instability. Perform local biomechanical testing for the hip and gluteal musculature, including manual muscle testing and treat what you find. Proceed to the ankle flowchart.

If half-kneeling is dysfunctional or painful, a stability dysfunction needs to be addressed.

SFMA

Further, reduce the stability requirements by taking the patient to a non–weight-bearing position, both prone and supine, and assess the rolling patterns. If rolling is dysfunctional, the patient has a fundamental hip or core stability and motor control dysfunction, or both. If rolling is functional, the patient has a weight-bearing stability problem. Proceed to quadruped diagonals.

Take the patient to quadruped to separate weight-bearing spine from a weight-bearing hip or core stability dysfunction. If quadruped diagonals are functional, there is a weight-bearing spine or hip or core stability or motor control dysfunction, or both. If quadruped diagonals is dysfunctional, there is a hip or core stability and motor control dysfunction, or both.

If rolling or quadruped testing creates pain, stop and treat the pain.

No matter the result, continue to the ankle flowchart to check for further dysfunction.

ANKLE DORSIFLEXION AND PLANTAR FLEXION PROBLEMS

Here we check active and passive dorsiflexion and plantar flexion. Have the patient attempt to heel and toe walk. If either movement is dysfunctional or painful, or both, perform prone passive range-of-motion testing to differentiate between a mobility and a stability problem.

If the movement is dysfunctional or painful, or both, when performed actively but functional and non-painful when tested passively, there is a dorsiflexion or plantar flexion stability or motor control dysfunction, or both. If dysfunction is consistent between active and passive testing, there is a joint mobility or tissue extensibility dysfunction, or both, limiting dorsiflexion or plantar flexion.

If passive testing is painful, stop and treat it.

ANKLE INVERSION AND EVERSION PROBLEMS

If ankle dorsiflexion and plantar flexion are both functional and non-painful, assess ankle inversion and eversion by having the patient perform the seated ankle inversion and eversion assessment. Since this test is active, if there is dysfunction or pain, or both, with inversion or eversion, or both, the problem can still be either mobility or stability. Perform local foot and ankle biomechanical testing to further assess the dysfunction.

PROPRIOCEPTIVE PROBLEMS

The only way to get a true diagnosis of proprioceptive dysfunction is to rule out other causes of poor single-leg balance. Therefore, if ankle inversion and eversion are FN and all other assessments in the single-leg-stance breakout are FN, you can assume proprioceptive dysfunction is the primary problem.

SINGLE-LEG STANCE BREAKOUTS

In the presence of a limited *single-leg stance assessment* in the top-tier testing, proceed with the following breakouts.

VESTIBULAR AND CORE BREAKOUTS

Vestibular Test—CTSIB
Half-Kneeling, Narrow Base
Rolling Patterns (see page 187)
Quadruped Diagonals

ANKLE BREAKOUTS

Heel Walks
Prone Passive Dorsiflexion
Toe Walks
Prone Passive Plantar Flexion
Seated Ankle Inversion and Eversion

VESTIBULAR AND CORE BREAKOUTS

See the *Vestibular and Core Breakouts flowchart*, page 349.

VESTIBULAR CLINICAL TEST FOR SENSORY INTERACTION ON BALANCE (CTSIB)

This test is only performed if there is a bilateral dysfunction with single-leg stance, eyes closed, during the top-tier assessment.

OBJECTIVE

The *Clinical Test for Sensory Interaction on Balance (CTSIB)* is a complex sensory strategy balance test. The CTSIB assesses the influence of vestibular, somatosensory and visual inputs on postural control.

DESCRIPTION

The CTSIB is administered by manipulating the support surface (firm versus foam) and visual conditions (eyes open versus eyes closed) while a patient is asked to maintain standing balance.

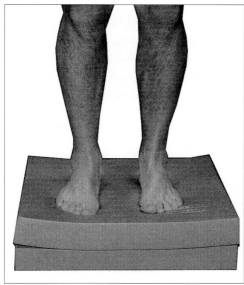

CTSIB

The CTSIB assesses the patient's ability to maintain upright posture under four progressively more difficult sensory conditions. Each of these trials is timed for 20 seconds as the examiner watches for excessive postural sway or loss of balance.

Condition 1: Normal base of support on a firm surface with eyes open

Condition 2: Normal base of support on a firm surface with eyes closed

Condition 3: Normal base of support on a foam surface such as two Airex pads with eyes open

Condition 4: Normal base of support on a foam surface such as two Airex pads with eyes closed

Condition 1 provides a baseline reference with which the other three situation-dependent conditions will be compared.

Condition 2 removes vision, which provides information about the patient's ability to use somatosensory input to maintain upright posture.

Condition 3 disadvantages somatosensory input, which indicates the ability to use visual inputs to maintain upright posture.

Condition 4 disadvantages somatosensory and removes vision, which indicates the ability to use vestibular input for maintaining upright posture.

If the patient passes Condition 4 of the CTSIB, the clinical sensitivity of the CTSIB can be enhanced by modifying the protocol to include dynamic head tilts. Requiring active rotation, flexion, extension and side bending head movements during upright stance generates visual and vestibular stimulation in addition to the sensory input created by postural sway. Altering the multisensory information increases the postural demands associated with maintaining upright stance. Consequently, the inclusion of head tilts can provide a more challenging balance task that could quantify subtle balance deficits.

Have the patient stand with a normal base of support on a foam surface with eyes closed. Instruct the patient as follows—

Tilt the left ear toward the left shoulder
Return the head upright
Tilt the right ear toward the right shoulder
Return head to upright
Tilt head forward
Return head to upright
Tilt head backward
Return the head to upright

These head movements are performed at one-second intervals; the complete sequence of head tilting should take eight seconds. A metronome set at 60 beats per minute may be used to pace the patient.

ADDITIONAL INFORMATION

Many times balance dysfunction can be due to vestibular system imbalances. This includes the inner ear's semicircular canals, which help indicate rotational movements, and the otoliths, which help indicate linear translations.

The vestibular system sends signals primarily to the neural structures that control eye movements, and to the muscles that keep us upright. The projections to the former provide the anatomical basis of the vestibulo-ocular reflex, which is required for clear vision, and the projections to the muscles that control our posture are necessary to stay upright.

NOTE

The CSTIB and modification with head tilting screens vestibular function as it relates to postural control. These tests do not screen for impairment in gaze stabilization, the vestibulo-ocular reflex.

POSSIBLE FINDINGS

- Ability to stabilize with eyes open and eyes closed on both the stable and the non-stable surface (FN)

- Inability to stabilize with eyes open and eyes closed on the stable or the non-stable surface (DN, DP or FP)

If the finding is FN, proceed to the *half-kneeling narrow base assessment.*

If the finding is DN, DP or FP, the patient has a vestibular dysfunction and should be diagnosed by an appropriate clinician.

HALF-KNEELING, NARROW BASE

OBJECTIVE

To reduce the mobility and stability needed to support a single-leg stance

DESCRIPTION

Instruct the patient to get into a half-kneeling position, with both feet and knee in a straight line.

ADDITIONAL INFORMATION

If the patient is uncomfortable, put an Airex pad or a foam pad under the knee.

Closely monitor the patient's breathing pattern. If unable to maintain a natural diaphragmatic breathing pattern, the test is dysfunctional.

POSSIBLE FINDINGS

- Ability to maintain balance (FN)

- Inability to maintain balance or the presence of pain while maintaining balance or trying to maintain balance (DN, FP or DP)

If the finding is FN and *dynamic leg swings* is DN or painful, perform local biomechanical testing for the hip and proceed to the *ankle flowchart.*

If the finding is FN and *dynamic leg swings* is FN, proceed to the *ankle flowchart.*

If the finding is DN, FP or DP, proceed to the *rolling patterns breakouts.* If the rolling patterns outcome is FN, proceed to the *quadruped diagonals assessment.* If the rolling patterns outcome is DN, there is a fundamental hip or core stability and motor control dysfunction, or both.

If the rolling patterns outcome is DP or FP, stop and treat the problem.

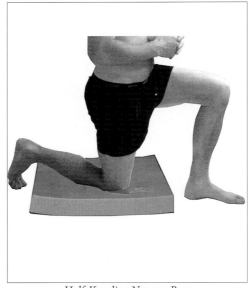

Half-Kneeling Narrow Base

QUADRUPED DIAGONALS

OBJECTIVE

To separate a weight-bearing spine from a weight-bearing hip or core stability dysfunction

DESCRIPTION

Instruct the patient to get into a quadruped position such that the arms and thighs are at 90-degree angles to the torso. Have the patient extend the right arm and the left leg, maintaining balance with the right leg and left arm. Repeat on the other side.

POSSIBLE FINDINGS

- Ability to perform the movement with no pain and maintaining balance (FN)

- An inability to perform the movement or pain while performing the move or while trying to perform the move (DN, FP or DP)

If the finding is FN, there is weight-bearing spine or hip or core stability or motor control dysfunction, or both. If hip extension is DN, treat it first. Proceed to the *single-leg stance ankle flowchart.*

If the finding is DN, there is a weight-bearing hip or core stability and motor control dysfunction, or both. If hip extension or shoulder flexion, or both, are DN, treat this first. Proceed to the *single-leg stance ankle flowchart.*

If the finding is DP or FP, stop and treat the problem.

ANKLE BREAKOUTS

See the *Ankle Breakouts flowchart,* page 350.

HEEL WALK

OBJECTIVE

To help identify gross limitations of dorsiflexion function in the sagittal plane

DESCRIPTION

Instruct the patient to take 10 steps forward with the toes off the ground, in full dorsiflexion.

ADDITIONAL INFORMATION

The first balance strategy used by the human body is closed-chain dorsiflexion with a concentric contraction of the plantar flexors. Look for either foot unable to maintain toes-up.

POSSIBLE FINDINGS

- Ability to perform the movement with no pain and maintaining raised toes (FN)

- An inability to perform the move or there is pain present when trying to perform the move (DN, DP or FP)

If the finding is FN, proceed to the *toe walk assessment.*

If the finding is FP, DP or DN, proceed to the *prone passive dorsiflexion assessment.*

Quadruped Diagonals

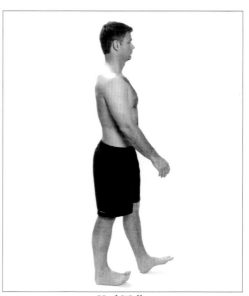

Heel Walks

PRONE PASSIVE DORSIFLEXION

OBJECTIVE

Helps differentiate between a true dorsiflexion stability problem verses a mobility restriction at the ankle

DESCRIPTION

Instruct the patient lie prone with the knee extended; measure the total passive dorsiflexion of the ankle. Then instruct the patient to flex the knee to 45 degrees and repeat the test. Take the average of these two results and consider that average as total dorsiflexion.

ADDITIONAL INFORMATION

Normal dorsiflexion is 20–30 degrees.

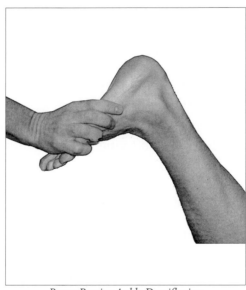

Prone Passive Ankle Dorsiflexion

POSSIBLE FINDINGS

- Ability to dorsiflex the ankle to within normal ranges (FN)

- The presence of pain while completing or attempting the movement (DP or FP)

- An inability to perform the movement (DN)

If the finding is FN, there is a dorsiflexion stability and motor control dysfunction. Proceed to the *toe walk assessment.*

If the finding is DN, there is a lower posterior-chain tissue extensibility dysfunction or joint mobility dysfunction, or both. Proceed to the *toe walk assessment.*

If the finding presents with FP or DP, stop and treat the problem.

TOE WALKS

OBJECTIVE

To help identify gross limitations of plantar flexion in the sagittal plane

DESCRIPTION

Instruct the patient to take 10 steps forward with the heels off the ground, in full plantar flexion.

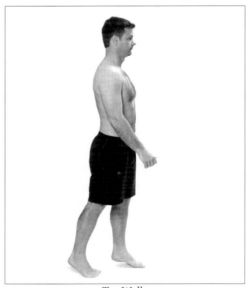

Toe Walks

ADDITIONAL INFORMATION

Look for either foot unable to maintain heels-up.

POSSIBLE FINDINGS

- Ability to perform the movement with no pain and maintaining raised heels (FN)

- An inability to perform the move or there is pain present when trying to perform the move (DN, DP or FP)

If the finding is FN, proceed to the *seated ankle inversion and eversion assessment*

If the finding presents with FP, DP or DN, proceed to the *prone passive plantar flexion assessment.*

PRONE PASSIVE PLANTAR FLEXION

OBJECTIVE

Helps differentiate between a true plantar flexion stability problem versus a mobility restriction at the ankle

DESCRIPTION

Instruct the patient to lie prone with the knee extended; measure the total passive plantar flexion of the ankle. Then have the patient flex the knee to 45 degrees and repeat the test. Take the average of these two results and consider that the total plantar flexion. Test both sides.

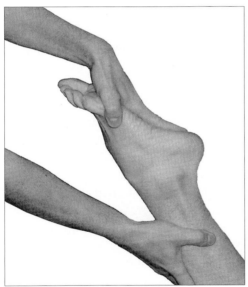

Prone Passive Ankle Plantar Flexion

ADDITIONAL INFORMATION

Normal plantar flexion is 30–40 degrees.

POSSIBLE FINDINGS

- Ability to plantar flex the ankle to within normal ranges (FN)

- The presence of pain while completing or attempting the movement (DP or FP)

- An inability to perform the movement (DN)

If the finding is FN, there is a plantarflexion stability and motor control dysfunction. Proceed to the *seated ankle inversion and eversion assessment.*

If the finding is DN, there is a lower anterior-chain tissue extensibility dysfunction or joint mobility dysfunction, or both. Proceed to the *seated ankle inversion and eversion assessment.*

If the finding presents with FP or DP, stop and treat the problem.

SEATED ANKLE INVERSION AND EVERSION

OBJECTIVE

To identify gross mobility limitations of the ankles in the frontal plane, eversion and inversion

DESCRIPTION

Instruct the patient to sit on a chair with the knees pelvic-width apart, feet flat on the ground, toes pointing forward. Then instruct the patient to invert and evert the feet. Have the patient move back and forth between inversion and eversion for 10 seconds.

ADDITIONAL INFORMATION

Because of ankle-joint limitations, many people create this motion at the hips. The patient should be able to do this without moving the hips or knees.

Seated Ankle Eversion

Seated Ankle Inversion

POSSIBLE FINDINGS

- Inability to evert the ankle

- Inability to invert the ankle

- The presence of pain while attempting the movements (DP or FP)

- Ability to do the movement, no pain (FN)

- An inability to do either movement (DN)

If the finding is dysfunction with eversion, there is an ankle eversion joint mobility or tissue extensibility dysfunction or stability and motor control dysfunction, or all three. Perform a local biomechanical foot and ankle exam.

If the finding is dysfunction with inversion, there is an ankle inversion joint mobility dysfunction or tissue extensibility dysfunction or stability and motor control dysfunction, or all three. Perform a local biomechanical foot and ankle exam.

If the finding is FN and there have been no other limitations or pains, treat for a proprioceptive deficit.

If the finding is DP or FP, stop and treat the problem.

If the finding is dysfunction both ways, there is an ankle joint mobility or tissue extensibility dysfunction or stability and motor control dysfunction, or all three. Perform a local biomechanical foot and ankle exam.

OVERHEAD DEEP SQUAT BREAKOUT RATIONALE

TRUNK EXTENSION AND SHOULDER FLEXION PROBLEM

The first step in the squat breakout is to rule out trunk extension dysfunction or shoulder flexion dysfunction, or both. Have the patient perform the interlocked fingers-behind-the-neck deep squat assessment. If that is functional and non-painful, proceed to the spine extension breakouts to further check the trunk and shoulder dysfunction.

If the patient still cannot squat, there is dysfunction lower in the chain. Proceed to the assisted squat assessment.

CORE STABILITY VERSUS LOWER-QUARTER PROBLEM

Next, have the patient perform an assisted squat. By allowing the patient to squat using assistance, you can determine if there is adequate bilateral, symmetrical ankle, knee, hip, spine and shoulder mobility to get into an overhead squat. If the patient still cannot perform a full deep squat, assume there is a hip, knee or ankle dysfunction, and continue the breakout.

If the patient can demonstrate a functional and non-painful squat, there is a core stability dysfunction or a motor-skill dysfunction, or both. The patient just proved there is mobility to squat, but has a lack of ability to stabilize during the squatting maneuver. Recheck the multi-segmental extension breakout, since active thoracic extension and shoulder flexion still haven't been cleared.

ANKLE MOBILITY PROBLEM

One of the most common limitations preventing functional squatting is limited ankle mobility. Use the half-kneeling or the standing dorsiflexion assessment to check ankle mobility. If the patient can demonstrate full dorsiflexion, the problem is in the knee, hip or core. If dorsiflexion is limited, there is a lower posterior-chain tissue extensibility dysfunction or an ankle joint mobility problem, or both. The patient can still have a thoracic extension and shoulder flexion dysfunction as well, so also check the multi-segmental extension breakout.

Regardless of how dorsiflexion turns out, proceed to the supine knees-to-chest assessments to further break out the knee, hip and core.

KNEE, HIP AND CORE PROBLEMS

The last steps in the overhead deep squat break-out are the supine knees-to-chest assessments. These tests are used to evaluate the knees and hips in a non–weight-bearing position. If the patient can show full hip flexion combined with full knee flexion while holding the shins, and dorsiflexion is normal, consider this a weight-bearing core, knee or hip stability dysfunction or a motor control dysfunction, or both.

If the patient can show full hip flexion combined with full knee flexion while holding the shins, and dorsiflexion is painful, stop and treat the ankle for pain.

If you see full hip flexion combined with full knee flexion while holding the shins, and dorsiflexion is DN, consider hips, knees and core normal, and attack the dorsiflexion dysfunction.

If the patient has limited hip or knee flexion while holding the shins, switch the hand placement to the thighs. This removes knee flexion from the challenge. If the patient can now show full hip flexion, consider knee joint mobility dysfunction or lower anterior-chain tissue extensibility dysfunction, or both, as the primary dysfunction.

If the patient still cannot show full hip flexion, consider hip joint mobility dysfunction or posterior chain tissue extensibility dysfunction, or both, as the primary dysfunction. This does not rule out knee involvement, but does suggest starting with the hip dysfunction in the treatment protocol. If the patient has pain with this assessment, stop and treat the area of pain.

No matter what outcome, the patient can still have a thoracic extension and shoulder flexion dysfunction. Check the multi-segmental extension breakout.

OVERHEAD DEEP SQUAT BREAKOUTS

See the *Overhead Deep Squat Breakouts flowchart*, page 351.

In the presence of a limited *overhead deep squat assessment* in the top-tier testing, proceed with the following breakouts.

Interlocked Fingers-Behind-Neck Deep Squat
Assisted Deep Squat
Half-Kneeling Dorsiflexion
Supine Knees-to-Chest Holding Shins
Supine Knees-to-Chest Holding Thighs

INTERLOCKED FINGERS-BEHIND-NECK DEEP SQUAT

OBJECTIVE

To remove the upper body component and re-duce the level of dynamic stability needed to per-form the squat

DESCRIPTION

Instruct the patient to place the hands behind the neck with the elbows facing forward. Then have the patient repeat the full deep squat, mak-ing sure to keep the heels on the ground and toes pointing forward throughout the move.

ADDITIONAL INFORMATION

This position lowers the difficulty because the shoulders are not flexed vertically and the thoracic spine doesn't need to extend.

Interlocked Fingers-Behind-Neck Deep Squat

SFMA

POSSIBLE FINDINGS

- Ability to do a full squat with no pain (FN)

- An inability to do the full squat or pain while attempting to do the squat (DP, DN or FP)

If the finding is FN, recheck all of the *extension breakouts*.

If the finding is FP, DP or DN, proceed to the *assisted deep squat assessment*.

ASSISTED DEEP SQUAT

OBJECTIVE

Allows a look at the true symmetrical mobility of the lower body joints without the requirement of dynamic stability—also allows greater mobility into the pattern to investigate undiscovered pain provocation

DESCRIPTION

Instruct the patient to reach out with both hands and grab your hands for support. Have the patient repeat the full deep squat; if the patient gets all the way down into the deep position, elevate the hands above the patient's head, and try to let go. Make sure the heels stay on the ground throughout the movement.

ADDITIONAL INFORMATION

Look for normal squat mechanics and full dorsiflexion of the ankles.

POSSIBLE FINDINGS

- Ability to do the squat with no pain (FN)

- An inability to do the movement or the presence of pain while trying to do the movement (DP, DN or FP)

If the finding is FN, there is a core stability dysfunction or a motor skill dysfunction, or both. Clear the *multi-segmental extension breakout*.

If the finding is dysfunctional or presents with pain (DN, DP or FP), proceed to the *half-kneeling dorsiflexion assessment*.

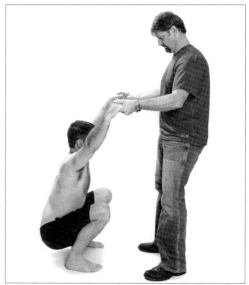

Assisted Deep Squat

HALF-KNEELING DORSIFLEXION

OBJECTIVE

To assess ankle mobility

DESCRIPTION

Instruct the patient to place one foot on a bench or stool or kneeling on one knee with the other foot in front. Have the patient lean forward onto the front foot as far as possible without the heel coming off the ground or bench. Look to see whether the knee can move forward past the toe a minimum of four inches. Repeat the test on the other side.

ADDITIONAL INFORMATION

Often a person will have a limited deep squat due to a mobility issue in the calf and ankle joint. Any restriction in the lower posterior chain can limit the closed-chain dorsiflexion and thereby limit the full deep squat. Normal dorsiflexion is 20–30 degrees.

POSSIBLE FINDINGS

- Ability to move the knee four inches in front of the toe with or without the presence of pain, or an inability to move the knee four inches in front of the toe due to pain (FN, FP or DP)

- An inability to move the knee four inches in front of the toe (DN)

If the finding is FN, FP or DP, proceed to the *supine knees-to-chest holding shins assessment.*

If the finding is DN, treat for a lower posterior-chain tissue extensibility dysfunction or an ankle joint mobility problem, or both. Make sure the *multi-segmental extension* and *single-leg stance breakouts* are clear.

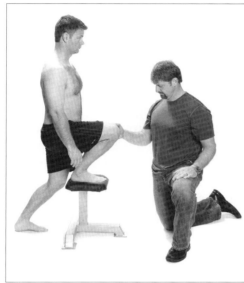

Half-Kneeling Dorsiflexion

SUPINE KNEES-TO-CHEST HOLDING SHINS

OBJECTIVE

To quickly check the mobility of the hips, knees and spine in a non–weight-bearing position

DESCRIPTION

Instruct the patient to lie supine and bring both knees up toward the chest. Next, have the patient grab both shins and try to touch the thighs to the lower rib cage and calves to hamstrings. If the patient cannot get the calves to touch the hamstrings because of knee tightness, have the patient grab the thighs instead of the shins and repeat the pull.

ADDITIONAL INFORMATION

The only lower-body joint not being checked is the ankle. If this is functional and the standing squat motion is not, suspect a weight-bearing sta-

bility problem. This move will help differentiate between hip and knee mobility issues.

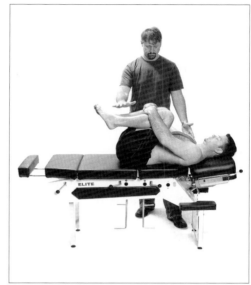

Supine Knees-to-Chest Holding Shins

POSSIBLE FINDINGS

- Ability to pull the knees to the chest with no pain (FN)

- An inability to pull the knees to the chest or can only do the move in the presence of pain (DN, DP or FP)

If the finding is FP, DN or DP, proceed to the *supine knees-to-chest holding thighs assessment.*

If the finding is FN *and* dorsiflexion is DN, consider knees, hips and core normal. Make sure you clear the *multi-segmental extension breakout.*

If the finding is FN and dorsiflexion is a DP or FP, treat the ankle for the pain. Clear the *multi-segmental extension breakout.*

If the finding is FN *and* dorsiflexion is FN, the patient has a weight-bearing core, knee or hip stability and motor control dysfunction. Make sure you clear the *multi-segmental extension breakout.*

SUPINE KNEES-TO-CHEST HOLDING THIGHS

Only perform this test if the patient could not perform the test correctly when holding the shins.

OBJECTIVE

To quickly determine if the mobility restriction in the squat is due to a hip or knee dysfunction

Description

Instruct the patient to lie supine and bring both knees up toward the chest. Have the patient grab both thighs and pull the thighs to the lower rib cage.

Additional Information

If the patient cannot get the thighs to touch the rib cage, there is a potential hip dysfunction. This test will help differentiate between hip and knee mobility issues.

Possible Findings

- Ability to pull the knees to the chest with no pain (FN)

- Inability to pull the knees to the chest (DN)

- Pain while pulling or trying to pull the knees to the chest (FP or DP)

If the finding is FN, there is a knee joint mobility dysfunction or a lower anterior-chain tissue extensibility dysfunction, or both. Clear the *multi-segmental extension breakout*.

If the finding is DN, there is a hip joint mobility dysfunction or a posterior-chain tissue extensibility dysfunction. Remember, the knee joint mobility has not been cleared yet. Proceed to the *multi-segmental extension breakouts*.

If the finding is DP or FP, stop and treat the pain.

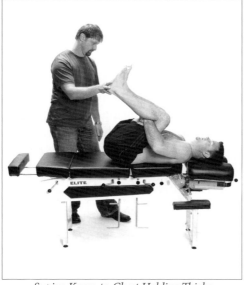

Supine Knees-to-Chest Holding Thighs

ROLLING PATTERNS BREAKOUT RATIONALE

Rolling is used as a terminal point in many of the breakouts. We'll comment on rolling in greater detail in the corrective section of the book, but rolling deserves some explanation here as well. Rolling can actually be a refreshing perspective if you allow it to be, however you need to take some time to appreciate some specific points regarding rolling.

First, the breakouts remove potential mobility limitations prior to rolling. This dictates and demonstrates that rolling should only be explored if mobility problems have been effectively managed. The positions, mobility and movements required for this type of rolling breakout are fundamental, and if these are compromised, rolling cannot provide the information it is designed to produce.

Second, consider rolling a fundamental representation of motor control, used to observe motor control from the bottom up and from the top down. It looks at patterns that move the subject from prone to supine and supine to pone. This provides an effective opportunity for the observation of left and right symmetry.

Third, rolling is an activity that at first glance does not appear to be a demonstration of stability. On closer inspection, rolling is the soil where stability is planted. Rolling is a basic demonstration of motor control and segmental sequencing. This sequencing is demonstrative of the timing and coordination that work behind the scenes in both static and dynamic demonstrations of stabilization.

It is easy to see the robust muscular stabilization in a bridge, side plank or quadruped diagonal exercise, but in the developmental sequence of events, rolling was ground zero. It preceded all other activities that move the body from one place to another.

As you observe rolling pay close attention to sticking points. These are instances where a patient simply cannot move any further into a rolling pattern without cheating or using momentum. This sticking point is a where the sequence of motor control cannot naturally flow. Don't look for a weak muscle to train or something to activate—the pattern is broken.

Make sure the patient is not sticking on an overlooked mobility restriction. If not, just identify the quadrant or quadrants where rolling is compromised. The rolling breakout has done its job, and you can now move into a rolling correction… but that is for later.

You might feel obligated to comment as to why the individual cannot roll, but there might be multiple reasons. By the time you have discussed them, the person can reset the broken pattern and start performing corrective exercises. From this point, it is a brain-body thing, and you need to allow these to reconnect. A dysfunctional postural habit or movement pattern has eroded the motor control pathway for this particular rolling pattern. All you need to do is provide the programming to rebuild the pathway.

Fourth, rolling is not something we normally practice or do, and there is no world championship. When we do roll, we surely don't use the goofy quadrants suggested in the breakouts. The quadrants are intended to disadvantage the patient and identify dysfunction. For the most part, we either can or cannot roll. We either have it or we don't—rolling seems to be all or none.

Big, beefy prime movers can't help here and our fitness level does not give us an advantage. A high threshold strategy will make attempts at rolling feel as if glued to the floor. The ability to automatically coordinate stabilizers is the key. Unfortunately, we cannot force them.

Breathing and relaxation are important keys to jump-starting the rolling patterns. Perceived exertion is another indicator. Some people can actually complete a rolling pattern, but the struggle is pronounced and very different from the other rolling patterns. This can also be indicative of a dysfunction. Don't look for perfection; just look for inability or struggle, and compare each pattern to its contra-lateral counterpart.

Fifth, upper body rolling creates more demand than lower body rolling. The legs simply provide more ballast. Upper body patterns also utilize head and neck movement patterns, and postural issues involving the head and neck can potentially make rolling more awkward in the upper quarter.

Don't coach rolling. Simply cue the patient to reach across the midline as far as possible with the legs in the lower quarter rolling patterns. Cue the patient to really lead with the eyes, head and neck with the upper quarter patterns, reaching across the midline with the arm.

Lastly, most people focus on the moving segments of the rolling patterns, but in fact, the non-moving segments are the key. The non-moving contra-lateral side must remain elongated and stable, but do not coach this. Initially, just instruct the individual to only reach with one segment and let the other segments remain motionless.

The coordination of one moving segment and three stable segments is the fundamental motor control you are attempting to observe. When it is not present, you have discovered the root of the motor control problem that caused a DN in the top-tier test.

Do what it takes to reset the rolling pattern and work your way back up to the top tier. If you are good with your corrective strategies, you might be surprised how quickly this pattern resets.

And remember, you do not need to teach rolling. It is a fundamental pattern and it is already on the hard drive. You simply need to dust off the program.

ROLLING BREAKOUTS

See the *Rolling Breakouts* flowchart, page 352.

Prone-To-Supine Rolling, Upper Body
Prone-To-Supine Rolling, Lower Body
Supine-To-Prone Rolling, Upper Body
Supine-To-Prone Rolling, Lower Body

PRONE-TO-SUPINE ROLLING UPPER BODY

OBJECTIVE

This pattern observes upper quarter, head and C-spine movements that coordinate rolling from a prone to a supine position. It is used primarily to observe motor control and symmetry.

DESCRIPTION

Instruct the patient to lie prone in an open space, with the legs extended and the arms flexed overhead. Tell the patient to roll the body to the supine position leading with the right arm. Evaluate, and repeat on the other side.

ADDITIONAL INFORMATION

Gross stability dysfunctions combined with extension maneuvers are often identified using rolling patterns. A good sequence of core stability and segmental loading of the entire body is required to perform this test well.

POSSIBLE FINDINGS

- Able to roll from the prone position to the supine position or an inability to roll (FN or DN)

- Able to or unable to roll from the prone position to the supine position, and experiences pain in the attempt (DP or FP)

If the finding is FN or DN, proceed to the *prone-to-supine rolling with the lower body.*

If the finding is DP or FP, use DP or FP in the *rolling pattern outcome* in the flowchart that suggested this assessment.

Prone to Supine Upper Body Rolling

PRONE-TO-SUPINE ROLLING LOWER BODY

OBJECTIVE

This pattern observes lower quarter and pelvic movements that coordinate rolling from a prone to a supine position. It is used primarily to observe motor control and symmetry.

DESCRIPTION

Instruct the patient to lie on the stomach with the legs extended and the arms flexed on the ground overhead. Tell the patient to roll the body to the supine position starting from the right leg. Evaluate, and repeat with the left leg.

ADDITIONAL INFORMATION

Many times a patient's sequence of coordination in active rotation is dysfunctional, and rolling patterns easily identify this. A good sequence of core stability and segmental loading of the entire body is required to perform this test well.

POSSIBLE FINDINGS

- Able to roll from the prone to the supine position or an inability to roll (FN or DN)

- Able to or unable to roll from the prone to the supine position, and experiences pain in the attempt (DP or FP)

If the finding is FN or DN, proceed to the *supine-to-prone rolling with the upper body.*

If the finding is DP or FP, use DP or FP in the *rolling pattern outcome* in the flowchart that suggested this assessment.

Prone to Supine Lower Body Rolling

SUPINE-TO-PRONE ROLLING UPPER BODY

OBJECTIVE

This pattern observes upper quarter, head and c-spine movements that coordinate rolling from a supine to a prone position. It is used primarily to observe motor control and symmetry.

DESCRIPTION

Instruct the patient to lie supine, on a flat surface with the legs extended and the arms flexed overhead. The patient should actively roll the body to the prone position starting with the right arm only. Evaluate, and repeat on the other side.

ADDITIONAL INFORMATION

Many times, gross stability dysfunctions combined with flexion maneuvers can be identified using rolling patterns. A good sequence of core stability and segmental loading of the entire body is required to perform this test correctly.

POSSIBLE FINDINGS

- An ability or inability to roll from the supine position to the prone position (FN or DN)

- Able to or unable to roll from the supine position to the prone position, and experiences pain in the attempt (DP or FP)

If the finding is FN or DN, proceed to the *supine to prone rolling, lower body.*

If the finding is DP or FP, use DP or FP in the *rolling pattern outcome* in the flowchart that suggested this assessment.

Supine to Prone Upper Body Rolling

SUPINE-TO-PRONE ROLLING LOWER BODY

OBJECTIVE

This pattern observes lower quarter and pelvic movements that coordinate rolling from a supine to a prone position. It is used primarily to observe motor control and symmetry.

DESCRIPTION

The patient should lie supine with the legs extended and the arms flexed overhead. Instruct the patient to roll the body to the prone position starting with the right leg. Evaluate, and repeat with the left leg.

ADDITIONAL INFORMATION

Often, gross stability dysfunctions combined with flexion maneuvers can be identified using rolling patterns. A good sequence of core stability and segmental loading of the entire body is required to perform this test acceptably.

POSSIBLE FINDINGS

- An ability or inability to roll from the supine position to the prone position (FN or DN)

- Able to or unable to roll from the supine position to the prone position, and experiences pain in the attempt (DP or FP)

If the finding is FN or DN and there were no DNs in the tests above, use FN for the rolling pattern outcome of the flowchart that suggested this assessment.

Supine to Prone Lower Body Rolling

If there were any DNs in the tests above, use DN for the rolling pattern outcome of the flowchart that suggested this assessment.

If the finding is DP or FP, use DP or FP in the rolling pattern outcome in the flowchart that suggested this assessment.

Important Instructions for Clinicians

The SFMA information provided in this book has parallel goals. One goal is to provide a movement profile that separates painful movement from non-painful movement and dysfunctional movement from functional movement. The other goal is to provide systematic structure to a corrective exercise prescription. To be more specific, the intent is to provide an exercise platform for dysfunctional, non-painful movement patterns (DN).

The clinician is advised to collect data for impairments for each DN pattern to create greater clarity and consistent impairment baselines whenever possible. These impairment measurements include but are not limited to testing of range of motion, strength, ligament integrity and segmental mobility. Movement patterns that do not involve pain can be considered mobility and stability problems, and manual therapy and exercise options can be considered as corrective options to be retested against movement and impairment baselines.

Our system does not generally suggest or recommend corrective exercise for patterns involving pain (DP and FP). The current best evidence suggests pain affects motor control with inconsistent and unpredictable outcomes. We recommend any attempts to exercise into painful patterns should be specific to individual situations and at the professional discretion of the clinician. Furthermore, it is recommended that the clinician reasonably and responsibly evaluate the movement pain behavior patterns (DP and FP) in greater detail than is provided by the SFMA logic and decision trees.

Before working with the flowcharts you'll find beginning on page 337, it's important for you to understand these are directional, not specific, and were created to demonstrate a possible path. The SFMA decision tree is only intended to produce an outcome suggestion for DN patterns regarding the nature of corrective exercise and manual therapies. This means the SFMA decision tree is not a terminal point for painful movement patterns (DP and FP). Suggestions for painful patterns in the SFMA decision tree are simply directional and require validation by specific testing and professional judgment appropriate to each patient.

It is the responsibility of the clinician to measure and report specific impairment findings in regions at, or associated with, the painful areas within painful movement patterns.

Please see www.movementbook.com/chapter8 for more information, videos and updates.

ANALYZING THE MOVEMENTS
IN SCREENS AND ASSESSMENTS

Before we discuss the key movements in the screens and assessments, note that the screens and assessments are, and should always be, used as collective tests. Do not single out a particular test and attempt to give it different a spotlight, extra weight or greater value. It is also inappropriate to attempt to modify screens and assessments simply to save time. As stated before, modifications should only be done on an individual basis, necessitated by physical limitations or restrictions imposed for safety or professional caution. Individualized professional discretion is expected and assumed when screening and assessing movement.

The Functional Movement Systems were designed specifically to be used as a combined group of movement tests viewed together to provide a fundamental movement map. Jon Torine, a name you may recognize from his foreword in this book, is a long-time friend and peer whose experience, opinions and innovation are well respected in the field. He was probably the first professional strength coach to jump in with both feet and apply a complete version of the Functional Movement Systems to his program. To say his perspectives have been valuable to us is like saying satellite images are somewhat helpful in battle—Jon's observations have been key to our corrective strategies. He's become an expert in the application of the systems because of the sheer amount of time in the trenches, not from reading biomechanics and motor learning books. His specific expertise is using the screens in large groups, something that many think is futile, impractical or impossible.

He is often approached by people working with groups and interested in screening. They are new to the concept and immediately want to shave time from the screening process—the think abbreviating the screen might be a solution. Jon's answer is classic and worth repeating, "I have time for fewer people, but I never have time for fewer tests."

Jon is telling us we save time by screening fewer people, not doing fewer tests per screen. Of course, he screens everyone under his care, but his point is not lost and his is message is clear. The screen should be used as it was intended, and it should be done correctly and completely each time it's done. Your data only has value if it is collected correctly. If you only have time for a few people, better to screen half your group correctly than have your entire group screened halfway.

What happens after a screen or assessment affects your reputation and ultimately it's your call. To date, no one who has used the Functional Movement Screen (FMS®) to improve statistics, get published or make a profound impact on individuals or groups has felt the need to abbreviate the screen. Only those who have done these things are actually qualified to suggest abbreviation, but they have not. All who suggest modification or abbreviation of the FMS are usually asking the question without achieving a level of expertise with the FMS. It's a freshman mistake and totally forgiven if the freshman gains perspective.

In this section, we'll analyze some of the possible results of each screen and discuss what you might see. The first portion covering the squat pattern will develop many concepts of screening application, not just to the squat, but also to the entire screening method. Carefully review this squat discussion even if the deep squat is not your primary interest.

THE DEEP SQUAT

In Olympic weightlifting, the snatch is a lift in which the athlete pulls the weight off the ground and accelerates in a vertical line in front of the body; at a certain instant, the athlete drops into an overhead squat position and quickly centers under the weight. The weight's momentum stops its vertical ascent and descends onto a well-aligned body—the shoulders in a flexed and abducted position and hips, knees and ankles aligned to catch the weight in the squatting movement. From the deep squat position, the athlete then stands up with the weight overhead.

An efficient pattern catches this weight, not a group of muscles. This amazing feat of power and skill cannot be done without near perfect flexibility and a perfect application of coordination, quickness and power. This is mobility and stability at its finest, working behind the scenes so the prime movers get all the usual credit.

You have been looking at the squatting motion in some form all your life. The neuro-developmental progression of humans demonstrates how we use the squatting pattern to achieve our initial standing position. Since that's the way we learn to move the first time, why not apply the model to sports rehabilitation and orthopedics?

New parents are amazed at watching a child develop mobility and stability and a command of balance by rehearsing the fundamental movements of rolling, crawling, squatting and eventually standing. A well-aligned and steady squat from the ground to standing will reward a toddler with a few seconds of control at the highest position. There are no sets and reps for reinforced strength. Instead, positive feedback reinforces perfect movement when the infant reaches the goal of movement exploration and expression.

As a new father, I watched as my first child came to standing, looked around the room, lost her balance and quickly fell, only to resume the activity and repeat her mission. If she did not stand from the squatting position with proper alignment, if she rushed or did not center her mass over her base of support, she dropped and her developing brain disregarded the less-effective pattern as not useful for her needs.

Trial and error of movement against gravity taught her how to manage her body through balance, coordination, mobility and stability. Her conscious attention and reflex behavior functioned as a matched set. At the time, I was a teacher, author, lecturer, physical therapist and strength coach. With all my credentials and with all my knowledge, I had nothing to add—there was nothing to remove, refine or modify; there was nothing at all to do, but sit and witness a miracle.

Many cultures assume the squatting position as way to rest for a moment, and it is used in many different ways to develop functional strength. Adherence to a squatting program with no upper body work whatsoever will yield upper body development. However, attention to an upper body strength-training program does not yield the same benefits in the lower body. That in itself represents how powerful the squat is as a developmental platform.

In the early days of weight training, trainees took time to learn to squat properly with full range, balance and control. They developed a strength platform built on a good squat pattern, but modern attitudes disregard the benefits of slow, steady development and consistent acquisition of the squat skill. In this, we made the fundamental mistake that created the need for a movement screen in the first place. We started thinking more of exercises than of movements.

That may seem confusing, but that is exactly what we did. We saw the obvious benefits of training the legs with the squatting movement, and went on to develop squatting exercises. Some people couldn't do those exercises, so we modified the squat purity to make it more generic for greater mass appeal. We modified the general exercise rather than correcting the flawed movement patterns displayed by the trainees. Moving weight became more important than moving.

Just use heel lifts.
If you can't go deep, just squat to parallel.
I can lift more in a back squat than a front squat,
so it must be better.

And then the big one—

If squatting hurts, do the leg press.

As more and more people migrated to training with weight, poor movement patterns followed them into the gym. They wanted a workout, but didn't realize good movement is a prerequisite for weighted exercise. Then people applied a typical modern solution to this growing problem: If we can't do the pure form of the exercise, modify it to fit our limitations. These modified exercises went from partial-pattern exercises to total muscle isolation. As a result, muscles grew and movement patterns atrophied. Modern equipment technology found a way to actually strengthen limitations and reinforce poor movement.

Some old-school coaches noticed this, and this caused them to warn against weight training altogether. They noted that the kids who lifted couldn't move well and voiced their opposition to weights because they thought it would make athletes stiff and slow. This is what they saw in many instances, but by the time the coaches observed the problems, the popularity of weight training was gaining momentum. It was too late—weights were here to stay.

These observant coaches didn't understand that weights weren't the problem; the programming was simply incomplete. Training partial patterns reinforces partial patterns. The weights were reinforcing whatever was put under them. When we put a bad squat under weight, we reinforce a bad squat. The trainee will get stronger initially, but will eventually experience problems.

The gym equipment industry offered us another solution. If a person couldn't squat but still wanted to work leg muscle development, they were there to help with a leg press, a leg extension and a leg curl machine. With these machines, we can work the leg musculature without ever performing the functional patterns these muscles support.

This is a big problem because the prime movers still get exercised while the stabilizers lag behind. The stabilizers do not have to work in a natural manner in a partial pattern, during isolation exercises and on most weight machines.

Own the movement before you do the exercise.

Instead of creating a workout program to reacquire a squatting pattern once fundamental to everything we did, we decided to make machines that worked the constituent parts. People dissected the squat into a series of prime movers, and today people still evaluate the exercise the same way. The stabilizers were forgotten in the evaluation and in the exercise, but upon inspection, the movement pattern reveals the truth.

If your client can't squat—that is, correctly use the squatting movement pattern—don't train the squat until it is fixed. Isolation and partial pattern exercises only complicate this problem.

MISTAKES ON SCREENING THE SQUAT

After we introduced the FMS, my colleagues and I saw other assessments begin to incorporate the overhead squat.[46] The overhead squat was conveniently turned into an assessment once it was popularized as a part of the FMS. It is unfortunate that the principle of screening movement patterns was converted into a testing method that represents the reductionist tendency discussed previously. See the appendix on page 387 for further discussion.

Obviously, the developers of other squat-based movement pattern tests appreciated the FMS deep squat pattern, but they knew many people could not perform the movement. To avoid the realistic perspective the true screen provides—that we might move poorly even when considered fit—they did not go for full squat depth.

Avoiding the end limits of the subject's range of motion, stability and control, they picked an arbitrary point in space to stop the squat, and in this ambiguous position proceeded to discuss misalignments as if they were medical diagnoses.

Knee valgus means nothing if you don't identify the cause. Pronation of the foot means nothing if you don't discuss what the hip could have done to prevent it. A tilting pelvis could actually be the result of a mobility or a stability problem.

This was probably done to create ownership of some form of assessment and appear more complex and thorough, but instead it showed a wealth of anatomical knowledge with little appreciation of functional movement patterns and motor control. If you truly understand movement, you understand that the complex chain of events in a multi-segmental pattern is rarely the result of one anatomical structure success or failure.

Mistake 1—They performed an assessment on the squat before screening it.

Mistake 2—They blamed the parts of a particular pattern without investigating other fundamental patterns.

Mistake 3—They did not identify the weakest link pattern to create a system to only focus on corrections within that pattern.

Mistake 4—They adopted an isolation approach, working on impairments without working on functional-pattern restoration.

Mistake 5—They neglected to look at the total full-range pattern and did not expose the limits of an individual's squat. They set an arbitrary limit and discussed imperfections at that specific point.

Bottom line: If you want to break down the squat with a more involved assessment, use the Selective Functional Movement Assessment (SFMA®). Anyone interested in the source or nature of a squat movement pattern problem can use the SFMA to break down movement. However, remember, the SFMA follows the same basic rules as the FMS: *Correct the fundamental patterns first, and address asymmetry first.*

If you do these two things before obsessing on the squat pattern, you will address most of the problems preventing optimal squat function. The SFMA can demonstrate a systematic breakout of the segments involved in the squat, but this is only after you have first established the functionality of the other patterns. Dysfunctional segments within the squat reveal mobility problems, whereas the absence of dysfunctional segments within the squat reveals a patterning or motor control problem. If you look closely, the FMS will tell you this.

The SFMA is for clinical movement pattern assessment in the presence of pain. If pain is not an issue, the SFMA only offers a pass or fail appraisal of movement. Consequently, the breakdown maneuvers in the SFMA are designed in parallel with the correctives in the FMS. This means the corrective exercises in the FMS are the breakdowns of each movement pattern in action.

Breaking down the squat for non-clinical purposes will often create confusion and cause a more narrow approach.

SYMMETRY VERSUS ASYMMETRY

Called a symmetrical pattern, the deep squat is one of two tests in the FMS that do not systematically separate the left and right sides of the body. The pushup is the other symmetrical pattern—it demonstrates movement where both sides of the body contribute to the push equally.

Do not tackle the deep squat or pushup with corrective activity if an asymmetry exists elsewhere in the screen. The asymmetry is probably creating the limitation or is compromising motor control. Reduced mobility or stability on one side of the body is almost certainly affecting the entire symmetrical pattern, causing inappropriate muscle contraction, inappropriate weight shifting and even torsion in the body.

First focus on the asymmetry and reduce its effect on the movement pattern in which you first saw it, and then once again recheck the symmetrical deep squat or pushup pattern. Removing the asymmetry may or may not completely change the symmetrical movement pattern, but it is the most responsible and logical approach because exercise progressions for the symmetrical pattern will not effectively address asymmetries.

The side of the body most limited by either a mobility or stability problem deserves special attention. The movement pattern in which you found the problem will continually serve as a baseline to recheck improvement in the limitation. Once you demonstrate an asymmetry, using the opposite pattern on the other side of the body as comparison creates a systematic baseline for your corrective exercise progressions.

STICK TO THE SCREEN

The screen will do its job if you let it. If you need to correct the squatting pattern, the screen will direct you to the squat. However, if the squat pattern is poor but you note an equal or greater limitation in a primitive or asymmetrical pattern, the squat pattern is not the problem, and it is not the weakest link. It displays a representation of the weakest link within the squatting pattern. The only time it is advisable to target the squat for corrective intervention is when it is the lowest scored test in the screen, and when no asymmetries are present.

You should never look at the squat as a test in and unto itself. Of all the tests in the FMS, the squat probably involves the greatest simultaneous display of control—*stability*—and range of motion—*mobility*. Because of this, we don't use the squat as a simple diagnostic tool. It can be a marker of underlying problems, but it is not a complete screen or assessment. The other six tests of the FMS often serve the fundamental purpose of describing why the squat is faulty in the first place.

We don't reduce the inability to squat to a single problem, like a tight muscle group, stiff joint or muscle weakness. A severely restricted ankle joint can most assuredly impede the deep squat, but that is the exception, not the rule. A limited squatting pattern is a disconnect between the body and the brain in the squatting pattern. The motor program that manages the squat has changed for some reason, perhaps from an earlier or current mobility or stability issue. Even if you reveal and alter the problem, the motor program has no reason for permanent change unless you develop corrective exercise programming to reinforce the original underlying pattern.

There are often mobility and stability problems, each playing off the other in an attempt to avoid unfamiliar territory. Tightness is often inappropriate muscle activity perceived as extra tension, when really it is extra contraction of one muscle group in the presence of poor control of another.

Describing a squatting problem as a solitary anatomical restriction or inability is shortsighted and assumes all else is normal. Address the obvious problems causing a limitation in the squat, and deal with isolated limitations as the situation dictates, but always address the pattern as soon as possible.

The deep squat has earned quite a bit of attention and controversy since we first introduced the FMS. The main reason is that people who consider themselves fit, strong and athletic often have difficulty in the deep squat movement pattern. This does not reduce their accomplishments in fitness or athletics—it suggests movement-pattern atrophy.

This atrophy cannot be observed on the surface, like muscle atrophy where a muscle withers when it is not worked. Movement-pattern atrophy is witnessed when an otherwise fit person has lost control of a fundamental movement pattern. Overtraining, inappropriate training, unresolved injury, compensation, muscle imbalance and combinations of these are often the underlying cause of movement pattern atrophy.

The squat movement pattern's central purpose is to move as much of the body's mass as low as possible. Think of it as a pattern, not as an exercise. It only becomes an exercise when we choose to rehearse it, but it is a pattern before it is an exercise.

Here are two examples.

EXAMPLE ONE

A person who is limited in the deep squat and cannot complete the pattern without moderate or serious compensation will often be viewed as having a dorsiflexion restriction. If you took a snapshot at the very bottom of this limited squat, the photo may show only 10 or 12 degrees of dorsiflexion at the ankle, and you might assume the limitation is at the ankle.

However, we can measure dorsiflexion in a closed-chain manner with one foot on a small stool, allowing the client to put a majority of weight on the rear foot and lean forward onto the forward foot, demonstrating unrestricted closed-chain dorsiflexion. In this position, the person will often exhibit significantly more dorsiflexion than available in the deep squat. In this case, the dorsiflexion limitation cannot be to blame for the limitation in the squat movement pattern.

When you find the limited pattern or patterns, pick the weakest link by identifying the lowest score, and use the primitive and asymmetrical rule to decide between comparable low scores. Once you've found the weakest pattern, you've captured the first problem. You'll then use a corrective strategy to tackle the most significant mobility and stability problems you found.

We refined the screen to address problems by working on the pattern or a portion of the pattern. Laser-like focus on one bodypart does not provide greater effectiveness with corrective exercise. Dorsiflexion may not be the problem or may be one of several problems. Don't take the bait by going for the obvious. The corrective exercises and progressions for each pattern will address key mobility issues within the targeted pattern.

EXAMPLE TWO

Tight lats are often the professional explanation for an individual's inability to keep the arms vertically centered overhead throughout the deep squat. In the deep squat test, the arms frequently pitch forward. The ability to hold arms overhead in the squat demonstrates they aren't needed to compensate, leaving them unrestricted. It is once again easy to oversimplify movement by identifying the arms as drifting forward due to lat tightness.

You can test this by having the same person lying supine with arms overhead, demonstrating full shoulder range of motion into forward flexion and bringing the arms back down. You then have the client flex the hips and knees towards the chest, keeping the pelvis and lumbar spine on the surface below, maximally flexed at the hip and knee. This will remove a significant amount of slack from the deep lumbar fascia, which is the opposite end of the latisumus, the proximal muscular attachment. If the client's arms can still hit the surface, demonstrating full forward flexion, you can assume there's no a flexibility restriction of the latisumus.

THE RESULTING COMPENSATIONS

If a person has full mobility at the shoulder or at the ankle, why would we see limitations in joint mobility in a deep squat pattern? My hypothesis is that people who do not effectively coordinate the core stabilizing muscles need to call in secondary muscles to help stabilize and maintain posture and balance. This compensation makes the pattern possible, but not optimal. Without effective stabilization, prime movers can impair joint alignment. This misalignment can produce the feeling of joint stiffness or even impingement.

Poor movement sequencing can also occur and further compound the difficulty within a pattern's full range of motion. Stretching the prime movers will not reset this system, and even if you note some improvement after stretching, it will most likely be temporary. When squatting without good trunk or pelvic control, muscles such as the lats, quadriceps and gastroc-soleus complex have more secondary activity as they contract against a body folding onto itself.

Even in the event of a flexibility problem in the lats, this would still not warrant attacking lat tightness initially because the screen may reveal a greater deficiency or asymmetry elsewhere. The lat may be responding to a poor movement pattern or poor stability elsewhere in the body.

Never blame a movement problem on one muscle. By blaming the tight muscle, you will neglect the weak muscle or poor stabilizing reaction. By blaming the weak muscle or poor stability, you will neglect the tight or compensating muscle.

STABILITY AND MOBILITY IN THE DEEP SQUAT

Occasionally we see people with significantly different squatting patterns who do not fit the mold. For example, an extremely strong individual with unbelievable capacity to squat weight may be unable to complete the overhead deep squat maneuver with only bodyweight. Then another person who obviously has significantly less ability to lift weight may demonstrate a near perfect overhead deep squat.

This is not to diminish the strength accomplishments of the first person. It simply means the individual developed extremity strength exceeding the stabilizing capacity of the core and trunk musculature. This provides for a very strong squat, as long as the person doesn't explore full ranges of motion. As long as the body stays at parallel or above, inappropriate muscle contraction can still be used to push through the movement. Once below parallel, a quad-dominant individual will not be able to maintain balance. This person will fall backward, unable to release tension in the quadriceps to allow the glutes, pelvic floor and core musculature to take over and stabilize.

In the deepest range of a squat, the glutes, stabilized by those deep core muscles, initiate the vertical movement out of the hole and up to parallel where the hamstrings and quadriceps can assist. But strengthening the core is not the answer. A person can go through numerous core-strengthening exercises, but it is only with developed core stabilization and movement pattern control in the squatting pattern that the deep squat will improve.

People often assume stability and strength are synonymous, when in fact they are not. It takes far less effort to stabilize a joint than it does to move that joint, especially under load. In effect, gaining more strength should give us more effective stability. However, *the motor program to move and the motor program to resist movement are two completely different pieces of software.* These are different programs altogether.

This confuses many, because they focus on the hardware—the muscle—but as with the computer hardware, it is the software that controls the action. Your laptop can be a music player, a photo editor or perform advanced accounting programs. You're using the same hardware, but different software. Similarly, moving a muscle through a range and hoping it learns to hold and stabilize makes as much sense as using a spring-loaded grip trainer to steady the hand of a sculptor, artist or musician.

Random muscle-building exercise will not significantly affect stabilization because proprioception and alignment are of greater importance than the mere development of tension. The secret to stabilization is timing, and the broad term is motor control. The stabilizers are smaller and weaker than the prime movers. The only chance they have to exert influence—to stabilize—is to fire first, because they cannot produce more overall tension. By firing first, they compress or hold the joint and create an instantaneous axis. The stabilizer reduces sheer and slide of the joint and actually increases prime-mover efficiency. Maintaining axis and alignment enhances the efficiency of the prime mover.

Training strength for stability is futile. Stabilizers are arranged close to the joint axis to offer a mechanical advantage for control and stability, but if they do not fire first, the mechanical advantage is useless.

Strength training without optimal movement patterns will only create prime-mover activity that overshadows the stabilizers. Strength will seem to go up, but the movement pattern screens will diminish in quality. Actually, true functional strength will diminish as well, but those who train strength in non-functional ways also test strength in non-functional ways. They witness a local improvement and assume a global one.

Good quality movement patterns in the screen indicate the stabilizers are still in the game. It is human nature to want to isolate and test stability, and it is possible as a secondary confirmation. However, a clean squatting pattern is a better representation of the stabilizer activity for squatting than, for example, a side plank. The side plank is an outstanding representation of static stability, but does not represent the dynamic stability needed in a full movement pattern like the deep squat.

SQUATTING:
THE PATTERN AND THE EXERCISE

Squatting is not an exercise; it is a movement pattern. The movement is part of growth and development as a transition from the floor to standing. Squatting can be used as an exercise, but is first and foremost a movement pattern.

The first time you squatted, you did it from the bottom up, not the top down. It is ironic that when we exercise the squat movement pattern with weight, we start at the top with a load and go down. That never happens in nature, whereas a movement like the deadlift actually does happen naturally. In essence, with the squat we are training a natural movement pattern in an unnatural way. In nature, we would squat down to get a load close to the ground, not get under the load conveniently hovering at shoulder level and then squat down.

Similarly, we didn't intend for the movement screen's deep squat test to mimic the squat weight training exercise. For example, the foot position of the deep squat test is straight forward, whereas an outturn is usually thought the most efficient foot position when squatting with weight.

The screen's purpose is to put the person performing the squatting test into one of three categories—

- Full capability (score of three)

- Moderate limitation (score of two)

- Significant, severe limitation (score of one)

The outturn assists a person with moderate limitation to compensate enough to be scored at full capability. If we allow an outturn of 10 to 15 degrees, it often reduces the deep squat's three-point scoring scale to a two-point scale. In this case, half in the middle category—those with moderate limitation, a score of two—would be considered perfect squatters and the other half would be categorized as significantly limited. It is important to make the delineation of three different categories, because it gives us greater ability to identify the weak link and address it with effective training practices.

By pointing the feet in the same direction on parallel lines slightly wider than shoulder width, the deep squat becomes a more difficult move.

Your client will only be able to do this with little or no mobility or stability problems in the squat pattern. This does not mean there are no mobility and stability problems. They just do not appear in the squatting pattern.

The outturn provides a mechanical advantage; it creates a wider base, and it does not challenge pelvic control or hip medial range of motion to maximum capability.

We often recommend an outturn in exercises and activities that require squatting or partial squatting, but that is a difference between testing and training. Testing should outline limitations, and effectively mark capability. Training should reinforce efficiency of the available capabilities.

Remember, stiffness hides a stability problem and causes a mobility problem. Those who lack that extra squat range for a score of three have probably created some degree of stiffness to gain support, control and stability. The stiffness you witness in the deep squat and other functional movement patterns is an active part of the individual's normal function.

Addressing the obvious mobility issues will probably not address the complete problem. If you return lost mobility to a region or segment of the body, you need to again review the movement pattern. If it has not improved, neuromuscular stability is not available and you will still have to address that. The screen shows the confirmation of appropriate stability only when the complete movement pattern is restored.

WHOLE PATTERN TRAINING

Since first introducing the screen, we have learned from those clients with limitations and difficulty in the deep squat pattern. My colleagues and I originally attempted the isolation approach by systematically stretching or strengthening all the muscles important to the squatting pattern, but this yielded little or no success.

We used countless pre-packaged protocols for improving the squat. When this proved ineffective, we threw away all preconceived notions of the squat as a mechanism and looked at the squat as a behavior, a pattern of moving behavior. Without awareness of the isolated parts of the squat, we observed the way a poor pattern compromised everything involved in the squat.

That is not to say we only worked on patterns in training and rehabilitation. We dealt with stiffness and poor flexibility one region at a time, but we did not assume we had accomplished anything. We caused a temporary change in the hardware, but wouldn't last if without resetting the software.

It was a huge relief when we realized it isn't necessary to *create* movement patterns and motor programs. We just have to reset the system—all these patterns were once within most people's capability. Some will never regain a previous level of function, but you will be surprised by how many improve beyond expectation.

When first presenting the screen, people criticized us because it appeared too hard for some athletes and fitness clients. Regularly, the audience asked if we would lower the bar, but here's the thing: The screen is not an absolute where everyone must achieve a perfect score. The goal is not a three on every test, and it's not to find a one or asymmetry on any test. We're merely trying to gauge major deficiencies, not making an unrealistic standard that forces everyone into corrective exercises.

We love training and want everyone to find activities where movement recharges, reshapes and reenergizes. Exercise and activity is not often managed in this way. We push through pain to get an activity dose. We take anti-inflammatory medicines to keep going. We use exercises we cannot perform with quality, hoping for correctness and never achieving it. Exercises and athletics should never cause problems or reinforce poor movement patterns, but they do—they do for every person who has an asymmetry or score of one on the movement screen.

Experienced users approach the screen from the bottom, while the novice focuses on the top of the screen, which often looks like an unreachable summit. People absolve themselves from working on the screen because they assume the three score is out of reach. They should not make that assumption, but even if it's correct, that's not even the goal.

The goal is to remove the one scores and resolve asymmetries.

If we achieve these goals, a slow methodical attack on greater movement quality in parallel with higher performance is the most responsible way to train. Great trainers do this intuitively, but the rest of us need a compass and a map. This methodology looks like a paradigm shift, but it's just a new perspective, with movement screening and assessment as the tools. This refreshing outlook was both enlightening and frustrating for our team, but it was the right way to go.

Whole patterning whenever possible is the rule. It is essentially the way we first learned to squat, and it is the most efficient path to relearning the squat when it's lost. Some will want this to be an absolute statement, but it clearly cannot be. There are actually very few absolute statements in the world and you will not find something so important in a screening and assessment text. The *whole pattern whenever possible* rule just means not to isolate if you don't absolutely have to.

Just the same, things will usually go smoother if you address inappropriate muscle tone or tightness before whole-pattern training. Joints with significant limitation or restriction will also have to be corrected, as will joints that display instability, but these are extreme problems and don't represent a large percentage of those who cannot squat.

As a group, we need to agree on what the problem is before we investigate the why. We all need to agree when a squat would be considered limited, and that the observed limitation is the initial problem. The rest of the screen will give perspective as to why.

Reread that last statement. If you are performing a movement screen, do not stop to investigate the squat or break down the squat. Just complete the screen. You may find an asymmetry in one of the other patterns and greater limitation in another. And you have already answered your question of why your client can't squat. He or she can't squat, but squatting is not the primary problem. The person cannot squat because there's asymmetry between the left and right sides. There's a limitation in the screen more fundamental than squatting. Screening rules say you must address this first. Sometimes it will drastically improve the squat and sometimes you'll see little improvement. The point is not to fix the squat, but to fix the problems *under* the squat.

Once you've done this, the corrective strategy designed for the squat will usually work. Each higher strategy in the corrective system is dependent on the preceding fundamental corrections.

As we developed our screening knowledge and gained new perspectives in movement, we tried to follow our rule to use whole patterns whenever possible. One unique way to create whole pattern exercise and training is by unloading or performing assisted exercise. Assisted exercise essentially allows the person to move with less than bodyweight when training a pattern. Usually assisted training allows more reps or greater range of motion, or both. It allows the individual to practice the squatting pattern with less than bodyweight to develop proficiency until able to handle bodyweight, at which time we begin to overload slightly, and then move into resistance training.

REVERSE PATTERNING IN THE DEEP SQUAT

We've also employed a very powerful technique called reverse patterning, developed further beginning on page 292, by which we train the squat from the bottom up. We think this works for two reasons.

One, there is no preconceived notion of how to do it, and therefore the individual learning the move has a nearly clean slate.

A person who cannot deep squat does not know what it feels like to be at the bottom of a squat, with arms overhead and heels flat. Without knowing the destination, how can the person arrive? When descending into the squat, the first automatic inclination is not to stabilize the pelvis with the core musculature, it is to use the quads to decelerate the descent of the body mass. In doing so, this person unnecessarily tightens the wrong muscles, limiting the range of motion.

The second reason reverse patterning works is because in the early days of our motor development, we did not start the squat in a standing position and then descend, we actually crawled and came up into a squatting, kneeling or half-kneeling position, and from that position ascended.

In my book *Athletic Body and Balance,* I discuss a deep squat progression in which a person bends over to touch the toes, and while maintaining a toe touch, drops into a deep squat. Bending over and touching the toes relaxes the low back musculature, and actually deactivates the core. We have seen in research that too much core activity is nearly as bad as not enough core activity.[26, 47-49] By rounding the back and flexing at the hips, we removed excessive core and thigh activity temporarily from the equation. The quadriceps or tight lats cannot dominate the pelvis in this position; removing the extra influence from the mix, the pelvis is free to find a neutral position.

We then ask the person, still bent forward in the toe-touch, to bend the knees, hips and ankles, and lower into the squat position while maintaining contact with fingertips to the toes—heels flat even if a using a slight heel lift. The client is now far below parallel in most cases. Now all the person needs to do is take the arms to the front or lift them overhead if possible, and stand back up.

This reverse patterning goes below the squat's sticking point, which is parallel for people with serious limitations. We've removed weeks or months of frustration instantly by just breaking an old motor program using this reverse patterning. Of course, as with other exercises, the client needs to repeat this if it is to change a behavior.

The most common question that arises when teaching this exercise reinforces the Functional Movement Systems philosophy—*What if my client cannot touch his toes?* Answer: You should not be working on his squat!

Most people who cannot touch their toes will demonstrate some degree of difficulty with the active straight-leg raise. *We always fix the active straight-leg raise first.* If the leg raise is perfect, but toe touching is still not possible, perform patterning for the toe touch.

REACTIVE NEUROMUSCULAR TRAINING FOR THE DEEP SQUAT

You can also apply reactive neuromuscular training (RNT) to the deep squat pattern with great success. This is another way to get the neurological system to do what you need it to do. This practice, developed more fully on page 294, magnifies or exaggerates the subtle imperfections in the squat to cause a proprioceptive balance reaction—*the don't fall* reflex. You can use elastic tubing to magnify or exaggerate an unintentional misalignment or poor balance strategy, and the automatic reflex-based mobility and stability reactions should correct the observed problem if applied correctly.

Since these problems are not at a conscious or intentional level, it is nearly useless to coach correct technique with verbal cues. You may see better form for the moment, but you will not be pleased with the lasting effect of your well-intentioned efforts. We don't learn to move by verbal cues; the language of movement is not recorded in mental words or pictures. The language of movement is feel. You must *cause a change*, not coach a change.

The tubing is a way to push the mistake a little further; it's not elastic resistance to load movement. You're not looking for a training effect for a certain muscle group, or trying to cause increased strength adaptation. You're trying to create the subconscious impulse to improve motor control and efficiency. Exaggerating the mistake is an attempt to change the sequence of stabilizer and prime mover activity in the movement pattern.

A common example of how we do this in the squat is one or both knees caving inward, usually accompanied by an outward turn of the feet twisting over the ground. Instead of coaching the client against that, we use our hands to press inward on the knees with just enough pressure to establish stability. We establish the pressure while the client is standing tall, and then cue to maintain the pressure all the way through the squat, going as deep as possible.

Instinctively, the client perceives the need to press outward much harder when descending. In reality, we're observing the lazy stabilizer showing itself. By enhancing the collapse, we actually correct it. Once we establish that the technique works with that person, we wrap tubing around the knees and slowly provide less and less resistance until the stabilization sequence has improved.

The same thing can be done when the shoulder blades round and the arms extend forward in the squat. While standing in front, place the tubing around the client's shoulder blades and pull forward with just enough force to activate better upper core stability. Once again, we establish standing stability before allowing the client to attempt the squat.

It is advisable to practice these techniques with someone who has had expert training. You can also review these techniques on our instructional videos[*] and can practice with a partner. Don't

---

[*] These videos, available at Perform Better, cover the following material. The Corrective Exercise and Movement Preparation DVD shows all our tubing drills-assisted, RNT, resisted for all patterns with tubing; Secrets of the Hip and Knee covers the squat, hurdle step and lunge tests; our Cable Bar DVD shows assisted exercise, RNT exercise and expanded examples of chopping and lifting; Secrets of the Core, the Backside covers the active straight-leg raise test; Secrets of the Shoulder covers the shoulder mobility test; Secrets of Primitive Patterns covers the pushup and rotary stability tests, as well as discussions of rolling.

attempt a new technique with someone who actually needs it. Become proficient, then implement the technique in your work.

THE DEEP SQUAT SHADOW

The deep squat casts a long shadow over the rest of the FMS. When you see a three on the deep squat, you're watching an individual who possesses good mobility and stability in the squat pattern, and you can expect to see higher overall scores in the other tests of the screen. The deep squat seems to be a biomarker for a well-functioning body, but it is not the only biomarker and you shouldn't consider it in isolation. Likewise, a score of one on the screen demands deliberation. A one score can have serious implications and will usually be accompanied by low score and asymmetry elsewhere in the screen.

There is a breakpoint in the movement screen where increased injury risk is noted. This can be in the form of the total score, but it is better to look at the minimum requirements for increased risk to be a score of two or better on every test with no asymmetry noted.

There also seems to be a breakpoint in the screen where using corrective exercise alone has reduced value, and exercise alone will not yield the same effective results as it would with a little higher score. This breakpoint seems to center around the squat. In preliminary investigations, an individual who scores a one on the squat does not respond to conventional corrective exercise as favorably as the person who scores above a one.[50] These people can still improve, but will probably require more intensive strategies.

In these cases, advanced soft tissue work from a qualified professional may improve results more than exercise alone. A consultation with a physical therapist, chiropractor or athletic trainer will also accelerate progress. Severe limitation in the deep squat can hide many unforeseen problems that could potentially worsen with conventional exercise programs.

Once again, this does not indicate a focus only on the deep squat. *You should address all other scores of one first.* Consider a one on the deep squat to be a warning sign. Corrective exercise approaches may take longer, outside perspective may be helpful, and advanced techniques may offer accelerated improvement.

Ultimately, the deep squat is just a part of seven tests in the FMS. Many times the quickest way to achieving an improved score in the deep squat is to focus elsewhere in the screen. This, beyond all others, is the greatest lesson that movement screening and assessment can teach. Keep an open mind and let the tools guide you. The system begs us to follow a sequential model for redevelopment of functional movement patterns.

If you follow the rules, you will find the weakest link, greatest limitation and most difficult pattern. It is paramount to first identify and then systematically and safely overcome the weakest link. If another emerges, address it then. Once you establish movement-pattern minimums, conditioning can resume while still pursuing greater movement quality. The ultimate goal is not to perform corrective exercise, but to avoid exercises, activities and lifestyles that dictate ongoing corrective efforts. Basically, we must strive to train in such a way as to maintain fitness and movement patterns in parallel.

THE HURDLE STEP AND SINGLE-LEG STANCE

The hurdle step screen in the FMS and the single-leg stance tests of the SFMA both comprise one movement pattern of seven in each system. In the FMS, the hurdle step is part of three tests that challenge the client in a narrow base. The lunge test and the rotary stability test are also challenging due to a narrow base of support.

The hurdle step and single-leg stance testing explore both mobility and stability simultaneously. We don't discuss one without the other; instead, get used to talking about primary and secondary problems. A significant mobility restriction in hip flexion on the un-weighted leg will automatically cause compensation in the stance-leg stability, and will compromise core stability.

You can picture the problem by imagining a barrier at the point of about 85 or 90 degrees of hip flexion on the un-weighted leg. The instant the brain recognizes a barrier, it will start to make changes in stability to get over the hurdle. Movement *quality* is compromised to gain movement *quantity*. Compensations in the un-weighted leg, weighted leg, torso or shoulders will appear spontaneously. Conversely, a stability problem will result in altered or limited movement that looks like a mobility problem, when in fact, it's poor motor control.

The body will always sacrifice movement quality for quantity, probably as a survival buffer. This is one reason why we never try to train a stability problem initially. Current methodologies in both fitness and rehabilitation have great ways to isolate and test stability, but this is a moot point until we have established basic mobility.

Most trainers assume that a stability deficit indicates a lack of fitness and the need for more exercise, but sometimes the opposite is true. Too much exercise or activity with poor technique or compensations can actually be a contributing factor to poor stability. Without specific direction, more activity or exercise may not only be ineffective, it may even cause further problems. It is more important to look for the reason for the sacrificed stability than to intervene with stability exercise following initial observation.

Studies indicate that a high-threshold strategy can actually cause poor motor control.[26,47,49] A high-threshold strategy is a hyper-protective core musculature response following injury or prolonged periods of inappropriate training.

The modern obsession with core strength has probably added fuel to the fire of dysfunction. The core doesn't always have to brace against the world. Sometimes this musculature needs to yield, stretch and reset. Nearly half the movements in yoga create mobility and movement through the core region, while the other half demand stability and control. This balance is most assuredly not an accident. For more than 4,000 years, humans have known and practiced equal amounts of spinal mobility and stability, intuitively demonstrating that one problem is often a lack of balance between the two.

Improvement in one dimension should not hinder development of another. In modern practice, we often witness a resistance program meet the goals for strength, but while it reduces flexibility and mobility measurements. Programs centered around mobility and stability often compromise ability when force production, load management and explosion are necessary. The demonstration of isolated ability, like the strength of one muscle group or mobility at a particular joint, does not guarantee movement competence. The ultimate representation of harmony within the body is symmetrical and unrestricted movement patterns.

As far back as the mid-1970s, Dr. Vladimir Janda recommended the single-leg stance as an effective way to view postural muscle activity, his way of suggesting a functional appraisal of stability. His basic contention was that we should not view posture as a static entity. Many in his day observed and evaluated posture against a grid or plumb line. They made assumptions about postural control and musculature contribution by documenting misalignment and asymmetry.

We should not discount this perspective. In fact, many skilled exercise professionals and clinicians have developed remarkable insight using static postural assessment. Janda was not offering a replacement to this, but a contrast. He contended we were intended to move, noting that the majority of the human gait cycle is predominantly in single-leg stance.

Single-leg stance offers a narrow base and an initial weight shift followed by static control. This is more functional and more dynamic than standing still with the feet shoulder-width apart.

Let's consider the different perspectives offered.

- **Scenario One**—If you note misalignments in static bipedal stance against a grid, but the introduction of single-leg stance on each side normalizes nearly all observable dysfunction, you may not be looking at a purely functional problem. What you have witnessed is a bad postural habit that may have little effect on function.

- **Scenario Two**—If static bipedal standing is observed to have nearly perfect alignment, but single-leg stance is limited to three to five seconds before losing balance, you have once again seen systematic disagreement. Bipedal standing posture seems to be hiding an underlying postural problem demonstrated by a dysfunctional single-leg stance.

- **Scenario Three**—This would be a scenario that does not demonstrate any conflict. Both tests agree; each test demonstrates consistent function or dysfunction.

Normal standing posture can simply be a habitual demonstration of misalignments that spontaneously disappear with activity. Let's look at how and why scenarios one and two disagree.

People who stand a lot adopt efficient static postures of the lowest possible energy expenditure, another sacrifice of quality in favor of quantity. This stance is efficient for long episodes of continued standing, but lacks the ability to respond quickly or to become dynamic. Single-leg stance requires a functional demand—people will always lose alignment before sacrificing balance. They compensate and contort their bodies to avoid the loss of balance and that, in nutshell, is the point.

Although normal gait cycles don't require three to five seconds of single-leg stance, asymmetry, poor alignment and limited single-leg stance ability can significantly reduce gait efficiency. Studies suggest 10–20 seconds of single-leg stance is considered normal. Obviously, the observed 10–20 seconds must have a certain degree of quality. The stance should be undisturbed by jerky movements, attempts at recovery and deviations from the central position.

Nearly all exercise and rehabilitation professionals practicing at the expert level migrate to dynamic movements or functional movements in contrast to static postural grids. And, in fact, studies suggest that movement patterns are a more reliable predictor of function than static measurements. This is one reason why the FMS and SFMA do not use a grid or static postural measurements. If we are going to make decisions largely based on dynamic movements, we can skip static assessments against a grid or use these only as a secondary evaluation.

EXERCISING THE PATTERNS

The FMS and SFMA offer contrasting levels of movement patterns and stability. It is not advisable to limit stability appraisal to one maneuver, such as a single-leg stance. Because the client is practicing the test, you'll see improvement in that specific stability exercise, but there will be no carryover into function.

The contrasting perspective of multiple movement patterns is the key to the FMS and SFMA. If you think a stability program has improved function, don't wonder, hope, argue or defend the unknown. Screen it, assess it, and objectively evaluate it across multiple functional patterns and you will know.

If fundamental and functional movement patterns are unchanged by the addition of a particular exercise or program, it is safe to say little functional carryover has occurred, and that motor control has not been affected or improved. This exercise did not accomplish its intended purpose, the improved movement quality and efficiency in patterns other than the movement rehearsed in the exercise.

The best exercises produce carryover—they improve movement capacity and movement quality in movement patterns not directly practiced.

When our team began considering our training programs within the perspective offered by the FMS and SFMA to check behind ourselves, as you might expect, the programs were not always changing what we thought they would change. At that point, we had a choice. We could trash our functional screens and assessments or we could trash our old views on corrective exercise.

We kept the screen, dumped our old protocols and started using the screen and assessment to check progress using different exercise strategies. This is where the pattern-specific stability approach first appeared in our work. Once we questioned ourselves, everything changed. Whenever possible we adopted a whole-pattern approach, working mobility and stability together the way movement happens authentically in nature.

However, we used special safeguards and proprioceptive strategies to enhance the ability of the weakest link. It is not enough to find a mobility or stability problem. Whenever possible, the problem also needs to have a movement-pattern classification. It needs to be the most dramatic example of movement dysfunction, and be the most significant asymmetry or most limited pattern. If multiple dysfunctions and asymmetries are noted, you must address the most primitive or basic pattern first. Once you find that weak link, your corrective strategy will address the situation by working mobility, static stability, dynamic stability and functional pattern retraining.

While viewing the hurdle step of the FMS or a single-leg stance in the SFMA, the novice will often try to implicate the moving leg as a mobility problem, or the static leg as a stability problem, when in reality the pattern is the problem.

If you note difficulty or limitation in the hurdle step or single-leg stance, it would be inappropriate to say anything other than the hurdle step or single-leg stance pattern is limited or dysfunctional. Attempting to implicate one side of the body over the other will only create neglect in training the other components of the faulty pattern.

Once again, corrective strategies should focus on the whole pattern. Even if the hurdle step or single-leg stance is compromised, there are still six other movement patterns to consider. During our workshops teaching the FMS and SFMA, we watch first-timers spend 10 minutes scrutinizing details of the hurdle step or single-leg stance. This isn't necessary—call it as you see it based on the criteria and move on.

COMPENSATIONS IN THE HURDLE STEP AND SINGLE-LEG STANCE

The hurdle step and single-leg stance patterns demonstrate the requisite mobility and stability needed for stepping, running and climbing. The high striding position involved in extreme climbing or technically correct running and sprinting is demonstrated in the hurdle step, and functional stepping and weight shifting is reviewed in single-leg stance.

Subtle compensations, loss of balance and limitations in mobility are indications that substitution and compensation are probably necessary in everyday function. Inappropriate movements are required by another bodypart to allow stepping and striding length and height to occur.

The hurdle step and single-leg stance patterns give us insight into dynamic movement, but also have implications in posture when we see a loss of height when shifting from double- to single-leg stance. A subtle loss in height between the stances is an indication of reduced postural quality. Dr. Janda taught that we could better observe posture in single-leg stance, because using both legs allows opportunities to stabilize and maintain posture through incorrect or sub-optimal means. Single-leg stance not well stabilized will often result in poor alignment or compensation the instant of transition from double- to single-leg.

When we see the lack of reflex stabilization during the stance transition, we know the transverse abdominis and core stabilizers are not firing appropriately. The transverse abdominis is a central corset around the mid-section, and when optimal, it is known to fire before the primary mover in almost any stabilization situation.

Research has shown the transverse abdominis fires before the deltoid in arm lifting, and before the hip flexor in lower extremity movements. This activation is delayed in people with low back pain, and it is delayed in athletes with long-standing groin pain.[51-55] Loss of height indicates the transverse abdominis has been temporarily overpowered or deactivated to allow a postural compensation because of inappropriate mobility or stability.

The core stabilizers function reflexively to help the body balance in postures and in the transitions between postures. Conscious thought is not a part of reflex stabilization. When a person transfers from double- to single-leg stance, this is often done without thinking. This is good, because in this you'll have a quick snapshot of the efficiency and ability of the stabilization reactions.

A loss of height, a lateral lean, a slight rotation or a sway will demonstrate lack of optimal stability and adequate motor control. This gives implications of faulty posture and function in single-leg stance, but this does not necessarily implicate the individual's standing leg—it could be part of a faulty motor control within a pattern.

Poor mobility can compromise stability and look like a stability problem, and poor stability can compromise movement and look like a mobility issue.

Call it as you see it; score the hurdle step or single-leg stance and move on. If it is not perfect and the other six tests are, there is a good chance you're seeing a stability problem in single-leg motor control and in weight shifting from double- to single-leg stance. We can assume this because the person will have demonstrated adequate mobility by perfection in the other six tests, and simple deduction points us toward a stability problem in the remaining movement pattern. This situation is rare, but it is entirely possible. Likewise, consistent limitation with hip flexion or hip extension in the other movement patterns will usually indicate a mobility problem.

Most of this discussion is not necessary since the corrective strategy for the hurdle step and single-leg stance will first establish mobility, and you will advance to static and dynamic stability work only after confirming adequate mobility. The corrective strategy will target the primary and secondary problem simultaneously. the hurdle step or single-leg stance corrective strategy will get to all the mobility, stability and motor control issues that need remedial efforts.

This strategy will only work to correct the hurdle step or single-leg stance patterns if these are the problem. If a lower or equal score in a more primitive pattern is present, it is inappropriate to first work on the hurdle step or single-leg stance.

The hurdle step and single-leg stance tests provide immediate insight into whether a person uses reflex stabilization or an alternate, less-efficient compensation when going from double- to single-leg stance. The formality of stepping over a hurdle or holding a leg up at or near the height of the tibia assures enough time in single-leg stance to see the necessary compensation representing a lack of reflex stabilization.

RUNNERS AND THE SINGLE-LEG STANCE

In clinical practice in sports physical therapy and sports medicine, we work with many runners and triathletes with various problems or injuries resulting from their sport. It's always surprising how many very accomplished runners have little or no reflex stabilization in single-leg stance on one or both sides.

This in no way diminishes their accomplishments in running or triathlons. But it does imply that wasted energy may be a factor, wasted energy that translates into poor efficiency. In running, efficiency is the name of the game, and yet to enhance efficiency we see mostly coaching of diet, cardiovascular performance and running techniques instead of these movement fundamentals.

Some runners assume their mechanics are not bad, but perhaps they're only looking at specific running mechanics with little consideration of fundamental mobility and stability. They assume a good running routine will keep them protected from mobility and stability problems. Others understand the importance of mobility and stability and become obsessed with supplementary work and isolation exercises. These two extremes are equally unfounded.

In the middle path, nothing in excess is good, including running. In running circles, this is blasphemy, but it is truth. The middle path—run in a correct volume to challenge the energy systems and yet not overstress the musculoskeletal system.

A balanced approach requires running enough to practice race strategy and new techniques without neglecting weight training and flexibility work. Runners should run, and should also maintain basic strength and a balanced and acceptable FMS score. When basic strength declines or the FMS score starts to go down, the focus on running has caused a neglect of the basic systems that support running.

Watching markers like basic strength and functional movement patterns in addition to activity-specific performance markers is the best way to keep a foundation under training. This is actually a perfect way to determine running and training volume.

We also use the FMS to determine lifting volume with strength training, looking for a strength gain with no reduction in FMS quality. Most lifters want to improve in their chosen activity, but we don't want to see this at the expense of reduced functional efficiency or increased injury risk. The FMS is a simple way to track the fine line between training, overtraining and misdirected training.

Excessive running on an incorrect platform can cause a lack of reflex stabilization, which we observe at a basic level in the hurdle step. Inappropriate muscle tightness from any source, including excessive or inappropriate weight training, can also result in some muscles working harder than others. Extremity muscle stiffening occurs because of inefficient core reactions. Repeated activity reinforces the stiffness because it is compensatory, and stretching will not fix a problem reinforced by the conditioning. The observation of a poor or asymmetrical hurdle step can imply both reflex stabilization and muscle tightness problems.

Inappropriate prime mover activity and poor reflex activity of the stabilizers reduces overall efficiency. Since both speed and endurance in running are paramount, and efficiency is absolutely the name of the game, we shouldn't have to oversell the hurdle step as the gold standard for baseline mobility and stability in a stepping or striding pattern.

Running without an adequate hurdle-step pattern will only strengthen compensation and a less-than-optimal stride, robbing efficiency from the activity of choice. Runners insist they cannot take time off from running to work on these patterns because they believe endurance will decline, but in fact, reduced efficiency is guaranteed when continuing to train and practice sub-optimal patterns with high volume training. The enhanced efficiency gained by two weeks of mobility and stability corrective exercises and calisthenics targeting weak links will far outweigh any microscopic loss in metabolic efficiency.

It is hard to deal with resistance to change from clients and athletes and patients, and it's hard to achieve compliance in rehabilitation, training or coaching. Still, recreational runners cannot seem comprehend that not running, even briefly, can actually improve running speeds.

The running analogy has dominated the hurdle-step section. This is to demonstrate how important maintaining a foundation is to all activities. We first build a movement base, then develop supportive energy systems, and then and only then can we build specific skill atop that complete base.

THE INLINE LUNGE

The lunge test in the FMS is the third test of seven. There is no lunge in the SFMA because the lunge may be too taxing in patient situations in the clinical setting. The SFMA does not require high-demand movement patterns to demonstrate pain and dysfunction. The lunge pattern is important, but is best screened before return to activity, not to provoke symptoms.

The lunge is similar to the hurdle step and single-leg stance in two ways: It is a narrow-base activity, as well as an asymmetrical activity. The lunge pattern actually exaggerates asymmetrical demand on the body since it creates opposing extremes for the upper and lower body, whereas the hurdle step only imposes an asymmetrical pattern on the lower body segments.

Picture the asymmetrical movement of upper and lower extremities in activities like crawling, walking and running. When things are working properly, the upper and lower extremities function as perfect counterbalances. As each segment moves, it also creates a point of dynamic stability. All this dynamic stability and counterbalance meets in the core and is redirected to conserve energy and to complement locomotion.

That is, if all is working properly. The contrasting upper and lower body movement patterns in the lunge test serve to push the limits of mobility, stability, motor control and dynamic balance. Here we get a glimpse of the individual's asymmetrical movement.

The lunge test looks at the hips and legs in a split or stride position with a very narrow base. The narrow base contributes to the difficulty and provides better distribution across the potential scores we assign to the lunge.

Workshop participants often complain the lunge is too difficult and ask us to modify it. This is the softhearted advocate fighting for everyone to get a perfect score—not by earning it, but through lowered standards. We don't help people by lowering difficult standards. In the FMS, a score of two without asymmetries is not bad. The FMS is about getting as many clients as possible above a one and about removing asymmetries, not presenting every athlete and client with a perfect three on the lunge.

When a field or court athlete has a significant asymmetry or limitation in the lunge test of the FMS, it is inappropriate to work on the parameters of speed, agility, quickness and power without first improving mobility and stability in the inline lunge pattern. As we train explosive athletes and as we prize safe and efficient deceleration and direction changes, a score of three on the left and right lunge should be our goal.

PATTERN-SPECIFIC CORE STABILITY

Sometimes confusion comes along with the administration of the lunge test because an individual may do poorly in the squat pattern, but do quite well in the lunge pattern. Alternately, a person might function well in the squat pattern and do poorly in the lunge pattern on one or both sides.

This is only confusing if you assume a mobility problem is the reason for the dysfunction. Many of the restrictions that compromise the squat can also affect the forward leg on the lunge. However, it is possible for a body to have an adequate lunge pattern and a faulty squat pattern, and vice versa. This has more to do with pattern-specific motor control and core stabilization, and this creates a second degree of confusion.

You might wonder why a client with core stability would pass one test and flunk another. The best answer is that stability is linked more to movement patterns than to anatomy. The muscles of the core may function perfectly well when the hips are in asymmetrical position, but poorly when in symmetrical positions.

The reverse is also true. In exercise and rehabilitation, people often under-think this or miss it altogether. The assumption that muscular control providing stability in one pattern will be equally present in all patterns is not consistent or reliable.

Dynamic stability and controlled mobility are pattern-specific. The abilities grow within each pattern naturally and don't seem to be altered unless a specific limitation is imposed.

Therefore, we view core stability as a specific behavior within a pattern and not an absolute demonstration of unquestionable multidimensional neuromuscular function. You can program six months of core training in one pattern and it may not benefit another that is limited or underdeveloped. If it is limited or deficient in some way, it will not develop unless you work with that specific pattern through varying degrees of mobility and stability training.

Here's the point: *Core stability is pattern-specific.* The ability to appropriately fire muscles in a coordinated effort is almost individual for each pattern. Once the patterns are established, work in one can complement the other. Working on one pattern can offer benefits toward the conditioning and training of another pattern, but patterns do not become perfected unless we train them as patterns.

Toddlers don't exercise; they explore patterns and establish them to a basic competency. As adults, we lose certain patterns because we have habitual and favored movements, and neglected or avoided movements. We have unresolved injuries and chronic problems that cause automatic compensation. Stress produces neuromuscular tension that interrupts the natural fluid movement. When the average adult starts to exercise with one or more limited fundamental patterns, the exercise programs rarely address the basic underlying problems. To add insult to the potential injury, the increased activity causes an elevation in the compensation behavior.

Establish basic patterns and put exercise and activity on top.

In the same way as discussed in the squat section, it is not appropriate to blame one structure as the focal point of the failure of one side of a lunge pattern. It is more suitable to assume that multiple problems exist within a faulty pattern, and that we can train these problems simultaneously and optimally within the pattern following basic flexibility and mobility movements. A truly integrated approach will focus on patterns as much as possible.

THE LUNGE AS A MOVEMENT PATTERN

With all this talk about the lunge, the inline lunge test in the FMS is not a lunge at all; it is actually a split squat, because we don't grade the step into the starting position. The step has already taken place in the pre-screen positioning, and only a descent and ascent must occur from the stride stance to complete the test. This is challenging because the base of support is extremely narrow. It is difficult to a further degree because the arms and legs are in alternate positions to replicate the reciprocal action of the extremities. This will expose mobility problems.

If poor stability is present, the lats, quads, hip flexors and muscles that attach to the iliotibial band will often struggle to function as both movers and stabilizers. This will fall short of a perfect pattern, and at best limit the score to a two, revealing compensation and loss of appropriate body angles.

The limitations collected here give evidence to faulty mechanics in the lunge pattern as well, and the lunge pattern is integral to deceleration and direction changes that require control with a low center of mass.

We have only discussed the lunge with respect to deceleration and direction-change activities, but the lunge is also important in situations that involve throwing, striking and some swinging movements.

Efficient throwing, striking and swinging are very often a result of properly timed weight shifting from the back foot to the front foot. This linear power transition turns into rotational power when the wave of energy generated in the lower body reaches the upper body, creating a throw or a punch. This can also occur in sports that involve a swing where one or both hands are on a stick or racquet, and the hips and feet are dynamic, in alternate positions. Other swings, such as the golf swing, occur in more static hip and foot positions, more like a squat stance. In baseball, the hitting movement involves a swing that sometimes looks like a squat and sometimes looks like a lunge.

Both the squat and the lunge give insight into movements where power is generated from the bottom up and the top down. An example of a top-down movement would be running or jumping, where the upper body begins a counterbalance and momentum initiator prior to motion in the lower body. We initiate bottom-up activities, such as striking, throwing and swinging, from the bottom to create motion, to reinforce stability and to add power to upper body movements.

Do not assume movement patterns like the lunge and squat only give implications for lower body foundation, function and mechanics. These can also create heavy influence on upper body movements that rely on core stability and lower body generation of power.

CORE STABILIZATION IN THE LUNGE

The inline lunge is an excellent test to demonstrate how people actually engage the prime movers of the upper and lower body prior to the stabilizing muscles of the shoulder girdle, hip complex and core. This bodyweight lunge is more about control than strength. When mobility and stability are not adequate, the prime movers go into action to help stabilize, but the result is not at all stable. The lunge looks shaky, awkward and difficult, and is taxing for those who lack adequate mobility and stability.

We use this test to demonstrate a point about efficiency—the lunge will be just as difficult with greater strength. It will only be easy if mobility and stability are adequate, and the lunge cannot be optimal if a person doesn't first initiate stabilizer activity. Inability to perform the inline lunge is a result of a prime mover contracting before a stabilizer has performed its role.

Once there is inhibition of natural activity, some form of facilitation must occur. The stabilizers must grow back into their roles by rehearsing the pattern where their activity is needed. Isolated mobility and stability testing is not necessary, and forcing the issue will only create compensation. If the movement pattern is not improving, the effect of each will not support the coordinated function.

It's not necessary for stabilizers to exceed the prime mover strength to function appropriately, they just need to fire first and endure. This flies in the face of training methods that attempt to strengthen stabilizers, assuming that the intricate timing involved in true stabilization will automatically reset. Human movement is amazingly resilient; it often resets itself in spite of rehabilitation and exercise, not because of it.

Stabilizer roles include joint centralization, joint tracking and alignment in coordination with movement. The basic stabilizer role is to create instantaneous control around a joint so the joint can find its appropriate axis throughout each degree of movement.

The prime movers often attach to anatomical segments that rely on stability so sufficient force can produce optimal leverage. When stability is not adequate, this advantage is not maintained or the axis of the joint is no longer optimal. In this case, the prime movers will appear weak when they may not be weak at all—the stabilizers simply have not performed their roles in a natural, coordinated fashion. Our typical reaction to weakness is to strengthen, but conventional strengthening exercises only will reinforce this problem.

It is more important to change the stabilizer timing. Strengthening stabilizers is of little use since they can never be stronger than the prime movers challenging their control and alignment. Stabilizers need to have greater endurance, better timing and quicker action than prime movers. In this, they will exert the instantaneous joint stability and create the platform for the large muscle contractions that provide movement patterns.

Mobility and stability gained by working on the lunge movement pattern will enhance body awareness and foster motor learning to gain better power, speed and agility in asymmetrical positions. Most direction changes, throwing, striking, swinging and deceleration movements do not take us into the extremes the inline lunge test explores. Having this buffer zone is the best guarantee that enough mobility and stability are present to handle higher levels of training.

Remember, the FMS is about movement, and once movement is established, performance and skill can rule the day.

THE SHOULDER MOBILITY REACHING TEST

The shoulder mobility reaching test in the FMS demonstrates alternate patterns of the upper extremities initiated by extension and internal rotation on one side, and flexion and external rotation on the other. It mimics the upper body position of the lunge test, but demands significantly more mobility in each shoulder and in the thoracic spine. The reciprocal pattern is demanding, because the opposing movements borrow extra mobility and stability from other segments.

You'll see reciprocal patterns throughout the FMS—in the hurdle step test, the inline lunge test and the active straight-leg raise test. We test shoulder mobility in the SFMA, however in that we appraise each shoulder independently, because in the SFMA we want to identify and locate sources of pain and dysfunction in each shoulder.

The reaching patterns require the thoracic spine, shoulder girdle, shoulder and elbow to wind and reach into extremes. This maneuver requires joint mobility, muscular flexibility and fascial, vascular and neurological extensibility. A similar upper body position is adopted for the FMS inline lunge test, but that does not require full range and is used not so much to check mobility, but to remove the hard-to-detect upper body compensation and poor posture seen in lunging.

The FMS and SFMA use the upper body movement pattern to detect and document gross asymmetry and limitation. The FMS shoulder mobility test has a higher degree of difficulty than the independent SFMA test, but each test is appropriate for the information needed.

In both shoulder mobility tests, we take the movement pattern to extreme. This may look like a glenohumeral joint mobility test, but that is just a surface view. The test implications far exceed a basic appraisal of shoulder flexibility.

Automatic patterns occur with both thoracic spine mobility and scapular stability that immediately precede the arm reaching movement patterns. When missing, these are great examples of movement-pattern atrophy. The movement is hard to detect, but extremely important to the complete functional movement pattern.

Most people slump forward when performing the alternate reach-back maneuver in the FMS, and to a lesser degree in the SFMA. This is ironic, because upright posture and some thoracic spine extension is the optimal position for both reaching patterns. The upright posture and thoracic extension are both reflex-driven and are not a conscious part of the pattern, but are imperative to a complete pattern.

COMPLEMENTARY MOVEMENTS IN SHOULDER MOBILITY

Lack of mobility and poor posture can limit the thoracic extension contribution and lower the quality of the pattern. Poor inner core function can allow a loss of the natural erect posture necessary for optimal shoulder movement and efficient breathing.

Shoulder girdle motor control is just as important, and is often reduced in an automatic compensatory attempt to make up for a lack of spine erectness and thoracic mobility. The stability of the scapular segment of the shoulder girdle is not static—it functions in a rhythm with the humerus. This means both move, but the scapula moves at a slower rate, giving stability to the muscles that move the shoulder joint.

In a normally functioning shoulder, the scapula automatically retracts and rotates to complement arm elevation. Excessive scapular movement is called scapular substitution, and is a common problem in shoulder rehabilitation. Limited movement of the scapula can also contribute to shoulder dysfunction. As with most muscular problems, the muscles controlling the scapula are not the problem; they are a representation of the problem. The problem is usually an alternating mix of restriction and poor motor control, each contributing to the reinforcement of the other.

Obvious causes of poor scapular stability or scapular substitution are restrictions and limitations of thoracic or glenohumeral mobility. The scapula stabilizers will give up their primary role of dynamic stabilization to accommodate an intended arm movement, even if that puts the shoulder at a biomechanical disadvantage. The shoulder is the most obvious example of this problem, but is not the only example. All segments of the body fall victim to this phenomenon.

It is part of human nature and survival to give up movement quality to gain temporary or immediate quantity. The brain assumes all tasks are of utmost importance and will reorganize segmental contribution of each part to complete the task at hand, even if that task sacrifices some degree of protection, efficiency or control.

Thoracic mobility and scapular stability are often the underlying problems in people who display both asymmetry and significant limitation in the shoulder mobility tests. Most of the corrective strategies we initially use to address this are targeted more at thoracic spine mobility and scapular stability than at glenohumeral shoulder mobility. This is not because they are more important, but because they create the platform for normal glenohumeral movement.

All the muscles that move the shoulder depend on scapular stability to create a point of proximal purchase to produce distal movement. The scapular stabilizers depend on thoracic mobility, and extension and rotation of the upper thoracic region reduces the need for excessive scapular protraction and elevation.

This should not imply that scapular mobility problems do not exist, because they do. Thoracic and glenohumeral mobility limitations just represent a common tendency in sedentary populations and populations with incomplete exercise practices.

The SFMA provides anatomical landmarks that represent normal range of motion parameters. The FMS shoulder mobility test grades the distance between the closest points of a clenched fist with the thumb inside, but you should still note the unnecessary compensation that may occur during this maneuver, specifically on one side compared with the other. In some cases you will see a loss of torso height, an immediate jutting forward of the head, and flexion of the cervical spine. The client will assume an anterior head posture, round the shoulders and flex the thoracic spine, all in an attempt to bring the hands closer together.

COACHING OR RETRAINING THE NATURAL PATTERN

The natural pattern for reaching backward is actually maintenance of a tall spine, a neutral cervical position, a slight extension of the thoracic spine with mobility throughout the rib cage segments, and instantaneous scapular stabilization. Together, each contributes in part to allow the glenohumeral musculature to exert influence and move the shoulders to the optimal position. Proper alignment allows for better anchoring of the important stabilizing musculature.

Do not discuss the observed faults with the person being screened or assessed, and don't attempt to improve the test score with corrective instructions like *stand up straight, pull your shoulders back before you move your arms* or *don't look down—keep your chin up and back.*

Let the screen or assessment do the job. This test was designed to demonstrate the automatic response of a person to a basic movement task, a chance to show maximal range in either test. The individual will automatically choose a particular path and pattern, giving you the opportunity to grade the quality. Coaching to a better score will not help matters—it actually hurts your chance of correcting the underlying problem. Verbal cueing and instruction may improve the test outcome slightly, but it will not change the automatic response that a real life situation will impose.

A true change in fundamental movement must come from an automatic and appropriate postural position, followed by an automatic and appropriate movement pattern. This is why an isolated segmental approach has limited functional affect on changing a pattern. To achieve an automatic postural and movement pattern response, we have to reacquire and rehearse developmental movement patterns with appropriate mobility.

The SFMA looks at shoulder mobility independently, and also considers the thoracic or cervical spine. The SFMA uses the standing rotation test to look at thoracic spine and hip rotation movement patterns. If we don't see adequate mobility or alignment, we use a breakout to investigate all the components of rotation.

Thoracic spine rotation is a combination of extension on one side and flexion on the other. If thoracic mobility is a significant problem, the SFMA breakout will assist the investigation. Sedentary life will compromise thoracic mobility, and this will likely be common in most of the people you screen.

Likewise, the SFMA also reviews the cervical spine. Movement patterns of the cervical spine are performed, and if found limited, the cervical spine is investigated with the shoulder girdle in different positions to discover dysfunctional relationships between the shoulder and cervical spine.

THE ACTIVE STRAIGHT-LEG RAISE

The active straight-leg raise is much more than a hamstring length test, and is one of the most misunderstood tests in the FMS. Even those who understand this will often contemplate a tight hamstring, forgetting they are grading a pattern involving two legs and a degree of core control. We look at a limitation in the active straight-leg raise as a pattern problem only.

Three specific, equal parts are required to allow this pattern to function normally.

- Adequate extension of the down leg

- Adequate mobility and flexibility of the elevated leg

- Appropriate pelvic stabilization prior to and during the leg raise

The SFMA also uses the active straight-leg raise, but as a systematic breakout for a dysfunctional forward bend or toe-touch movement pattern. The toe touch, sit and reach and active straight-leg raise are all independent yet interrelated movement patterns. This means someone can be dysfunctional in one pattern without equal amounts of dysfunction or even any dysfunction in others. The differences might be varying amounts of loading, symmetry versus asymmetry and whether the movement is initiated from the top down or bottom up. The same body segments are bending and stretching, but in different postures and positions and in different directions.

THE HIPS AND THE ACTIVE LEG RAISE

Many single factors or a combination of problems can limit this pattern. To start, let's look at a lack of hip extension on the leg remaining on the surface below. If this leg lacks hip extension, the pelvis will have to adopt an anterior-rotated position to allow the leg to lie flat. This puts the lumbar spine in a lordotic position and creates a faulty alignment.

In this case, when the lifted leg moves toward the high position in the leg raise, its hamstring becomes taut at an earlier point because the muscle is pre-stretched by the anterior pelvic position. Even though there is tightness in the hamstring, this may not be a hamstring problem.

The hips are a window to the core. Hip strength is usually weak in the same direction that spine stability is poor. When hip flexion strength is weak, we see spine problems associated with poor anterior or flexion stability. When hip extension is weak, there will be spine problems associated with poor posterior or extension stability. Hip problems between internal and external rotation will be associated with trunk rotation stability, and hip adduction and abduction problems will relate to trunk stability in lateral flexion—side bending.

Back dysfunction is associated with hip asymmetry. We can identify asymmetry by a strength test, a range-of-motion test or even a movement-pattern test, and the active straight-leg raise test is one such test. The asymmetry could be a cause for compensation and poor back mechanics and explain the pain, or the asymmetry could be a result of dysfunction in the core and spine. Regardless of the explanation, we cannot ignore the relationship.

When we use mobilization, muscle work and corrective exercise targeting the asymmetry, we'll observe improvements in the previously observed back dysfunction. If flexion is limited, we'll see it improve without directly working on flexion of the spine. We don't see the same response when the limitation is rotation or extension. Creating symmetry does not correct or resolve all spinal problems, but it definitely has a strong effect on restoration of function and on relief of pain.

People often use the leg raise as a diagnostic test for the presence of sciatica, but it really only serves as a practical symmetry and function screen. If it is limited or asymmetrical, you should address it with corrective strategies that promote mobility and stability, which are usually deficient in some combination.

CORRECTIVE PROGRESS

The active straight-leg raise provides one of the best examples of a new perspective on progress. Other tests in the FMS can display this, but the active straight-leg raise provides the best and most common example.

Suppose your client initially scores a two on one side and a three on the other, and after two weeks of corrective work is rescreened only to display a pair of twos. One would expect progress to look like matched threes, but the goal is first and foremost symmetry.

Perhaps the initial three was scored with some overlooked or undetected compensation, and you recorded a false three, a false three being one given in the presence of asymmetry. Single-side dominance, hyper-mobility and pelvis and spine oriented to favor asymmetry can all play a role in a false three. That three score will not stand the test of corrective exercise, because corrective strategies performed well will often rob from the three and give to the two, a Robin Hood effect.

Often the pelvis will orient to provide more mobility to one hip and leg than the other. This can also be observed in the thoracic spine in the shoulder mobility test. The three score comes not so much by appropriate mobility and stability, but by a posturing bias or orientation toward one side. Habitual movement patterns, unilateral sport dominance and compensation can all produce this behavior.

People challenge the utility and simplicity of the active straight-leg raise, questioning the point other than for activities such as running hurdles or dancing in a Vegas show. Here's the point: The test demonstrates freedom of movement.

MOBILITY BEFORE STABILITY

The two most primitive and fundamental patterns are the shoulder mobility pattern and the active straight-leg raise pattern. These are primitive patterns because in the developmental sequence, mobility is present before stability. Even though a small degree of stability is required to perform each of these patterns, they are a representation of functional mobility as well as left and right symmetry within the patterns.

The FMS hierarchy forces us to consider the principle of mobility before stability. Many see the FMS and feel that the tests might be too challenging for some people, but this thinking demands a review of the screen implications. If all scores on the FMS were one, the individual would only be required to work on the active straight-leg raise and shoulder mobility. This person would be required to stay at this level of corrective exercise until improvement was noted. This forces a change in mobility before attempts to improve stability are brought into the corrective program.

If an elderly person displaying poor FMS scores and poor fitness scores were to start on the corrective strategies for shoulder mobility and the active straight-leg raise, we would suggest thoracic spine mobility and scapular stability exercises for the one, and hip mobility and core stability exercises for the other in a very simple and controlled situation. The person would be working at the appropriate functional and fitness level. What better place to start someone who needs to gain mobility before stability, as well as ease into fitness?

In a study by Yamamoto et al,[56] it was discovered that among people 40 years and older, performance in the sit-and-reach test could be used to assess arterial flexibility. Arterial stiffness often precedes cardiovascular disease, and the results suggested that a simple movement test could be a quick measure of a person's risk for early mortality from a heart attack or stroke. The study, called *Poor Trunk Flexibility is Associated with Arterial Stiffening*, essentially uses a movement pattern to predict a problem.

Although it is not the intent of the active straight-leg raise test to provide anything other than movement information, a recurrent theme is starting to emerge. Patterns can be predictive if we know how to use them. It would also follow a line of logic that as muscle, joint and movement patterns stiffen and become dysfunctional, other structures in the body consistently diminish in functional capacity as well. The takeaway here is that simple patterns can be predictive. These aren't diagnostic; they are indicators that function may be compromised and greater investigation might be prudent and protective.

Mobility and stability problems co-exist. Intense focus in one region causes unintentional neglect in another. A central mistake in both rehabilitation and functional exercise is the attempt to create stabilization with inappropriate mobility.

Revisit and attempt to maximize mobility whenever possible. True functional stabilization cannot occur in the presence of inappropriate mobility, because the instant a mobility restriction comes into play, reflex stabilization is inhibited or compromised and becomes a less valuable factor in function.

Once we've maximized mobility, stability is the next function to target—greater amounts of mobility will require greater stability. Many people think it takes too long to gain appropriate mobility, and since stabilization is of utmost importance, they go for stabilization exercises and programs, neglecting the requisite mobility necessary for spontaneous stability.

When all things are limited in the screen, start with the shoulder mobility and the active leg raise. It is a safe and effective path, and the improvements gained in these two patterns will lay a better foundation for restoring optimal function.

THE PUSHUP

You must address problems noted in the active straight-leg raise or in shoulder mobility before moving on to the next two screens. Once those are managed, the pushup and rotary stability become the next priorities for movement correction.

It would be inappropriate to work on any of the big three tests in the FMS if a lower score exists in the pushup test. This would indicate that even if functional stabilization in the three primary foot positions is not optimal, a significant lack of basic core reflex stabilization is still present at a developmental level.

The trunk stability pushup does not measure strength of the pushing movement, but instead looks at reflex core stabilization. The abdominal musculature does not so much flex the trunk, as it avoids unnecessary extension of the trunk. In doing so, it also helps transfer energy from the lower body to the upper body, and the upper body to the lower body.

The test requires movement as a stiff, rigid segment from the floor to the top of the pushup position, with no sag, sway or twist in the low back, and no rotation occurring at the pelvic region indicating a loss of stabilization. When the hips and shoulders do not remain in the same plane, the pushup is dysfunctional.

If you see compensations or dysfunctions in the pushup, you should address each at that developmental level. This means to not just work on strength—look for problems in other primitive postures. If pushup corrections are not working as quickly as you would like, return to the active straight-leg raise pattern, shoulder mobility pattern or rotary pattern and work toward any three scores not already attained. These three tests work to provide improvements in timing and alignment that may help improve the pushup test score.

Other corrective options are reaches from prone-on-elbows and pushup positions, reaches from a plank position that cause weight shifting and reflex stabilization. Moving from prone-on-elbows to the pushup start position also has deep developmental roots from a sensory standpoint.

Yet another option is to perform a downward dog position yoga movement and add a pushup or half-pushup between each downward dog. This helps produce natural stabilization through better perception in the core and shoulder girdle.

All these suggestions keep the symmetrical positioning and follow the developmental sequence to improve reflex trunk stabilization. A few more examples to increase the challenge include eccentric pushups, tall-kneeling chops and lifts and side planks for symmetry, which can all help magnify this problem and offer insight as to the best corrective exercise options.

There is no SFMA representation of the FMS pushup test for a few reasons.

- The SFMA assumes a painful situation is present, and the pushup requires greater muscular loads than the other patterns in the SFMA.

- The position of the pushup is not practical in clinical evaluation situations because the risk-to-reward ratio is not good.

- The pushup does not specifically represent a movement pattern as much as it shows stabilization capability and capacity.

- The pushup is advantageous in a screen for active clients not in pain because it demonstrates the stabilization necessary for many exercise programs and sports activities. This level of stabilization is not necessary for the needs of the SFMA.

The pushup test in the FMS is the only test in the entire screen where we altered a criterion to reflect the strength differences between men and women. Because of a distribution in lean body mass of the upper body, there's a necessary adjustment to create an equal platform for both sexes.

MASTERING PUSHING SYMMETRY

A book called *The Naked Warrior*, written by Pavel Tsatsouline, completely reinforces confidence in having the pushup as a test in the FMS. The book's subtitle makes the topic clear: *Master the secrets of the super-strong using bodyweight exercises only.*

In the book, Pavel commits exercise book heresy by only discussing two exercises, the single-arm pushup and the single-leg squat, also known as

the pistol. Since most people cannot do these two moves, readers soon learn the book is about progression. These two exercises almost serve as a screen because the central premise is symmetry and movement competency with bodyweight.

I had the opportunity to perform an FMS on the author, who at the time had never before heard of or seen the screen. His score was nearly perfect. Many people try to become as strong as Pavel, but spend little or no time trying to be as flexible. The subtle theme suggested by his collective work is that good movement patterns are a precursor to good strength.

Obviously, one may have greater difficulty developing the pushup or pistol on one side compared with the other, and that is the point. Sometimes the best way to gain symmetry is to follow a nonsymmetrical path. Certain aspects of movement demand greater work, and others may come naturally.

ROTARY STABILITY

People new to the screen are confused when they witness the difficulty of fit people performing the rotary stability test and they question its value and efficacy. Rotary stability testing helps create clarity when we see dysfunction in other less-primitive functional movements like squatting, lunging and the hurdle step. The pushup tests one primitive movement pattern, looking at frontal and transverse plane stability, while the rotary stability test looks at sagittal and transverse plane stability.

The rotary stability test is representative of the first efficient form of locomotion for most humans, the creeping and crawling patterns seen in early development. These patterns demonstrate the same reciprocal movements of the arms and legs used in climbing, walking and running. Rotary stability is more often used as an exercise instead of an evaluation, which is unfortunate because you can learn a lot when you view these patterns in your clients.

We use two different movement patterns to observe stability and movement competence in the quadruped position, the first a unilateral movement and the second a diagonal. The unilateral movement is more difficult since it imposes greater demands on proprioception and reflex stabilization. It provides moderate difficulty for those who depend on strength in the extremities when core stability would be more efficient and effective.

Once you have described the task to your client, no further cueing or suggestions are appropriate. Because the unilateral pattern in quadruped is unfamiliar, it requires body knowledge to solve the weight shifting and coordination required to execute the movement effectively.

This is not an exercise; it is a test of a motor task. The person can either complete the task or not. If your client cannot comply, move to the less-complicated maneuver, the quadruped diagonal movement pattern, which is the next level of coordination and motor control to be tested. In either pattern, consider asymmetries both problematic and dysfunctional.

When teaching workshops, we remind those new to movement screening that it is not necessary to perform this movement perfectly. It's an indication of a problem when the client does the movement perfectly on one side and is significantly limited on the other. This identifies an asymmetry at a very primitive and fundamental level. You should rectify this whenever possible, assuming it represents the lowest score and greatest asymmetry, or lowest score outside of the shoulder mobility and active straight-leg raise tests.

The quadruped diagonal movement is common in both back rehabilitation and core stabilization training. From the all-fours position, the client will lift one arm into flexion and the opposite leg into extension. Remember to watch the movement as a pattern and don't dissect its parts. This means if you see a difficulty in lifting and extending the left leg, don't consider the faulty movement merely a result of weak left extensors. Poor support from the right hip bearing weight could be causing or equally contributing to the awkward movement. There may also be significantly different core stabilization between each pattern.

The quadruped position offers a posture, position and pattern performed less often than walking. When an ambulatory adult has difficulty with full extension in a quadruped diagonal movement, the problem is coordinated movement. In this position, the load against hip extension is far less than the loads during walking or climbing stairs. This should quickly squelch discussions of the need to perform bridges to improve hip-extension strength. The problem is motor control, timing and proper reflex stabilization. The problem is not strength, and basic strengthening will rarely change the pattern.

Adult walking patterns are complicated by years and miles of compensation, substitution, habitual patterns and unique rhythms. In gait, there is so much going on that we can't separate pure mechanics and mobility and stability from the walker's tendencies. Returning to the quadruped position offers insight into the pattern that preceded walking as primary locomotion—its relevance is demonstrated by its position in the developmental sequence. Joint loading, reciprocal movements of the arms and legs, head and neck control and reflex trunk stabilization all provide rehearsal of attributes necessary in walking and running. Asymmetry, limitation and compensation are more obvious and less complicated in the quadruped position.

ROLLING PATTERNS

In the SFMA, we use rolling patterns in breakouts for rotation, flexion and extension. Rolling patterns offer a low-load opportunity to review symmetry and motor control, and are even less taxing than quadruped movements.

A study on adult rolling patterns called *Description of Adult Rolling Movements and Hypothesis of Developmental Sequences*[57] demonstrated that adults with no neurological or physical problems do not display a dominant way to roll from a supine to a prone position.

This suggests it would be hard to standardize a single accepted movement pattern for rolling. Some people lead with head and neck movements, while others lead with upper extremity or lower extremity movements.

It was disappointing that a standard did not emerge, but in fact, too many variables were uncontrolled. I had used rolling for some time in my sports medicine and orthopedic patient practice, and used certain controlling criteria to get the most accurate possible feedback.

First, those with obvious range-of-motion limitations or flexibility problems weren't tested in rolling, because they would certainly use a non-authentic rolling pattern due to compensation for limited mobility.

Second, I separated my observation into upper body or lower body quadrants. Each patient rolled from supine to prone, but with a still lower body, using only the head, neck and arms to do the rolling patterns. The clients rolled to the left and right, and then performed the rolling patterns only using the lower body. Restricting these options exposed their deficiencies. Noting differences in rolling to the left and right, and in initiating from upper and lower, allows us to observe subtle asymmetries.

Third, to gather even more information, I looked at the opposite pattern of rolling from prone to supine. Many of the asymmetries were consistent in patients who had observable asymmetries in other, more complex movement patterns. Here we capture motor control and stabilization at a fundamental level by ruling out mobility as a contributor to the dysfunction.

We use quadruped diagonals in the same way, by separating the upper and lower body. We watch the patient on all-fours perform an arm lift on each side, and then check a leg extension on each side. Finally, we examine the reciprocal movement.

These movements provide unique perspectives for both clients and patients since they are not commonly practiced movement patterns. Moreover, these patterns do not require strength in the prime movers. In fact, those who try to muscle through the patterns are the least efficient and find the activity extremely taxing.

Rolling and quadruped movement patterns are unique observations of fundamental reflex stabilization. When correcting these problems, don't assume practice with sets and reps is the way to affect these. Repetition and practice are important, but in this case, there's nothing to strengthen or condition. The problem is sequence, timing, high-threshold strategies, poor breathing patterns and motor control. Your role is to first help make the patterns possible, and then promote relaxation and effortless control. You can do this using breathing drills, assistance, movement patterning and facilitation techniques.

The primitive patterns observed in rolling and quadruped movements speak to movement intuition. We can ascertain command of movement by watching a person get up from the floor. A developing toddler, a feeble old person or an injured athlete will each forecast movement deficiency in a matter of seconds on the way up off the ground.

This is why the kettlebell get-up, known as the Turkish get-up, and the yoga sun salutation are fundamental exercises that should precede other exercise endeavors. These movements serve as progressions to maintain and build basic movement skills across the lifespan.

Dr. Ed Thomas, a noted Indian club expert, gives a stirring presentation on our successes and failures regarding physical culture. Dr. Thomas discusses the American gymnasium before the invention of basketball, and shows picture after picture of open space with mats, gymnastic equipment, rings, ropes, pegboards, medicine balls, kettlebells and Indian clubs.

One interesting fact obvious in his discussions and pictures is the use of climbing as a form of general exercise and training. Today only climbers climb. Every now and then you might see children on the playground enjoying the freedom of climbing, but liability and paranoia will soon make that risky and irresponsible. Climbing has always been a large component of physical development. In its simplest form, climbing patterns replicate and reinforce rolling and crawling patterns.

We have used this section to elaborate on movement patterns used in the FMS and SFMA, to provide insight and explanation about the unique way each movement pattern contributes to the movement map. We did not go into lengthy discussions of anatomy regarding each pattern, and challenge you to do the same. Do not simply discuss the value of a pattern by the anatomy it engages. Discuss it as a fundamental piece of a complete functional platform—the stuff and authentic movement is built upon.

*Please see www.movementbook.com/chapter9
for more information, videos and updates.*

UNDERSTANDING CORRECTIVE STRATEGIES

COMMON CORRECTIVE EXERCISE MISTAKES

Corrective exercise is applied across many conditioning and rehabilitation programs, sometimes without set rules or systematic control. This is unfortunate, because lack of structure or logical thought processes can be counterproductive. As you scrutinize exercise and rehabilitation programming, consider the following examples commonly employed without an effective movement standard. When caught up in this type of thinking, it's best to learn from the mistake and move on.

These can be broken into four common categorical mistakes.

The protocol approach
The basic kinesiological approach
The appearance-of-function approach
The pre-habilitation approach

The following are some descriptions and examples of each.

The protocol approach is demonstrated when exercise is prescribed based on a general category and does not incorporate the individual appraisal of movement dysfunction. Examples would include general weight-loss programs, sport-specific training and medical protocols for issues like low back pain and other one-size-fits-all programming.

In each instance, the person was put in a category that identified a group or activity, but did not identify the individual level of movement competency. It is possible for two obese people to move differently and have different levels of risk associated with exercise. It is also possible for two soccer players to require completely different exercise programs to best address their respective functional movement needs for soccer. They play the same game, but have different movement-pattern profiles.

Two people suffering from low back pain probably have different mobility and stability needs. Without a basic understanding of the movement patterns associated with each, generalized back pain protocols could potentially put one or both individuals at even greater risk. Back pain is a symptom and should not imply the region of dysfunction. A movement dysfunction should be discovered by a professional rather than be inferred by the region of pain.

The basic kinesiological approach is also used with little regard for movement patterns or motor control. It follows a tidy map targeting mostly prime movers and a few popular stabilizers. If the legs are thought to be weak, strengthening exercises for the glutes, hamstrings, quads and calves are provided. If core weakness is assumed, side planks, crunches, prone extensions and leg lifts might be considered an appropriate remedy.

This approach can provide a general strength base, but fails to incorporate timing, motor control, stability and a full array of movement patterns. It has no system of checks and balances—it only assumes weakness, assigns concentric sets and reps, and waits for some arbitrary change in movement or performance.

The appearance of function approach might be one of the ways functional exercise got a bad reputation among serious strength coaches. The mistake here is less obvious. The appearance of function approach looks at movement, but just uses bands or some other form of resistance, load or challenge to the movement pattern in the hope that light loading will improve both the quality and the quantity.

Some performance examples of this mistake would include tying bands to a baseball to mimic the throwing motion or securing a load to the end of a bat or club and to perform a swing. The mistake is also perpetuated in rehabilitation. When poor single-leg stance quality is observed, it is not uncommon to see exercise done on a wobble board or other unstable surface. The instability does create a challenge to single-leg balance, but if the pattern was already identified as poor in a stable environment, the challenge may cause even greater levels of compensation. The single-leg work on an unstable surface may look functional, but will likely produce compensation in this scenario.

The hallmark of this mistake is the lack of standardized pre- and post-testing. If no movement standards are used, it is possible that exercises are inconsistently imposed. If the exercises are not difficult, it is assumed the exercise is not needed; if the exercises are difficult, it is assumed they are needed. The exercise becomes the test, baseline and confirmation. This situation uses poor logic and creates a slippery slope to irresponsibility.

Clinicians are cautioned not to let the treatment confirm the diagnosis, but to formulate a complete diagnosis before recommending treatment. In this scenario, the exercise professional should be cautioned not to let exercise difficulty confirm itself as the remedy to restore a functional movement pattern. Numerous factors come into play, so baselines without exercise bias should be standard. Standard baselines will reveal the best approaches, just as expert diagnostic testing will reveal the best possible treatments.

The pre-habilitation approach repackages rehabilitation exercises and introduces these into conditioning programs as preventative measures to reduce injury risk. The exercises are not based on actual movement risk factors—they're based on injuries common to particular activities.

An example is the use of rotator cuff rehabilitation exercises in throwing sports. Shoulder injuries in throwing sports can be caused by poor mechanics, as well as mobility and stability problems. They can come from overuse and inappropriate technique. The rotator cuff is often the victim, not the culprit. A little added cuff strength is not likely to change bad throwing habits, exercise habits, poor hip mobility or poor movement screens. Arbitrary strengthening of the rotator cuff musculature should not be assumed corrective or preventive unless a true weakness has been identified. This

work can often perpetuate an unfounded sense of insurance or protection without validation. Many other regional factors can contribute to shoulder injuries, including the dysfunction in the cervical spine, thoracic spine, AC and SC joints and scapular stability.

GETTING SOMETHING FOR NOTHING—MOVEMENT ALCHEMY

There is a hint of alchemy—profound transformation, getting something for nothing—in the proper application of corrective exercise strategy. It occurs when you put focus on one movement pattern, but changes are noted in other movement patterns not prioritized, essentially unaddressed. This is why it is necessary to perform a complete rescreen or reassessment when significant improvement is noted in the pattern of corrective focus. It almost seems as if the brain and body skip a step and repair or reacquire a degree of other patterns without a direct corrective nudge.

The secret to getting this to work is to work on the most fundamental problem. You can easily identify the fundamental problem when you use the Functional Movement Screen (FMS®) and Selective Functional Movement Assessment (SFMA®) rules of prioritization. In most cases, you can correct the basic problem if you use the proper corrective framework. The following sections will discuss corrective prioritization and framework.

SCREEN AND ASSESSMENT CORRECTNESS

To establish effective minimums, you must perform your screens and assessments correctly. If you take shortcuts or are distracted in your observations, they won't reliably direct corrective exercise choices. Make sure you don't simply agree with and support the FMS and SFMA as your movement systems—you need to own them. It's simple: Practice and practice some more.

You need to see movement, lots of movement. We recommend a minimum of 20 screens or assessments before you use these on your clients or patients, depending on your professional credentials and scope of practice. But don't stop there, practice until you feel smooth. You have no obligation to correct anything—it's just practice. Remember, when you do the FMS or SFMA, your first goal is to be a consistent technician.

The three biggest and most frequent mistakes in the FMS and SFMA are—

- Trying to convert movement dysfunction into singular anatomical problems, such as discussing isolated muscle weakness or tightness

- Obsessing over imperfections in each test instead of using the test to identify the most significant limitation or asymmetry

- Attempting to link corrective solutions to movement problems prematurely during basic data collection

Your first goal is to profile the movement of your subject without responsibility to change anything, because that burden will compromise your technical skills as you learn to screen and assess movement. That responsibility will come once you can produce consistent and reliable data from your movement tests. Be a competent technician before you try to be the master. Until you understand how the screens and assessments compartmentalize movement patterns, you will not have a gauge for corrective exercise success or failure.

A rush to correction causes another popular mistake seen in corrective exercise. This is usually the novice trying to correct a movement pattern that was incorrectly scored or erroneously deemed dysfunctional. It all goes back to movement.

The FMS and SFMA mentors with a firm hand. Even though our team developed these screening and assessment methods, we had no way of knowing every possible outcome when corrective exercise was used. Some exercises worked; some did not. We needed to refine our corrective exercise rules against the movement standard. Once we set a standard and work to it, we set ourselves up for feedback—both good or bad.

Learn to take aim; learn to take focused and deliberate aim. Learn to reproduce your ability to aim—then shoot. Don't randomly shoot your corrective exercises and hope your aim was true. Know your aim is correct; it is your professional responsibility. The screens and the assessments should run smoothly, and you should be correct and confident in your collection of the information.

Don't fall victim to the *ready—shoot—aim* approach to exercise demonstrated by pre-packaged programs. Those are based on anatomical parts and regions. You are now considering exercise and rehabilitation based on movement patterns as well as anatomical structure, and your command of movement patterns must be as good as your command of anatomy.

BEGINNING WITH EXERCISE BASICS

We've reviewed the wrong corrective exercise approaches, so what's the right one? Now that you have a basic understanding of screening and assessment and have hopefully run through the movement tests with colleagues, what do you do with the information? The movement pattern tests identified dysfunction and build the movement map, and now it's time to develop effective corrective strategies.

Using movement screening and assessment, some readers will reach a dilemma about previous exercise choices. When screening and assessment reveal that a movement pattern is dysfunctional, that pattern should not be exercised, practiced or rehearsed—especially with loading, impact and under resistance. The movement should first be broken down and reconstructed with corrective exercises designed to work on the aspects of mobility and stability that support the complete movement pattern.

This may interfere with some exercise programs and rehabilitation protocols, but you cannot achieve the full benefit of either on a platform of dysfunctional movement patterns. A temporary detour can bring out resistance and frustration in those individuals focused on fitness or performance goals, but most dysfunctional movement patterns can change in a week or two. Sometimes it will take longer; it takes as long as it takes, and it is what it is.

It requires professional diligence to take a stand and explain that these movement conflicts will not just work themselves out. Merely exercising is not good enough. It may have once been, but we are generations away from the lifestyle and movement base that could handle the increased activity load without breakdown or elevated risk of injury.

These corrective exercise concepts make all exercise options and choices subordinate to fundamental movement patterns. Screening the most fundamental problematic movement pattern will confirm or refute your corrective choices. This does not suggest that you should discard your programs, but instead to incorporate these new ideas into your existing programs.

The one thing we do not want to do is to put strength or fitness on dysfunction. In the end, this is actually more time-consuming than taking a break from strength and conditioning in favor of corrective exercises. Putting fitness on dysfunction can potentially increase the chance of injury. It also could contribute to an injury's severity once it happens, and it will probably slow down the corrective process.

Common training programs assume general kinesiological principles apply to all trainees. They target a specific movement or fitness attribute and operate as if a general universal movement platform supports the agenda, but it doesn't. The same exercises can impose completely different and unpredictable stresses on people if no screening or assessment first qualifies or disqualifies them for each movement.

This section will provide the foundation to hold off contemplation of the effectiveness of an exercise until first clarifying the situation, individual or group using it. This means using data to identify movement proficiency, risk factors and performance baselines when applicable.

Once this clarity is present, divide all exercise into three distinct categories—

- Exercises designed to restore movement patterns and remove movement-related risks

- Exercises designed to improve physical capacity and performance

- Exercises designed to improve skills

These categories are interrelated; each will influence the other. For the sake of argument, it is possible to improve your flexibility, grip strength and swing speed by practicing the golf swing. Swinging a golf club is considered specific-skill training, but it can create positive changes in movement and performance. Unfortunately, it can also create opportunities for compensation and cause problems as well.

Baselines on each level of function will make all exercise choices more objective. This means you can work on aspects of movement patterns, performance and skill simultaneously, but you need to be constantly aware of problems and deficiencies. To simply, just apply this basic rule: Only work on performance and skills that use movement patterns free from dysfunction or limitation.

Next we'll look at how those categories put structure to our training programs.

THE PERFORMANCE PYRAMID

The performance pyramid is a diagram constructed to offer a mental image and understanding of human movement and movement patterns. There are three rectangles of diminishing size, one rectangle building upon another, each representing a certain movement type. The quality pyramid must always be constructed from the bottom up and must always have a tapered appearance—a broad base and a narrow top.

The first rectangular pillar is the foundation, and represents the ability to move through fundamental patterns such as squatting, lunging and stepping, disregarding performance and physical capacity. The only focus is movement quality.

The second rectangular pillar depicts functional performance. Once we've established the client or athlete's ability to move, we look at movement efficiency. This efficiency represents power in this movement image.

In active groups, we can measure this type of power through activities such as pushups, situps or even the vertical leap, as well as other norms for running efficiency and weight lifting. The tests should always create a baseline of the physical capacity needed for activities, occupations or athletics.

These tests assess a person's generalized physical performance level compared with norms previously established. These are not intended to appraise specific skills, only physical capacity against other individuals in the same group or participating at the same activity levels. Examples of these would be tests for endurance, strength, speed, power, quickness, agility and coordination.

It is very important from a training standpoint to be able to compare individuals of different performance areas in a general format. The first two rectangular pillars allow us to make this comparison of functional movement ability and power, so active people can learn from each other concerning different training regimes. Moreover, we don't get task-specific with testing at this level of the performance pyramid. Task-specificity at this point reduces the ability to compare individuals with one another, as well as the ability to learn from these comparisons.

It is also important not to perform too many tests at this level. Statistically, this concept is known as the *Law of at Least One.* This refers to the statistical fact that the more diagnostic tests you perform looking for the same target disorder, the more likely you are to obtain a false finding. Simply perform a limited number of appropriate tests in each category. The more tests you do, the more you can overanalyze or even bias the person's performance in one direction or another.

The purpose is to get an overall assessment of the individual's abilities. Use a few simple movements to illustrate a person's efficiency and physical capacity in the categories you need—power, speed, strength, endurance and agility.

The last pillar of the pyramid represents functional skill. This pillar constitutes a battery of tests to assess the ability to perform certain functional skills.

In industry, this may be the performance of job-specific tasks, or in sports, it might be reviewing the attributes of a particular position. The idea is to compare normative data related to the specific skill tests.

The performance pyramid is only a map, designed to give a direction with which to categorize and identify areas of weakness. Examine the four basic appearances of the pyramids on the following page. These are simple generalizations, but each represents how the pyramid can guide the entire evaluation, and eventually even the conditioning program.

THE OPTIMUM PERFORMANCE PYRAMID

Functional Movement   Functional Performance   Functional Skill   Buffer Zone

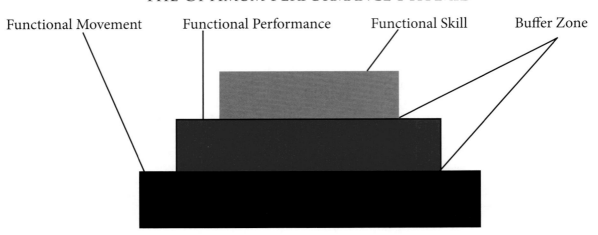

THE OVER-POWERED PERFORMANCE PYRAMID

Functional Movement   Functional Performance   Functional Skill   Buffer Zone

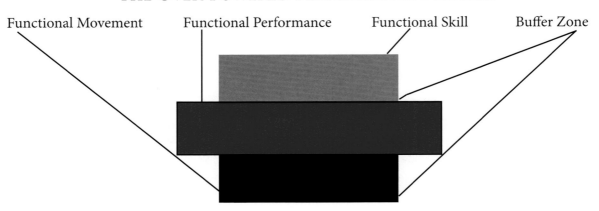

THE UNDER-POWERED PERFORMANCE PYRAMID

Functional Movement   Functional Performance   Functional Skill   Buffer Zone

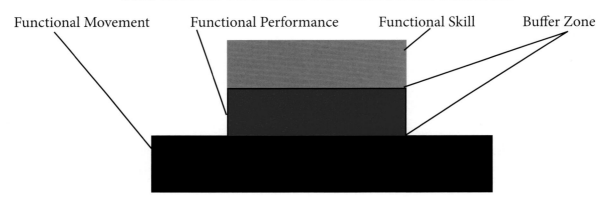

THE UNDER-SKILLED PERFORMANCE PYRAMID

Functional Movement   Functional Performance   Functional Skill   Buffer Zone

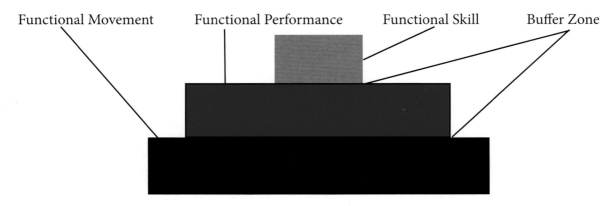

THE OPTIMUM PERFORMANCE PYRAMID

The first pyramid is the optimal pyramid, which represents a person whose functional movement patterns (demonstrated by the FMS), functional performance (demonstrated by performance testing) and functional skill (demonstrated by functional skill testing, sport-specific testing and production statistics) are balanced and adequate.

This does not mean the individual cannot improve, however, any improvement should not upset the balance of the performance pyramid. This broad-base representation demonstrates appropriate or optimal functional movement, the ability to explore a full range of movement and demonstration of body control and movement awareness throughout numerous positions. The first level of function—movement—is appropriate to support the other levels of function.

On the next level of an optimal pyramid, the person demonstrates a requisite amount of physical capacity. When compared with normative data, you should see average or above-average general performance capacity. This means well-coordinated linking movements or kinetic linking, and implies during a test such as the vertical leap, the individual loads the body in a crouched position, throws the arms, slightly extends the trunk and finally explodes through legs in a well-timed, well-coordinated effort presenting optimal efficiency.

This displays the potential to learn other kinetic-linking and power-production movements with appropriate time, practice and analysis. Efficiency can also be considered in activities less concerned with power and more concerned with endurance.

The third rectangular pillar of functional skill demonstrates an average or optimal amount of task- or activity-specific skill.

Note how the broad base creates a buffer zone for the second pillar, and the second pillar creates a buffer zone for the top pillar. This zone is extremely important; it implies that the individual exceeds the necessary mobility and stability needed to perform the specific tasks. Without the buffer, there may be potential for injury or for compromised power and efficiency. This buffer shows that the functional movements are more than adequate to handle the power generated. Between the middle and the top of the pyramid, the power generated can more than control the skill possessed.

THE OVER-POWERED PERFORMANCE PYRAMID

The second pyramid is a graphic demonstration of people who are over-powered. This does not mean they are too strong—it only means their ability to generate power exceeds the ability to move freely within fundamental movement patterns.

This pyramid provides a visual representation of a person who scores very poorly on mobility and stability tests, but very high on the second pillar of power production and adequately in skill, the third pillar. The way to rectify this problem is to improve movement patterns while maintaining the current power.

An individual with these characteristics lacks the ability to move freely because of limited flexibility or stability in some of the movement patterns. This produces a less-than-optimal functional movement score that would appear as a smaller rectangular pillar at the base.

This performance does not really offer the appearance of a pyramid, because of the functional movement base and the power block are inverted. This person is generating a significant amount of power with many restrictions and limitations in functional movement. The profile can easily demonstrate less-than-optimal efficiency where valuable energy is wasted overcoming physical stiffness and inflexibility.

Many highly skilled and well-trained people will replicate this performance pyramid when their performance is evaluated. An example of this would be an individual who displays tremendous strength and power in traditional weight training movements such as the bench press and squat, yet is unable to perform functional movements without compensations. This person may have never experienced an injury and may be performing at a high level, but the best focus for training would still be on functional movement patterns.

This focus would remove the limitations to functional movement, which would provide a broader base to the pyramid and create a greater buffer zone. There may not be an immediate tangible improvement in performance. In fact, task-specific performance and power production might remain the same or decrease slightly as mobility and stability improve. However, it is unlikely this person would improve in general power production or task-specific skills to any large degree without first improving general basic movement patterns. Whether you're targeting functional movement patterns for injury prevention or as a way to realize untapped performance through higher levels of efficiency, your client or athlete will eventually see improvements.

THE UNDER-POWERED PERFORMANCE PYRAMID

The third pyramid represents a graphic of an underpowered person who has excellent freedom of movement, but whose physical capacity is poor and needs improvement. Here we would plan training and conditioning to improve efficiency, endurance and power without negatively affecting the movement patterns.

This person demonstrates a broad base and optimal movement patterns with very poor power production at the second level, while showing optimal or above average skill in a specific movement. The individual has the requisite movement patterns to perform multiple tasks, activities and skills, but lacks the ability to produce power in simple movement patterns. The most beneficial program for this classification is power, plyometric or weight training.

It is very important to maintain functional movement patterns while working on strength, power, endurance and speed. This power reserve will create the buffer zone for task-specific skills, while still improving efficiency.

Consider the example of a firefighter who has extremely good mobility and stability, and has honed work skills through practice and expert instruction. This person must use very high energy expenditure in order to perform at high levels for a short period, but does not need to be on a mobility or stability program. The firefighter probably does not need to work on improving certain tasks specific to firefighting, but instead should create better strength, power and endurance reserves, thereby improving overall physical capacity.

Such interventions would create a buffer zone between the pyramid's second and third pillars. This would allow performance at the same level of effectiveness with higher efficiency or lower energy expenditure. This should provide greater effectiveness and durability during performance of specific tasks.

The profile displayed by the underpowered performance pyramid probably explains a second level of injury risk. Movement screen scores below a cut point are associated with greater risk of injury, however higher movement screen scores don't seem to suggest lesser risk. This seems to indicate that once favorable movement quality has been achieved, physical capacity and technical skills might also play a factor in risk.

It also seems to indicate that when poor movement quality is present, physical capacity and technical skills are of little influence. This supports the suggestion that the performance must be constructed from bottom to top.

THE UNDER-SKILLED PERFORMANCE PYRAMID

The last pyramid is a graphic demonstration of people who are under-skilled. This is a situation of adequacy in the first two blocks of the pyramid— movement pattern and efficiency or power generation. However, skill analysis demonstrates an overall inability to produce the desired outcome or mastery of skills needed to perform the tasks. People in this category are well conditioned, but are not appropriately skilled.

In this case, a training program specifically designed around skill fundamentals and techniques would be the best time investment. This would develop a greater awareness of the movements needed to perform the skill with more efficiency since a fundamental base has been proven.

USING THE PERFORMANCE PYRAMIDS

During training, the performance pyramid will continually change for some people; for others, it will remain the same. Some people will have the natural ability to generate power, but must consistently work on functional movement patterns to maintain optimal freedom of movement. Others will have excellent natural freedom of movement and movement patterns, but need supplementary training to maintain power production. Finally, some need consistent work on the fundamentals and on task-specific skills, while others naturally gifted with certain skills should invest their time in other types of conditioning.

The testing proposed in this book provides a method to acquire information to construct this performance pyramid. Memorizing this graphic representation will assist in identification of weak areas on which to focus training.

The performance pyramid explains why replicating the program of one person will not consistently yield the results it does for another. Many physical therapists, athletic trainers, coaches and athletes intuitively use this approach to identify the performance area of greatest weakness. For those who don't possess this intuition, the performance pyramid is a simple and effective graphic to evaluate and illustrate body balance or lack thereof. The graphics will also aid in communicating this thinking process with your clients, athletes or patients.

SHORT-TERM RESPONSE VERSUS LONG-TERM ADAPTATION

A response is a reaction or the sum total of reactions to a stimulus, training or treatment. For our purposes, consider it a temporary adjustment in physiological function or movement function brought on by a single exposure to exercise.

A physiological example of a response is the general cardiovascular warm-up that increases cardio-respiratory activity. A movement example might be a movement pattern made less awkward and more fluid following a few mobility, stability or patterning exercises.

An adaptation is an adjustment of an organism to a change in the environment. This might mean a persistent change in structure or function caused by repeated bouts of exercise, training or rehabilitation.

Most of us exercise with a focus on our quantities. We look at times, distances and weights. We rarely think about the difference in responses and adaptations. We do the work and automatically anticipate the adaptation—the change in the body—and miss the opportunity to see the immediate responses, especially in movement quality. We don't think to look at the responses because these are only temporary. The things they represent are short-lived and will fade quickly after the training session.

We hold the adaptation in much higher regard: better flexibility, muscular hypertrophy, fat loss, improved VO_2 max and a reduced resting heart rate. These are important and should always be tracked and measured, but they should not overshadow the responses. We consider responses in temporarily metabolic terms, but in neurological terms they mean everything—positive movement-pattern responses are favorable indicators of motor learning. Exercises that produce poor movement-pattern responses reinforce poor movement-pattern adaptations.

Focus on technical quality creates a greater neurological demand. It forces better and faster connections. If we exercise against qualitative minimums and push performance quantities, we should see improved movement responses in a single session. This will lay the groundwork for refined and efficient movement-pattern adaptation.

How could favorable adaptations be developed from unfavorable responses? In reality, adaptations are built on repeated positive responses. This is the hallmark of corrective exercise. Without multiple positive movement responses, it is unlikely for adaptations to occur.

Exercise science neglects this because of the biased emphasis on physiological responses and adaptations to exercise. Generalized physiological capacity seems to improve as an automatic response to repetitive exercise sessions. Movement pattern improvement is a little more technical and is unreliable without a movement standard because compensation can and will occur when training exceeds movement potential.

Screening reveals that not all exercise positively affects functional movement patterns. And it shows that balance, mobility, stability and movement patterns can be unchanged or might even worsen following a training session. This demonstrates how we're able to put fitness on dysfunction. It also explains how physiological capacity can improve, while at the same time functional movement patterns regress.

Let's look at how I restructured my thinking about functional movement-pattern correction because of that revelation. Breaking down movements with basic kinesiology after uncovering poor movement patterns in conventional exercises identified the prime muscles responsible for the movement in order to exercise them—usually in isolation.

For instance, seeing poor single-leg-stance quality, the initial response was to look toward a weak glute medius. As per my training, my clients and patients did hip abductions against gravity, in side lying or against an elastic band.

At the time, we assumed motor programs would reinstall themselves, that we could strengthen muscles, and could thereby change their timing and coordination. In PT school, it was taught but not directly stated that poor timing in the deep stabilizers could require superficial prime movers to work excessively, unnaturally producing increased tone. Observing the tone as tightness implied it would diminish after a few minutes of static or dynamic stretching.

Mobility and flexibility work is necessary to break poor movement patterns, but mobility and flexibility must be reworked or they will quickly reset to their default setting. When we observe a change in mobility or flexibility, we create progressive situations where the new range is used. This was a tricky step because half the time this works and half the time it doesn't.

That the corrective strategies used weren't working consistently was demonstrated by checking movement patterns at the end of exercise and rehabilitation sessions. This was not about inventing a new system—it was about getting feedback.

The difference between the clients and patients who improved a movement pattern compared with those who did not pointed out the lack of monitoring the challenge factor of the exercises. Exercises that presented too much difficulty forced the people to revert to the compensation pattern. Exercises people could do yet still presented a challenge produced much better outcomes.

CORRECTIVE EXERCISE PROGRESSIONS

The first set of exercises following a change in mobility tells us all we need know, producing one of three responses—

It is too easy: The person can perform the movement more than 30 reps with good quality.

It is challenging, but possible: The person can perform the movement 8 to 15 times with good quality of movement and with no sign of stress breathing. Between 5-15 reps, however, there's a sharp decline in quality demonstrated by limited ability to maintain full range of motion, balance, stabilize, stay coordinated or just becomes mentally or physically fatigued.

It is too difficult: The person has sloppy, stressful, poorly coordinated movement from the beginning, and it only gets worse. It's nearly impossible for the person to breathe, to relax and move.

Using this as a corrective exercise base, you can now observe the response and act accordingly. If the initial exercise choice is too easy, increase the difficulty, observe the response to the next set and repeat the process.

If the initial exercise choice is too difficult, decrease the difficulty, observe the response to the next set and repeat the process.

If you get lucky and the initial exercise choice is challenging but possible, try not to act surprised.

INCREASING DIFFICULTY

Increased difficulty rarely means increased resistance when discussing corrective exercises. A more advanced posture, a smaller base or a more complex or involved movement pattern usually indicates increased difficulty. A typical example would be some form of rolling movement pattern moving to a quadruped exercise, then going to a half-kneeling activity and finally using a single-leg stance movement.

When you look at human performance and expertise, you'll realize the best of the best practice things differently. They are masters at creating opportunities for deliberate practice. In *Talent is Overrated,* Geoff Colvin describes deliberate practice as much more than simple repetition and rehearsal. Deliberate practice is the act of repeating actions subject to consistent, specific and objective feedback.

Colvin establishes that people we commonly refer to as talented are not born that way. Their talent is knowing how to practice correctly. These people are known for their talent, but we overlook their unique style of preparation, the most important aspect of their development. They simply create opportunities for better feedback and they expose themselves to the subtle mistakes that most of us disregard.

The human central nervous system requires deliberate practice. The wobbliness that an individual might experience in half-kneeling with a narrow base will require a blend of conscious and reflex activity. The vestibular, visual and proprioceptive systems will quickly work to avoid a loss of balance. If balance is maintained, the feedback is positive, obvious and immediate. If balance is not maintained, the previous sequence and stabilization strategy will be modified.

The central nervous system seems to like deliberate practice, a model infants apply better than some coaches, trainers and rehabilitation professionals. Infants don't just rehearse movements, they tinker with them until they achieve a set desired effect.

ASKING THE RIGHT QUESTIONS

What do you do for speed development?
What are your favorite exercises for low back pain?

Much of our confusion and professional disagreement could be traced back to attempting to answer poorly formulated questions. How can we effectively answer those questions when we do not know the person who will receive the exercises? After all, many things limit speed, and initially removing the barriers to speed is more important than generalized speed development exercises. Data from three different athletes' movement screens might suggest three different plans to develop speed performance.

Likewise, back pain is a symptom caused by issues far more complex than a weak or stiff back. If there was a set program for all low back pain suffers, we would have found it by now and low back pain would be rare.

If you consider the questions, you'll realize the problem is the incorrect focal point. A lack of speed or the presence of back pain is only a situation. It should not be the focal point—the one experiencing the situation is the focal point. Both questions concern a general situation and assume it is applicable to all people within the category.

These two situations cannot be fixed directly or reliably with general exercise, but their presence can lead us to screens, assessments and other measurements that will reveal the cause, source and complications. Once we identify the complications, we can address them. The answer is not in the exercise. The answer is in the system that ruled out all but a few exercise options that fit the individual's needs.

This is the fundamental concept in movement screening and assessment often overlooked on the first pass. The role of the screen or assessment is not to pinpoint a single corrective exercise—the role is to remove all potential corrective considerations and narrow the choices to a select few. The select few will have varying levels of challenge within a specific movement pattern continuum.

The system mentors the novice professional, and as the professional becomes expert, exercise selection becomes second nature. The system is always there to check the work, but it's not paint-by-numbers since multiple correct approaches can effectively be applied. The approaches may be technically different, but are categorically the same.

It would be impossible to cover the specifics and special interests of each exercising population. The questions will continue forever, but the problem is the questions, rather than the lack of answers. To discuss exercise, the responsible professional needs to know the situation, specific goals and problems, as well as the person's current functionality.

Instead of attempting the impossible, we present the common foundation that all exercise, rehabilitation and physical training should integrate before specialization or special interests. These foundations and principles are authentic and naturally occur before specialized skill.

WHAT EXERCISES TO REMOVE FROM THE PROGRAM

In all cases, a movement screen or movement assessment makes two suggestions. It first suggests what you should remove from the current exercise program or list of activities. This is simply because it could be counterproductive or even causative of the current movement dysfunction. The second suggests what should be added to the program in the form of corrective exercise.

If screening or assessment have helped you identify dysfunctional movement patterns, we must assume they were caused, overlooked or allowed to develop for some reason. Regardless of the cause, they are present now, were probably not present in the first five, 10 or even 20 years of life, and must now be addressed.

The first step is to stop activities most likely connected to the dysfunctional pattern. If single-leg stance and hurdle stepping are dysfunctional, discontinue running, jogging and single-leg work until you see improvement in the basic functional pattern. Do not twist this message. It does not imply that these activities are unfavorable or even that they contributed to the dysfunction. It simply means any attempt at training a dysfunctional pattern with exercise geared toward conditioning would be counterproductive in regaining movement pattern integrity, and may actually increase risk of injury.

Unfortunately it is likely that favorite activities and exercises might be temporarily restricted for some clients, athletes and patients. The beloved activities and exercises can resume as soon as the movement pattern becomes acceptable.

How quickly the patterns change is usually up to the individual. How much will a client want to work on the pattern? How dedicated and focused is the athlete going to be? Is the patient willing to do some homework? The question is not how long will we have to stop their favorite activities; rather it is how diligently will they work on their weakest links so these can be resumed.

Use the screen and the assessment data to decide which activities to delete. Stop all loading and heavy resistance work with the shoulders and arms if the shoulder mobility patterns are severely limited or asymmetrical. If the squatting pattern is limited, do not allow squatting with load. That means no partial range-of-motion work, and no leg pressing. It sounds simple, elementary even, but instead of stopping exercises, many of your clients and patients have fought through the pain, only to get beaten down by another injury.

Coaches and trainers often assume attention to detail and technical suggestions will always work, but good coaching cannot overcome fundamental movement dysfunction. Technical instruction in exercise and weight training should only be used when basic mobility and stability are proven to be present and available. Once established, we can use and coach these attributes into new movements and exercises, but they will not spontaneously appear in a complex movement if they are not present in a basic movement.

This is why both screening and assessments use regressions to redevelop movement. In the FMS corrective strategy, basic mobility is established and followed by unloaded or assisted movements, and then by static and dynamic stability. This provides the body's movement systems a chance to redevelop lost or inactive connections.

The screen and assessments create a two-pronged suggestion. They suggest corrective strategy for movement patterns that are dysfunctional and not painful, and identify and allow continued activities in movement patterns not compromised by pain or dysfunction.

As you read this corrective exercise section, remember the order of importance. The first priority is the movement pattern exercises to avoid. Remove all related exercises from your conditioning programs temporarily as you rebuild the dysfunctional pattern. Both adding exercises and techniques and removing problem patterns are required in a complete corrective strategy.

In most cases, you should be able to see some degree of improvement in a dysfunctional movement pattern after the initial exercise session. Some will not become completely functional, and a few will not change at all, but as long as slow change is noted, do not stop. Most movement dysfunctions develop over time, and we have to give them time to change.

CORRECTIVE EXERCISE IS SUPPLEMENTAL

We should always consider corrective exercises as supplemental and temporary. The end goal is to restore movement to acceptable levels and to design exercise programs that maintain movement quality while addressing performance levels without needing ongoing corrective strategies. Obviously, they are indicated if chronic problems are present or if ongoing problems persist, but these should be mostly supplemental. They should produce changes in movement patterns, but if their continuous use is required to maintain movement patterns, consider the saying by Ben Franklin, "An ounce of prevention is worth a pound of cure."

Don't let corrective exercises become the pound of cure when correct approaches could essentially remove the need for continuous correction. Poor exercise choices, as well as exercises performed poorly, could be the major contributing factor to movement dysfunction in the first place.

Yoga and kettlebell training are examples of movement forms that require expert instruction, patience and personal investment. These and other forms of training that require mindful movements are the extreme opposite of mindless cardio-circuits and the eight-minute-ab mentality. These high technique forms seem slow at first. They can be frustrating and may lack the vigorous appeal of a high-intensity exercise class or a fitness boot camp, however the investment pays dividends in the end.

By learning movement skills that require a mental and physical demand, fitness is a byproduct of learning higher quality movement. Some of the fittest people in the world don't obsess about their exercise time slot—they don't require loud music or mirrors to motivate them. They simply practice movement skills, knowing they will never master them. They use exercise correctly and they stay in touch with movement. Exercise correctness is not a popular topic, but is a much needed perspective.

EXERCISES: CORRECT VERSUS CORRECTIVE

Corrective exercise becomes a popular topic once you are familiar with the concepts of movement pattern dysfunction as demonstrated by screening and assessment. Corrective exercise is probably the best remedy for movement pattern dysfunction, but it is not the best preventive measure. If we constructed and taught better exercise techniques, we could help prevent much of the need for corrective exercises and reserve corrective concepts to situations where rehabilitation and post-rehabilitation are necessary.

Corrective exercise is used also to undo the effects of poor exercise choices and premature physical decline due to the lack of or even excess of physical activity. Injuries and imbalances will always be present, but we can minimize the need for corrective exercises in all other forms of physical training. By designing exercises around outcomes that yield performance and adequate functional movement patterns, we'll see higher function and simultaneously enjoy reduced injury risk.

To that end, we should first develop guidelines for exercise and athletic conditioning that meet the goals of performance while managing risk. This way we can take the information that functional movement screening and assessments provide and become proactive in influencing future screens through better initial program design.

Before we review corrective exercise measures to resolve movement pattern problems, let's make steps toward a better exercise model that will not create movement problems in the first place.

CORRECT EXERCISE

Search for quick fitness fixes and you will find a strategically priced product, a fad or a temporary trend. All of us have seen them; we've shaken our heads at the absurdities, and sadly enough, some of us find ways to endorse them. Seek super athletic performance and you will find the drills and activities professional and elite athletes perform, but this requires everything else the professionals do if your clients want to attempt the same.

Correct exercise is a concept dedicated to the elimination of exercises that are not good choices to meet the goals of both performance and durability. It is not just doing a fancy high-tech exercise you picked up at the last professional meeting, convention or off the internet. The drive to find the right exercises first strips away the unnecessary ones. Correct exercise choices do not need to be supported by supplemental work. They are designed to target weak links and biomarkers that establish reduced risk and higher performance.

Historically, exercise supported endeavors for aggressive and defensive activities or survival situations. These activities eventually evolved into sports and hobbies that require exercise and training. Today, exercise has become as a ritual unto itself; it's no longer solely a path to athletic perfection or superior physical culture. People do it as a pastime without defined goals, to offset the sedentary lifestyle and perhaps to generate a few random endorphins while mostly keeping guilt and gut at bay. The goals, if there are any, are often based on one-dimensional markers of physiological or athletic performance. When exercise is done as a sport or for fun, one can only gauge performance against other forms or levels of exercise.

Training and exercise produce both tangible and intangible qualities. Physical training has classically been used to improve a physical skill set. History has also observed practices such as yoga, dance and even distance running as rites of passage and forms of meditation, escape and reconnection with one's self or with a group.

In one instance, the exercise investment has a specific desired outcome: We want to run faster. If we do not improve speed at an efficient rate, we abandon the exercise choices and seek better training. In another case, the benefits of exercise may have tremendous personal value.

But these cannot be quantified against a physical standard. Exercise should demonstrate higher physical perfection and durability, or foster well-being and physical recovery.

Ancient warriors quickly decided upon the best training methods, and friendly competition sprang from their preparation. Early practitioners of yoga did not ask about the flexibility or slimming effects of the movements—they focused on the breathing and enjoyed the benefits of the daily moving meditation. Those who ran did not need a stopwatch to connect with the accomplishment—they ran to survive, compete or reconnect.

Today, we simultaneously expect reconnection, great results and healthy competition every time we train. We do not know how to make the best training choices for health, fitness and athletic goals or effectively measure the outcomes. We also don't know how to enjoy exercise for the sake of personal reconnection and well-being. Usually there's only time for one exercise endeavor, so we let our guilt and time constraints push us into a single specific activity, forgetting that it should be pleasurable or productive and offer some variety. We have tried to make the ultimate hybrid activity by combining these, and we have done so to the detriment of both.

The way out of this trap is to first set physical goals and design the training toward those goals, and second, to make time for a healthy diet of movement that will re-energize and reconnect us. We must separate the two to maximize the benefit of each, but once we fully understand each, they can often be merged.

For our purposes, we'll focus on the objective exercise standards designed around the physical goals of enhanced performance and durability. The next great step in exercise and training will be in the field of recovery. We can only maximize the best exercise choices if they are paired with the most effective recovery. For now, as you review your programs against your knowledge of movement screening and exercises goals, recognize that the best forms of recovery are practices that yield normalization of physiological parameters and movement patterns.

Hard training and competition can lower movement screen scores temporarily because these endeavors cause us to push our limits. Without resetting our movement system, we can compound the negative effects of hard training and competition without being aware of a decline in movement-pattern quality. Appropriate recovery can accelerate a return of motor control, mobility, posture and muscle and tissue extensibility.

SELF-LIMITING EXERCISE— NATURALLY CORRECT EXERCISE

Self-limiting exercises make us think, and even make us feel more connected to exercise and to movement. They demand greater engagement and produce greater physical awareness. Self-limiting exercises do not offer the easy confidence or quick mastery provided by a fitness machine.

The earliest exercise forms were self-limiting— they required mindfulness and technique. Idiot-proof equipment and the conditioning equivalent of training wheels did not exist. Great lifters learned to lift great; great fighters learned to fight great; great runners learned to run great. Their qualities and quantities were intertwined.

Self-limiting exercise demands mindfulness and an awareness of movement, alignment, balance and control. In self-limiting exercise, a person cannot just pop on the headphones and walk or run on the treadmill, fingering the playlist or watching the news on a well-placed monitor. Self-limiting exercise demands engagement.

The clearest example of self-limiting exercise is barefoot running. While running barefoot, the first runners connected with the sensory information in the soles of their feet. This works perfectly—this is the very reason the soles of the feet have such a uniquely dense distribution of sensory nerves. This provides a window to our environment, like the nerves in our hands, eyes and ears. The information provided by sensory nerves in the soles help all who walk on two feet continually adjust their movement, stride, rhythm, posture and breathing to meet changes in the terrain.

The modern running shoe allows us to ignore a sensory perspective of running that is only second to vision, and, as you know, the increase in running-related injuries paralleled running shoe development. When running barefoot, over-striding and heel striking is not an option—it produces jarring, discomfort and pain because it is not authentic. Is it not a bit peculiar that the quick twinges of pain refine the barefoot runner's stride to help avoid running injuries, while the comfort of the modern running shoe later exchanged those friendly twinges for debilitating pain?

The modern runner uses braces to cover a weakness, often not taking responsibility to rehabilitate a problem, or dissatisfied with the rehabilitation process and its incomplete outcome. Christopher McDougall reveals this concept in an amazing story in his book *Born to Run: A Hidden Tribe, Superathletes, and the Greatest Race the World Has Never Seen,* a story that reminds us to temper all technologic advancements against historical facts and time-tested principles. He touches on medical and biomechanical issues, prehistoric man, exercise concepts and a detachment from the joy of movement we exchange for superficial results.

This book is highly recommended for trainers, coaches and rehabilitation professionals to help them see their respective professions through the eyes of the inquisitive, chronically injured runner. Christopher's investigation and story connects important dots we can all appreciate. In his journey, he discovered rehabilitation and coaching wisdom that is logical and simple. The problem is that he had to dig to find it. Part of his digging was caused by our incomplete practices of movement assessment, exercise and rehabilitation.

Examples of other natural, self-limiting categories are governed by breathing, grip strength, balance, correct posture and coordination. Some exercises combine two or more self-limiting activities, and each has natural selective and developmental benefits. These exercises produce form and function while positioning the entire movement matrix for multiple benefits. As we train movement, anatomical structures model themselves around natural stresses.

Self-limiting activities should become the cornerstone of your training programs, not as preventive maintenance and risk management, but as movement authentication—to keep it real. The limitations these exercises impose keep us honest and allow our weakest links to hold us back, as they should.

Used correctly, self-limiting exercises improve poor movements and maintain functional movement quality. These exercises are challenging and produce a high neural load, which is to say they require engagement and increased levels of motor control at the conscious and reflexive level.

Anytime we don't acknowledge our weakest links or confront them in training, we demonstrate the same behavior that caused our collective functional movement patterns to erode in the first place. Embedded in each workout, the self-limiting activities continually whisper the message that we cannot become stronger than our weakest links.

EXAMPLES OF SELF-LIMITING ACTIVITIES

These are only a few suggestions provided to spark your own imagination.
Each of the examples should be performed for technical correctness, not to failure and not even to fatigue.

PAIN OR DISCOMFORT TO LEARN BODY MANAGEMENT		
Balance Beam Walking	Climbing Activities	
Barefoot Running and Training—Pose, Chi or Evolution Running	Farmer's Carry	
BREATHING		
Crocodile Breathing (yoga)	Rolling Patterns	Classic Yoga Instruction
Classic Martial Arts Instruction	Pressurized Breathing for Power	SeeSaw Breathing (Feldenkrais)
GRIP / SHOULDER / CORE / CONTROL		
Goblet Squat, to Overhead Lift	Bottom-Up Clean, Bottom-Up Press	Bottom-Up C&P, Tall-Kneeling
Bottom-Up Press, Tall-Kneeling	Climbing Activities	Heavy Rope Work (Brookfield)
BALANCE AND SMALL BASE CONTROL		
Trail Running	Bottom-Up Press, Half-Kneeling	Single-Leg Deadlifting
Single-Leg Med Ball Catch	Half-Kneeling Kettlebell Halos	Tall-Kneeling Kettlebell Halos
Goblet Squat to Halos	Medicine Ball Throws, Half-Kneeling and Tall-Kneeling	Single or Alternate Leg Jump Rope
POSTURE AND COORDINATION		
Jump Rope	Indian Club Swinging	Turkish Get-Up
Kettlebell Overhead Walking	Farmer's Carry	Surfing & Stand-Up Paddleboarding
COMBINATIONS		
Cross Country Skiing	Trail Running	Single-Leg Squat
Single-Arm Pushup	Chop and Lift, Half-and Tall-Kneeling	Press—Bottom-Up, Half-Kneeling
Double Press, Tall-Kneeling	Single Bottom-Up, Clean/Squat/Press	Double Bottom-Up, Clean/Squat/Press
Yoga	Pilates Mat Work	Martial Arts Movements
Climbing Activities	Surfing & Stand-Up Paddleboarding	Obstacle Courses
Sparring	Running Uphill	Running Downhill
Compressed Athletic Activities—meaning smaller areas, quicker play, increased one-on-one contract and disadvantaged activities		

A word of caution: These activities are not magic. They don't automatically install movement quality. They simply provide the opportunity should the individual be up to the challenge. Each of these activities imposes natural obstacles and requires technical attention. There is usually a coordination of attributes not often used together, such as balance and strength or quickness and alignment. These activities usually require instruction to provide safety and maximize benefits. If you do not respect them, they can impose risk.

However, patience, attention to detail and expert instruction will provide a natural balancing of movement abilities. These do not have to make up the entire exercise program. Instead, they offer mental and physical challenges against natural limitations and technical standards. These activities will not only provide variety, but should ultimately produce physical poise, confidence and higher levels of movement competence.

CHALLENGING VERSUS DIFFICULT

We use exercise as a challenge, but it's important to make sure the challenge addresses quality as much as quantity. Rarely do recreational lifters or runners prize technique and quality equal to bigger weights or shaved times, but the elite athletes who own the records know that quality and quantity are a delicate balance they must continually monitor, and that balance is elemental.

People often consider high-intensity exercise the most extreme or radical form of self-imposed physical punishment for performance gains. To illustrate the error of this point, consider an example of the test drive and the tune-up.

A test drive is the exercise equivalent of pushing to the extremes and noting the limits of physical capacity. It is a necessary step to mark times and set goals as one strives for improved performance.

In contrast, a tune-up is the exercise equivalent of deliberate attention to any part of the physical system not functioning optimally or normally. Movement screening and assessment offer an additional piece of information to the tune-up. We optimize and normalize movement patterns in this tune-up process. In most cases, the application of the *fresh, frequent and flawless* mentality is the recipe of choice as we tune movement performance.

High-intensity work reinforces movement patterns. If the movement patterns are optimal, hard work supports them, but if the patterns are limited, those limitations will be fortified.

Approach high-intensity exercise with equal parts of preparation and conservancy. Strength coaches in professional athletics understand this. Ancient warriors understood this. It is okay to test the limits, but usually we need to leave something in the tank. The test-drive workout is used to measure improvements and set new goals, but those goals will not be met with more test drives—they're met with a batch of strategic tune-ups. The tune-ups are the key to having better test drives.

It has become popular and common practice in the exercise world to make nearly every workout a test drive. This focus on high-intensity exercise may be contributing to the current high number of training-related injuries. It is nothing new to make a workout a competition, and it's become commonplace to post a workout of the day on the internet. This is more like a daily test drive for the masses, and it turns exercise into competition, entertainment and exhibition. As an exercise professional, you should recognize the difference between sound, intelligent exercise programming and social neediness in the exercise world.

We must remind those we teach, train and advise that random selection of an exercise or an exercise program will never yield optimal results even when performed at high intensity. Any competition in exercise should always be measured against an individualized goal, a goal set using intelligent and objective principle-based practices.

True training and conditioning is as much about learning as it is about energy expenditure. Learning opportunities are all around: better mechanics, better angles, better coordination, better breathing, better strategy, better emotional management, better alignment and better recovery between bouts of exercise.

The pursuit of random difficult acts of exercise will give way to intelligently constructed correction and conditioning challenges. These challenges will be produced by informed professionals using systems that support principles and consistently produce safe and effective results.

THE GOALS OF CORRECTIVE EXERCISE

The goal of corrective exercise is to resolve or reduce measurable dysfunction within fundamental and functional movement patterns. Sometimes this will require a breakdown of the supporting mobility and stability within a pattern and then reconstruction of the pattern. Other times much of the necessary supporting mobility and stability will actually be present, and you can direct your focus on corrective exercise for movement patterning and motor control.

Regardless of the specific nature of the corrective needs of the individual or group, all corrective exercise in the Functional Movement System follows a simple but very specific path.

First—

Corrective exercise is driven by a functional screen or assessment that produces a movement baseline. The process of screening and assessment will rate and rank patterns. It will display valuable information about movement pattern dysfunctions such as asymmetry, difficulty and pain. Screening and assessment will therefore identify faulty movement patterns that should not be exercised or trained until corrected. They will also identify movement patterns that produce pain and should therefore be examined by a healthcare professional.

Second—

The corrective exercise framework will assist you in the best possible choices for corrective categories and exercises. There is no single best exercise for a movement problem, but there is an appropriate category of correctives to select from. Don't look at screens and assessments as a tool to pick the single best corrective strategy or exercise. The first order is to remove all counterproductive exercise options and to identify an obvious movement path with favorable exercise options.

Third—

Following an initial session using the corrective exercises, recheck the movement pattern for changes against the original baseline. Note any positive or negative changes. Use this information to modify approaches in future sessions.

Fourth—

Once you note an obvious change in the key pattern, perform a repeat screen or assessment to survey other movement changes and to identify the next priority. The alchemy of correcting the key fundamental pattern is positive changes in other movement patterns that were not the focus of the initial corrective approach. By resetting the most fundamental pattern, it is possible to see other positive changes. If no measurable improvement is noted in other patterns, at least know the next corrective approach will have a more favorable foundation.

These four steps provide the framework that makes corrective exercise successful in this system.

- The order of the rules of screening and assessment direct you to the most fundamental movement dysfunction.

- Choose and apply one or two of the most practical corrective exercises from the appropriate category.

- Once you've taught the exercise and seen it performed correctly, check for improvement against the basic movement-pattern dysfunction revealed by a specific test within the screen or assessment.

- If no change is noted, recheck the screening or assessment protocol, the corrective exercise choice and the instruction and execution of the exercise. This recheck minimizes time wasted on ineffective and incorrect exercise choices.

YOUR DEVELOPMENT

The screening system is designed to mentor you in your work. It is the best way to practice the professional skill of corrective exercise dosage and program design, and fits the definition of deliberate practice. The immediate feedback from repeated baseline comparisons will confirm or refute each corrective choice. This is the mark of professional development, and it is the best way to become proficient in the effective use of corrective exercise.

The system provides the structure—the actions you take next should be based on the feedback provided by the structure. The system demands strict adherence to the screening or assessment protocol, but produces a systematic, individualized approach to corrective strategies.

The term *systematic, individualized approach* sounds like an oxymoron, but it is exactly what the system produces. The system will generate the best possible exercise category to address the dysfunction you've identified, and will connect you to the most successful group of exercises for the dysfunction the screen or assessment revealed.

These exercises will first be defined under the broad categories of mobility and stability. Once you have grouped all the exercises you currently use into mobility and stability categories, you can create sub-categories based on movement patterns. Most exercises will use a predominant movement pattern or combination of two. Before you seek new exercises, take the time to categorize those you are currently using and are familiar teaching.

From that point, you'll take appropriate action based on the first corrective exercise outcome—your work is scripted until you observe the first response. That response determines the next step.

We have each become skilled in particular exercise methods. These methods should possess exercises that target mobility and stability. Stay in your professional comfort zone at first and identify the methods that produce positive changes and the ones that do not. Then seek alternate methods to adapt or build upon your deficits.

Please see www.movementbook.com/chapter10
for more information, videos and updates.

11

DEVELOPING CORRECTIVE STRATEGIES

Developing corrective strategies is where most of the major mistakes will occur, so it's time to slow down in your study. Our greatest teaching disappointments occur when people skim over the screening and assessment information in a hurry to get to the corrective exercise information. These people are usually looking for general catchall—quick fixes—but corrective exercises are not paint by number.

Success with corrective exercise requires intellectual and professional investment. Using movement-pattern screens and assessments as a starting point forces a paradigm shift from the conventional exercise thought process. Kinesiology 101 gives the impression that all we need to do is exercise a muscle group in isolation and it will spontaneously become functional and efficient in all of its roles. Most exercise theory is based on basic kinesiology and physiology using mechanical responses, and often overlooks the complex systems and behavioral responses involved in motor control. We now know better.

People with the same isolated mechanical limitation might demonstrate very different movement patterns, while people with different mechanical limitations can demonstrate similar patterns. Movement-pattern behavior represents mechanical limitations inconsistently. Once you understand this, you should always base your exercise choices on movement patterns because they represent the sum of movement systems.

Your screens and assessments will reveal dysfunctional patterns, and subsequent improvements in dysfunctional patterns will confirm your exercise choices and progressions. Once you're working on the right pattern, you will make other decisions based on the pattern breakdown.

Both the Functional Movement Screen (FMS®) and the Selective Functional Movement Assessment (SFMA®) provide a systematic form of movement pattern breakdown, but the methods and specificity are different. The FMS uses movement pattern hierarchy and corrective exercise categories to classify problems. The information gained in the FMS directly links the problem to the most practical and appropriate corrective category, and that category provides a general description of the problem influencing the pattern.

The FMS is not a diagnostic tool, and it's inappropriate to infer a specific cause for movement dysfunction without further testing. The role of the screen is to minimize risk with activity and to assist with the development of corrective exercise. Specific diagnosis is less important than risk management and suitable corrective exercise selection.

The SFMA employs specific breakout tests if pain or dysfunction is noted on a top-tier test. The tests are geared to assist in the formulation of a functional diagnosis and guide the user to the regions requiring testing for impairments. The information is also useful for corrective exercise development within clinical situations.

Either way, in the FMS or SFMA, movement patterns are somehow broken down, and ultimately you'll make specific decisions based on information collected on the breakdown of a pattern.

IT'S ALL ABOUT TENSION AND TONE

Human posture and movement are created by the conscious and unconscious tension in our muscular system acting on the skeletal, fascial and articular support and lever system. The tension is produced by muscular tone created by nerve impulses that cause the muscles to behave in a partially contracted state. This partially contracted state provides the baseline for posture and movement.

Muscles do not rest in a flaccid state waiting for a command from the brain. They are partially active at varying levels throughout the day. When we change posture, position or move in any way, some muscles relax and lengthen. Some contract and shorten, and others have no change in length but increase or decrease tension as they perform a supporting or stability role. Conscious activity and reflex activity act together to increase and decrease nerve impulses providing the framework for the muscular activity that gives us movement patterns and many of our postures.

When we perform certain movement patterns and postures repeatedly, we refine the neurological component of control at the same time as we develop better circulation, glycogen storage and tissue strength in the muscular system. We effectively model and remodel bodily hardware and our mental software.

At the most basic level, all exercise is designed to positively influence muscle tone and tension, producing efficiency in movement and motor control. Correct muscle tone and tension allow us to stand and move with proper alignment, allowing us to react quickly and to move efficiently with no energy wasted on non-productive muscle activity.

Exercise science has treated the muscles like a big meaty furnace. We engage in muscular activity to burn calories and produce a cardio-respiratory demand. This has proved to be effective for caloric expenditure and maybe even improved cardiovascular health, but it has not left us moving well. We didn't set quality movement as a goal—we focused on quantity. If we compound our incomplete exercise platform with a predominance of sedentary activities, we are left with extremely poor movement patterns.

We can also argue the case that movement-pattern dysfunction precedes musculoskeletal degeneration. Once pain is present, it becomes a driver of muscle tone and tension, compounding the problem with distorted motor control. Do we now move poorly because we have arthritis, or did years of dysfunctional movement, poor alignment and shear force produce the degeneration? The question is worth asking even if it cannot be completely answered.

Physical and emotional stress can cause problems with muscle tone and tension as well. When mobility and flexibility is limited, muscle tone is usually higher than necessary in certain positions and movements. It can limit movement directly by restricting joint movement or by imposing sub-optimal alignment in the joints, forcing them to bind or impinge.

When balance, stability and motor control are limited, muscle tone and tension can be low, producing delayed reactions and poor timing and coordination. We must always consider the autonomic nervous system with its sympathetic and parasympathetic drivers as we look at movement.

The central point of exercise is to improve tone and tension in the neuromuscular system in general and specific situations. The fundamental goal is efficiency with movement, both general and specialized. Some people will use the efficiency for endurance activities; some will use it for athletic speed and quickness. Some will use it to become stronger and more powerful, and some will use it as they rehabilitate to gain a level of function that was lost. But the general goal is always efficiency.

Exercises can be geared toward relaxation or high performance. Regardless of the activity, exercise can be simply defined as repeated movement packages designed to create greater amounts of control and efficiency within general and specific movement patterns.

As we begin to discuss the tenets of corrective exercise, keep in mind the neuromuscular basics of tone and tension. When screening and assessment expose movement pattern dysfunction, be prepared to consider the tone and tension as the underlying cause, and also as the solution. Is the system protecting itself in a sympathetic survival mode, or is it learning to become more efficient managing output and repair? The only way to answer the question is to keep checking movement-pattern quality for improvement or decline.

DYSFUNCTIONAL PATTERNS

In the simplest terms, it is best to break a dysfunctional pattern down before attempting to recreate or restore a functional pattern. The FMS will use a progressive exercise model to demonstrate a movement pattern breakdown, whereas the SFMA will use a breakdown system since pain must also be considered.

We must also acknowledge that the dysfunctional patterns we identify are now serving a purpose. Pain, compensation, unresolved injuries, de-conditioning, poor lifestyle, autonomic nervous system stressors, poor movement and postural habits are all contributing factors, but most people develop the dysfunctions at one level to remain functional at another. They sacrifice movement quality to maintain a level of movement quantity.

This is easily seen when an individual walks with a limp but has no pain. A gait specialist could coach, teach and train the individual with biofeedback, force plates and video, but a simple breakdown of movement may reveal a compromised control in single-leg stance. If the fundamental building blocks are present for gait but gait is compromised, it should be trained. If they are not present, gait is no longer the problem—it simply represents the problem and the foundation must be trained.

Systematic approaches to exercise and testing can also reveal the primary problem region. Regardless of the body segment—foot, ankle, knee, hip or core—a mobility or stability problem could be assigned and managed at the appropriate region. Corrective exercise should target the appropriate region with the appropriate training.

However, this is not isolation; it is managed with effective mobility work. Stability is managed developmentally with movements like rolling and in postures like quadruped, kneeling and half-kneeling. Once adequate mobility is present and the developmental approach is sound, feedback can be gained by a recheck of single-leg stance. Improvement would warrant a reappraisal of gait, and lack of improvement warrants deeper investigation, alternative training or the need for more training.

This approach is the most efficient strategy we know—it's imprinted it in our brains. The language is simple: progressive postural demands placed on a smaller and smaller base of support. All the system requires is appropriate mobility and lots of sensory input to develop perception and assign appropriate behaviors. Erect, smooth, rhythmical gait is imprinted in us, but when it is temporarily lost, we should not force it—we should allow it to reemerge naturally.

Our scientific advancements should nudge the natural system, not force it into submission. Attempts to force correct gait patterns on a dysfunctional system with poor fundamentals is like manually wagging the tail of an angry dog to make him happy. You are working from the wrong end—make the dog happy and the tail will wag automatically. Don't work directly on gait or other functional patterns until the developmental fundamentals are demonstrated.

As you strive to remove a dysfunction and replace it with function, note the steps that produce change and those that do not. You must adequately replace the dysfunction, or it will return as a default operating system designed to meet a basic need.

Corrective exercise is both art and science. Here we'll review the logic, basic science and practical steps toward efficient and effective corrective exercise strategies. It is up to you to develop that art and skill within the specific scope of your training and practice.

For a minute, let go of any specialization or professional bias. Start fresh, and be assured the strategies discussed here are designed to repurpose much of what you already know into a different way of thinking. Your particular exercise preferences do not need to be exchanged for something new or improved. Anything you're doing that has objective value will find a place within the strategy.

If you cannot find objective merit, discard it and move on. In the book *The Dip*, Seth Godin tells us that winners quit all the time. Most people quit things; they just don't quit successfully. Quitting unproductive practices early and moving on to something better is a hallmark of successful people. If you are accurate in your screening and assessments and your best corrective efforts do not change movement patterns, let them go.

THE FMS AND CORRECTIVE EXERCISE

For all practical purposes, the FMS is refined or even broken down by the corrective exercise progressions. This means exercises associated with each movement pattern in the FMS are actually a continued part of the screen. There's no need to score these, but we still pay close attention. The corrective exercises are not simply performed with blind confidence—they should be used as a gauge to identify proficiency or deficiency for each exercise task.

FMS corrective exercises are grouped two different ways. First, we group the exercises by the seven movement pattern tests within the screen. Second, the exercises follow a linear path from basic mobility to basic stability to movement-pattern retraining.

Specifically, each of the seven movement patterns is linked to—

1. **Mobility exercises**—focus on joint range of motion, tissue length and muscle flexibility

   These demonstrate the basic mobility required within each moving segment of a specific movement pattern. The mobility category includes any form of stretching or joint mobility work within the movement pattern. Exercises in this category need to explore and eventually demonstrate all the available mobility required for that pattern.

2. **Stability exercises**—focus on basic sequencing of movement

   These exercises target postural control of starting and ending positions within each movement pattern. The stability category includes any form of postural control work, with a particular focus on starting and end-range postural control. Don't think strength, think timing. Timing is a quick tap of the brakes, whereas strength is a force that locks the wheels. Stability is about fine-tuned control, not force. These exercises need to demonstrate appropriate postural control without verbal coaching or visual cues.

3. **Movement pattern retraining**—incorporates the use of fundamental mobility and stability into specific movement patterns to reinforce coordination and timing

   These exercises reinforce confidence through repetition and reactive drills and should explore the entire movement pattern in order for improved mobility and stability to interact and become coordinated.

The corrective exercise progression always starts with mobility exercises. These exercises are performed bilaterally to confirm mobility limitation and asymmetry. Never assume you know the mobility restriction location or side of the mobility restriction. Always check both sides and always clear mobility by performing all the mobility exercises.

If these exercises reveal limitation or asymmetry, you have confirmed a mobility problem within the pattern and it should be the primary focus of the corrective exercise session. If no change in mobility is appreciated, do not proceed to stability work. Use the exercises to prove mobility is present or continue working on all mobility problems until you note an appreciable, measurable change. Mobility does not need to become full or normal, but an improvement must be noted. You can proceed to a stability corrective exercise only if the increased mobility allows the person to successfully get into the appropriate exercise posture and position.

If there is any question about compromised mobility, always return to mobility exercises at the beginning of each exercise session before moving to stability exercises. This will assure that proper tissue length and joint alignment are available for stabilization exercises. The mobility exercise will remove stiffness or muscle tone that is performing the role of stability. If optimal mobility is achieved, it is appropriate to move directly to stability, but periodically reconfirm mobility just to be sure.

The stability exercises demand posture, alignment, balance and control of forces within the newly available range and without the support of compensatory stiffness or muscle tone. Consider stability exercises as challenges to posture and position rather than being conventional strength exercises.

When no limitation or asymmetry is present in the mobility corrective exercises, move directly to stability corrective exercises. Seeing no limitation or asymmetry indicates the mobility required for the movement pattern is present, but is not responding to efficient motor control.

Using the idea of motor control will help you think beyond weakness as the only explanation for poor stabilization. Motor control is a broad category that includes mobility, alignment, balance, timing, sub-maximal muscle quickness, coordination and efficient co-activation. The absence of efficient motor control looks like weakness, but strength training the stabilizers is not the solution.

Stability can be separated from strength by improved motor control demonstrated by rigidity and firmness at end ranges. This is why many of the stability exercises use a light load, good posture and a hold or a movement into the end ranges. Quick firmness and adjustments to changes in load are more important than force generation.

Stability in the middle ranges is also important, but particular interest must be taken to assure end-range function. If end-range function is present, mid-range function is usually acceptable, but the reverse is not necessarily true. Look at good mid-range tension as strength, and good end-range tension as stability, timing and integrity. This is the main reason why mobility is important. You must make sure you are testing end-range motor control.

When improved stability is noted, it is possible to progress to movement pattern retraining. Movement pattern retraining should always follow proper attainment and demonstrations of mobility and stability within corrective exercises. Perfection is not necessary and is rarely possible, but do not attempt to retrain a movement pattern if the mobility and stability to support the pattern is not available.

Many forms of assistance facilitation can be provided to reduce compensation and allow quality practice within movement patterns. The general rule is to only use techniques that improve form and quality of the movement. Overload is not an effective corrective tool at this level of training.

Quick progressions with load and intensity will usually cause a default to a more limited or dysfunctional movement pattern. Descriptions and definitions of movement pattern retraining will be discussed in the Chapter 12.

THE SFMA AND CORRECTIVE EXERCISE

The SFMA must not simply rely on corrective exercises to serve the purposes of a break down of movement-pattern dysfunction. The SFMA top-tier tests are supported by logical, systematic breakouts. These breakouts look at movements within each movement pattern in various ways. Movements are looked at loaded, unloaded, symmetrically and asymmetrically, and when possible, they are looked at passively.

The systematic breakouts are necessary in the SFMA simply because movement classification can change during the breakdown. The added complication of painful movement is the major consideration.

The SFMA carries higher levels of responsibility in corrective exercise choices. Rehabilitation must deal with pain and debilitating problems that require added investigations. The SFMA must consider the two variables of pain and dysfunction, whereas the FMS needs only to consider dysfunction.

With the SFMA, it is possible to see a dysfunctional, non-painful pattern on one of the top-tier tests, only to find it is painful when broken down. An example would be dysfunctional, non-painful right single-leg stance and dysfunctional, non-painful squatting, but when broken down, functional and painful dorsiflexion is noted. This potential problem negates the simple assignment of squatting and single-leg stance exercises. Most likely, mobility correctives would catch the problem, but in a medical environment, all potential pain-provoking movements should be reviewed before corrective exercises are prescribed.

The SFMA does not require the user to follow a linear path for mobility, stability and movement pattern retraining. The SFMA forecasts whether the problem is mobility or stability, and the user is required to confirm the forecasted problem with appropriate impairment measurements. Once the forecasted problem is confirmed, the clinician will provide mobility treatment and corrective exercise or stability treatment and corrective exercise as indicated by the assessment and impairment information.

In clinical cases, it is often necessary to facilitate corrective exercise with treatments to promote either mobility or stability. To clarify the terminology, a treatment would be performed on the patient; in this criterion, the patient role is mostly passive. Whereas the patient would need to participate in or perform the corrective exercises—in these, the patient role is mostly active.

Treatments that promote mobility can involve manual therapy, such as soft-tissue and joint mobilization and manipulation. Treatments for mobility might also include any modality that improves tissue pliability or freedom of movement. These techniques can potentially complement mobility corrective exercise and often improve the effects when measured against corrective exercises alone.

Treatments that promote stability can involve any form of facilitation or inhibition taping, or use of a functional orthotic or brace or other support. Treatments for stability can also include any modality that improves or facilitates motor control prior to stabilization exercises.

Other information will also come into play, the attributes or limitations of the person or group you're guiding. This means the SFMA will guide your treatment and corrective exercise, but you must still make specific decisions particular to certain situations.

It is better to follow a standardized system when developing a treatment and corrective exercise strategy, but in some cases, you'll have to make unique decisions to fit different situations. You'll classify each person categorically following the appropriate movement system criteria, but within the classification and category there is latitude for special considerations and for individualization.

The system will point out the problematic pattern and even set functional levels of mobility and stability, but ultimately you will have to choose the best treatment methods, exercise approaches and initial program from the toolbox you have studied and are skilled at implementing.

This is the best possible way for the system to work. The structure provides a reliable platform to help develop professional consistency, but it also fosters autonomy and professional development with its feedback loop. It is a standard operating procedure with professional discretion. It forces us to think initially in categories, but as we refine the information, it provides autonomy within its consistent structure. If you choose the wrong exercises or treatments initially, you will know quickly, because reappraisal is built into the package.

Categorically, corrective exercises for the SFMA closely resemble those used for the FMS, and in many cases the exercises are the same. The definition of corrective exercise should simply indicate an exercise focused on poor movement quality.

Poor movement quality is problematic in both non-painful non-patient populations and in patient populations. Therefore, corrective exercise is a tool used in clinical and non-clinical situations. The purpose of the SFMA and FMS is to govern the use of corrective exercise in different professional settings. Each has rules and guides for the use of corrective exercise, and each has a hierarchy of priority, as will be discussed later in this chapter.

Exercise and rehabilitation professionals have long debated the utility of one corrective exercise over another without a standard gauge to identify the most fundamental movement problem. Without the gauge, who is to say the best two possible choices are being compared?

This new perspective will most likely be different from most discussions on corrective exercises or even discussions of exercise in general. This system relies on movement-pattern classification, although conventional exercise perspectives are categorized in kinesiological terms.

The old paradigm was intended to cause a positive adaptation in an anatomical structure. The new perspective considers movement patterns and motor control in addition to anatomy, because attributes of movement can change without any change in anatomy.

You now have functional standards and your exercise choices are subordinate to them.

CORRECTIVE EXERCISE AND FUNCTIONAL EXERCISE

Corrective exercise falls into three basic categories.

Mobility—
targets basic freedom of movement

Stability—
targets basic motor control

Movement Pattern Retraining—
targets functional movement patterns utilizing both mobility and stability

Functional exercise generally falls into two basic categories.

General—*exercise that complements fundamental human movement patterns*

Specific—*exercise that complements specific skills or specialized activities or athletics*

There are many misconceptions about corrective and functional exercise. In physical therapy, corrective exercise is often referred to as therapeutic exercise and is defined as exercises specifically developed to maintain or restore function. In fitness and athletics, the word *functional* is often used to denote realistic purpose and physical preparation.

Functional exercise is simply purposeful exercise that displays a certain amount of carryover into other activities. Performing a functional exercise should not only improve the ability to perform that exercise, but also improve some other physical movement not directly practiced. If weighted squats improve the ability and performance with weighted squats and as a side effect the vertical leap also improves without being practiced, squats are functional for jumping. If abilities with weighted squats improve, but vertical leap heights do not, weighted squats are not functional for jumping for this particular scenario.

If you test this theory, you will find it may be true for some individuals and not true for others. The issue is not if squats are functional training to improve the vertical leap, the questions is if squats provide improvement in the physical attribute negatively affecting the vertical leap in that individual.

To summarize the terms for corrective exercise and functional exercise, it would be appropriate to say *corrective exercises* are performed to provide a functional base when dysfunction, limitation or asymmetry is noted. *Functional exercise* maintains that base while enhancing physical capacity.

PUTTING THE FUNCTION IN FUNCTIONAL EXERCISE

Functional does not just mean possible—functional means without pain, limitation, asymmetry or any other form of dysfunction.

Functional should mean authentic, the most basic aspects of human movement potential. Some strength coaches refer to this authentic fundamental capability as Tarzan strength. Tarzan can run, climb, swim and fight—he can move. He may not be highly specialized, but he is good at everything. He knows if he focuses only on running, he will lose some valuable swimming and climbing mobility. If he focuses only on swimming and climbing, he will not have the hardness and quickness that running and fighting provide. He does not have to supplement his exercise because he is always moving and always mixing it up. He has body knowledge.

Other trainers prize these capabilities in their ideas of farm-boy or wrestler strength. The typical athlete who fits this profile may not be the most impressive individual in the weight room, but intuitively knows how to use leverage, angles and the entire body to execute every task.

Many approach the question *What is functional exercise?* by discussing the equipment used. Others describe exercises that mimic movements in daily activities, occupational activities and athletics. Some just discuss natural forms of exercise void of machines and complex training devices.

We can all appreciate these explanations because they might all produce functional outcomes. However, these explanations all fail to create the best possible definition. We don't define functional exercise by how it appears, but by what it produces. The definition is practical and clear: Functional exercise promotes acceptable function against a standard, and if it doesn't—it's not.

Now we must define acceptable function, which takes us back to the performance pyramid. First, functional exercise must promote or maintain basic functional movement patterns. Second, functional exercise must promote or maintain basic physical capacity. Lastly, functional exercise must promote or maintain specific skills associated with athletics and activities. This is a big order, because it suggests that functional exercise choices must promote or maintain one level of function without compromising another.

Functional exercise has historically been seen three different ways. It can create—

- General functional movement quality—this is actually corrective exercise or functional movement maintenance

- General functional movement quantity, improved volume and capacity for general functional movements—this is general functional conditioning

- Specific functional movement quality and quantity, improved volume and capacity specific to a particular skill, activity or athletics—this is activity or sport-specific conditioning

Some will consider corrective exercise a subcategory of functional exercise simply targeting quality. Others may want to have corrective exercise in a category of its own. As long as a corrective approach is used to address poor movement quality, and functional exercises are used to reinforce and develop better functional movement quality and capacity, this is probably not a worthwhile argument.

The functional perspective can still be confusing, though. People will always define function through the image of their own particular interests or specific agendas. Just remember that general movement function provides a sound platform for functional performance and functional capacity. Functional capacity provides a sound platform to practice and perfect specific functional skills.

The better part of this book has been concerned with the philosophy and practice of setting functional movement baselines. We have refined our ideas about corrective exercise by constant scrutiny against these same standards.

The FMS and SFMA were not designed to prove opinions and preferred exercises, but to reliably confirm or deny the effectiveness of the endless lineup of exercises called functional or corrective. The FMS and SFMA actually revealed much of what we've presented here, and the systems deserve the credit for the corrective strategies we discuss. The systems provided a consistent standard and we kept checking our corrective ideas and exercises against the systematic structure.

CONDITIONING EXERCISE, CORRECTIVE EXERCISE AND MOVEMENT PREP

CONDITIONING EXERCISE

Conditioning exercise has a focus on positive neuro-physiological adaptations in structural integrity or performance over periods greater than a single exercise session. To effectively set goals, pre-testing is necessary to document adaptation with post-testing. Examples are skeletal muscle hypertrophy, strength gain, increased lean body mass, improvement in endurance and work capacity.

Conditioning exercises are usually progressed in cycles. Periodization models help athletes peak physical capacity and performance at important competitive times. Improved physical capacity is a blend of improved neuromuscular efficiency and metabolic efficiency. At any given time, one system will take the load and get the challenge.

When neuromuscular efficiency is optimal, metabolic system gets the load. It gets the stress as it tries to keep transportation of oxygen, nutrients and waste products flowing the correct directions at the correct ratios. When metabolic efficiency is optimal, the neuromuscular system gets the load. It get the stress as it tries to continuously coordinate stable segments, moving segments, sensory input, as well as conscious and sub-conscious management of posture and movement patterns. On any given day, the load can switch back and forth, which is the way nature intended it to be.

When we introduce new activities, the neuromuscular system gets the load. This can be called high technical volume or high neuromuscular load. When activities are routine and familiar, the metabolic system usually gets the load. This can be called high metabolic volume or high physiological load.

If either load is too great, compensation will most likely occur, and when compensation occurs, a degradation of movement patterns can result. This is why it is practical and prudent to perform movement screening periodically when training loads are high. This assures that conditioning is improving movement quantities without compromising movement qualities.

CORRECTIVE EXERCISE

Corrective exercise has a focus on positive neuro-physiological responses in the quality of mobility, stability and movement patterns within a single exercise session. The goal is not structural change or performance, but normalization of muscle tone, length, tension and freedom of movement. It is based on stability, proprioception, timing and motor control, and on the normalization of limitations and asymmetries within basic and functional movement patterns.

To effectively set goals, pretesting is necessary to document a positive response within a single session. The end goal is normalization of the movement screen or movement assessment, however smaller goals of mobility and stability may be necessary before addressing movement patterns.

This all depends on the severity and complexity of the movement dysfunction. Perfection is not necessarily the goal—symmetry and adequate function is the goal.

Corrective exercise should be flexible and dynamic. This means regimented programming can potentially limit progress. In many cases, change happens within a single session, so you must be prepared to progress activities and exercises on the fly.

Don't get stuck on a program—be dynamic. Expect change and be prepared to use it or loose it. If you gain mobility, use it in a static stability drill. If you see symmetry and competence in static stability, advance to something more dynamic. But don't assume the change will last to the next session. Be prepared to repeat the process to enjoy slow, steady progress.

Be willing to provide homework for clients and patients, but don't expect large gains at home. They should maintain at home what was gained within the last session. This is to say that home exercise should be maintenance of possible and proven gains. Large corrective gains should be made as you execute the corrective strategy and dynamically progress and regress the corrective exercises to create appropriate challenges.

MOVEMENT PREPARATION

In contrast to a physiological warm-up with a focus on cardiovascular preparation before activity, movement prep is directed at the patterns that will be used in activities, athletics or training. Movement preparation can also encompass movements that originally required corrective exercise. The exercises can replicate the corrective exercises used to achieve normalization of a movement pattern, but they are not employed to correct anything. These are now used to reinforce and maintain symmetry and function of movement patterns.

Movement prep should first address newly gained functional patterns with preparation activity. Once movement prep is used to address individual movement-pattern needs, sport- and activity-specific movements can be performed. This creates the best possible system to address movement-related needs.

First, the focus is on the individual, and then on the specific activity. It also provides routine appraisal of the status of the last movement pattern correction.

Movement prep should also be performed for recovery, and repeated even more often in times of higher stress like competition or a hard workout cycle. Movement prep can also replace a workout when time does not allow a full workout.

It promotes activity-specific readiness.

SKILL TRAINING, CONDITIONING AND CORRECTIVE EXERCISE

The training process begins with the basic understanding that we separate exercise into three distinct groups to improve professional communication and to facilitate problem-solving. This book will discuss corrective strategies and a bit of conditioning exercises and leave skill training to the experts.

Skill Training—*drills and exercises specifically designed to improve a particular skill related to a sport or activity*

Conditioning Exercises—*training, habituation and reinforcement of movement behavior to increase physical performance or physical capacity*

Corrective Exercises—*specific mobility and stability exercises used to improve the quality of a fundamental and functional movement pattern*

We'll consider corrective work in greater depth than conditioning, but be aware that your conditioning exercises should target functional movement patterns. Functional movement patterns should respond favorably to conditioning, whereas dysfunctional movement patterns will require corrective exercises. Likewise, faulty conditioning exercises can compromise functional movement patterns.

Let's review examples of why we approach conditioning with a functional perspective using the front squat as an example of a fundamental conditioning exercise.

To demonstrate the point, consider three athletes. They all play the same sport and the same position, and all have the same performance goals, including optimal leg strength in the squatting pattern. For the sake of the example, consider that Athlete One has a score of one on the FMS deep squat test; Athlete Two has a two score on the deep squat; and Athlete Three has a score of three.

After a trial lift, they all squat to parallel in the exercise. Does this fact alone make the front squat a good conditioning choice for all three athletes? Not exactly. Let's look at this at a deeper level.

Athlete One can squat deeper with weight than without weight, but the weight does not optimize or reinforce proper lifting mechanics even if it appears to. Tendons, fascia, joints and ligaments disproportionately carry the load, and optimal alignment is lost or compromised. The FMS revealed the actual motor control capacity of One's squat. Any greater depth in the squat would load and stress non-contractile tissues, and this is not advisable. The athlete may actually look good in the lift by some standards, and looks stronger than the other two athletes at first glance. He may seem the least fatigued of the three after the first set, because the loaded non-contracting tissues have relied on their elasticity and integrity to help control the load.

The entire point of weight training is to stress the neuromuscular system, to demand it to overcome and control load within and throughout a functional movement pattern. This stress on the neuromuscular will demand greater coordination and timing. In the first two to four weeks of serious strength training, significant gains will be made in strength without any appreciable hypertrophy. The improved ability is attributed to neural factors and improved motor control, meaning motor control has been optimized. Once motor control is optimized, muscular hypertrophy is stimulated since motor control cannot be further improved.

However, this natural process does not occur with Athlete One. He cannot optimize motor control because he is going beyond it. He is controlling the squat by compressing stiff hips and ankles and by collapsing his feet into excessive pronation. He is collapsing his knees into valgus or twisting his feet into greater amounts of external rotation. He is rounding his back to avoid pitching the weight outside his base of support.

Perhaps you can see the rounded back and valgus knees and even cue better form, but that's not good enough. What if the athlete only has a few degrees of poor alignment at each segment, just three or four degrees of poor alignment at each joint across all the loaded joints? You can't see all that; you're not that good—no one is that good.

And even if you were that good, you could not coach it away because these things must occur if Athlete One is to reach parallel in a loaded front squat. Remember, this person can't squat bodyweight to parallel holding an eight-ounce dowel overhead. Dropping his arms into the front squat position will allow an increase in depth because the shoulders and spine can now round to compensate, but do you really want to reinforce the poorest movement patterns under load?

Athlete Two with a two on the FMS squat test could improve the movement pattern by taking time with corrective exercises, but may still want to front squat. The optimal situation is to improve the movement pattern before heavy loading, however if this person chooses to perform the front squat for training strength, you can monitor the FMS squat test to make sure the movement pattern does not become more dysfunctional.

Athlete Three with a three on the FMS squat test could participate in strength training and pursue strength and performance goals using the squat pattern, since he's demonstrated full functional movement capability within the movement pattern. However, it is still advisable to monitor the FMS periodically during periods of heavy training even for people with a high FMS score.

In the Certified Kettlebell FMS (CK-FMS) certification workshop, which you can investigate at the Dragon Door website at *dragondoor.com,* we came up with a way to communicate these three examples by using the analogy of a traffic light.

- **Athlete One** receives a red light for conditioning exercises using the squat until the movement pattern improves to an acceptable level, at least a score of two. The red light simply means stop—do not proceed until something changes.

- **Athlete Two** gets a yellow light for conditioning exercises using the squat,—proceed with caution because this isn't an optimal situation. Supervision and periodic screening would be prudent.

- **Athlete Three** receives the green light for conditioning exercises with the squat—move forward with squat conditioning. The green light does not imply the athlete will be a great squatter or perform well in variations of the lift. It just indicates basic movement is not compromised. Problems can arise with technique and performance, but you have removed the basic issues involving poor mobility and stability.

The red-yellow-green examples work well with kettlebell and free-weight training. They provide a reasonable application of the Functional Movement System in fitness and strength training environments. If an over-enthusiastic client or athlete is upset by the red light implications, reinforce that the red light is only necessary until the movement screen indicates that corrective exercises have improved the pattern to an acceptable level.

The system is effective because it encourages movement quality at the same time other fitness and performance goals are attended. It also demonstrates that the practice of certain weight training exercises is earned by movement competency.

MOVEMENT PREPARATION VERSUS MOVEMENT CORRECTION

Movement preparation is not corrective exercise, even though it is based on corrective concepts. The exercises performed in movement prep may actually be the exact same movements used in corrective exercise, but the difference is in the intended outcome.

As long as mobility, stability and movement-pattern gains are being made, the exercise is defined as corrective. Once the corrective goal has been achieved, the goal shifts to maintenance. This maintenance is best achieved with movement prep if it cannot be maintained by the exercise program alone.

Movement prep is first defined by a successful movement screen, and can be further defined as a corrective exercise rehearsal specific to the mobility, stability and movement work used to achieve acceptable functional movement patterns.

Movement preparation is done not with the intention to achieve greater levels of functional movement, but to practice and demonstrate an acceptable level of functional movement before exercise and activity. This rehearsal is key to sound training because it's a quick recheck of movement-pattern capability and quality, and it provides a trial run for the mobility and motor control prior to more vigorous forms of exercise and activity.

It is important to remember that most people don't initially seek a movement screen. They're looking for some level of fitness, conditioning or sports performance. Likewise, patients don't usually seek a movement assessment; they want pain relief. The initial evaluation is an opportunity to teach the patient that both dysfunctional and painful movements must be considered to optimize rehabilitation from a musculoskeletal injury. The screen and assessment establish a movement-pattern map and a fundamental base to build fitness, conditioning, durability and rehabilitation.

Corrective exercises are used to reconstruct faulty movement patterns, but they must also be reinforced and maintained. When these individuals seek higher levels of fitness and conditioning or a return to an active lifestyle following an injury, they can easily resume poor training or lifestyle habits that erode freshly restored movement patterns. This is where a movement prep program is key—it's a way to focus attention on the weakest link so things will not get out of hand.

FMS CORRECTIVE PRIORITIZATION

The FMS squat, hurdle step and lunge are visually representative of function. People often see the remaining four tests as ancillary, when really they are fundamental. Although the squat, hurdle step and lunge are obviously the most functional movement patterns in the screen with their combinations of upper- and lower-body movement patterns, the mobility and stability tests that make up the remainder of the screen are basic and supportive and you should always reconstruct them first.

As you work through the corrective strategies, you'll move to the next tests when a minimum score of two has been attained in the previous one. Attempt to establish a score of three in the fundamental test before attempting to establish scores of three in other, more advanced movement patterns. If scores of two seem to be an obvious plateau, construct frequent movement preparation around these movements.

The shoulder mobility and active straight-leg raise should always be the first considerations. If either test has a score of one or displays asymmetry, it is a red flag in screening, and should be the first priority.

The next pattern to address is rotary stability. This precedes the pushup for two basic reasons: It offers a left-to-right appraisal and it uses low-threshold or "soft core" stability, which naturally and fundamentally must precede high-threshold or "hard core" stability.

The pushup we look at next is the last of the fundamental tests and represents high threshold or "hard core" stability. It demonstrates proper reflex bracing and integrity in high-load situations.

The next pattern to look at, the lunge, offers two points of stability in asymmetrical stance and precedes the hurdle step pattern because it offers a larger base of support.

The next pattern to address is the hurdle step. This pattern offers the smallest base of any FMS test and only precedes the lunge in the FMS corrective hierarchy.

The squat is the final pattern in the FMS hierarchy. Asymmetry is often a complicating factor for squatting dysfunction. Squatting also involves the greatest display of range of motion of all the seven tests. For these two reasons, it is placed last in the hierarchy to assure that all other issues have been addressed before approaching it.

Although it is possible to score a three on the deep squat without having threes elsewhere in the FMS, it is not advisable to use correction to achieve a three on the squat without first having threes on all other tests. This is because all the other tests provide complementary building blocks integral to the safe and complete restoration of the authentic squatting pattern.

If a three score on the squat naturally emerges as a result of the other work, that's great, but don't aggressively attack it with correctives if the foundation is not solid. Every other movement pattern in the screen contributes a part to the squat pattern.

At each pattern follow corrective sequence priority, which is based on the rules of mobility work, stability work and movement pattern retraining. Demonstrate adequate mobility and symmetry before moving exclusively to stability training. Demonstrate static and dynamic stability before completely moving to exclusive movement-pattern retraining.

SFMA CORRECTIVE PRIORITIZATION

The SFMA is organized to allow ease of movement from one test to another. It is also organized in order of corrective priority.

- Dysfunctional, non-painful patterns in the C-spine should be addressed first

- Next, consider dysfunctional, non-painful shoulder patterns

- Then, look at forward bending and backward bending patterns

- Next, review rotation dysfunctional, non-painful patterns

- Then, consider single-leg stance patterns

- Finally, address the squat pattern

This means if everything is equally limited and dysfunctional, corrective focus should follow a natural progression. This model also offers added efficiency since successful management following prioritization can reduce dysfunction at multiple levels.

It is more likely for removal of cervical spine dysfunction to influence forward and backward bending than the reverse. It is also more likely for removal of shoulder pattern dysfunction to influence rotation than the reverse.

These rules are not absolute; they are simply probable and practical. The C-spine and shoulders are often incorporated in other corrective strategies and they should be managed prior to exercise incorporation. Each pattern in the succession promotes a level of movement quality that supports the next level.

Note how forward bending and backward bending incorporate anterior and posterior weight shifting, while rotation and single-leg stance incorporate lateral weight shifting. Also note how rotation creates a subtle weight shift from one side to another, whereas single-leg stance requires a drastic weight shift. It is always efficient to manage a problem at its most fundamental levels and not at its most impressive presentation.

Ultimately, the clinician must prioritize treatment and corrective strategy, however it is always prudent to have a systematic logical approach as a starting point. It is entirely possible to locate a dysfunctional, non-painful pattern that does not easily represent the primary symptomatic complaint. These areas can also be treated, managed and exercised without neglecting the region of primary complaint.

The SFMA strategy does not imply painful areas cannot be treated; it only contends that movement patterns that provoke pain should not be actively managed with exercise for movement correction purposes. In some special cases, it is necessary to actively move painful regions after injury and post-surgically, but these practices are done to maintain a degree of mobility or basic function in the presence of inflammation and not to enhance motor control, timing or coordination. They are temporary measures best complemented with pain-free exercise practices once inflammation and pain have been successfully managed.

SFMA categories should be addressed in the following way.

DN patterns—
manual therapy and corrective exercise

DP patterns—
manual therapy and modalities

FP patterns—
modalities and manual therapy

FN patterns—
general exercise for the purpose of metabolism, circulation and so on

A professional recommendation that will improve critical thinking is this: Attempt to manage symptoms by only addressing the DN pattern initially. This way the clinician can gauge the direct relationship of the DN pattern to the painful patterns. Regardless of the level of influence, the DN should be managed as efficiently and effectively as possible. Of course, pain control treatments can be rendered if no change is noted

DEVELOPING YOUR CORRECTIVE TOOLBOX FOR THE FMS AND SFMA

Don't make the mistake of rushing out to adopt entirely new corrective exercises. If you have not been using your corrective exercises systematically, the problem may not be the exercises. Take the exercises you already use to teach, train and rehabilitate, and organize them in general categories.

First, group your exercises under two broad headings: *Corrective Exercises and Conditioning Exercises.*

To be absolutely clear, corrective exercises are those exercises you use to improve movement quality. These exercises can potentially show movement-quality changes in a single session, and are used to improve mobility and stability and clean up movement pattern deficiency.

Conditioning exercises build better physical capacity on a sound movement base. If this seems a little confusing, don't worry—it was designed to do just that. We have all come to expect too much from a single exercise. We load movement, hoping quality will improve. We foam roll for 10 minutes and assume movement patterns are improved without checking for changes. We blend and mix activities, hoping quality and quantity will both improve and honor our unique programs.

Review your exercise list and try to find what you would use if you absolutely had to improve movement quality disregarding all other goals. Pretend you only get paid if you positively change movement on the FMS or SFMA in a single session. You have no responsibility for fitness, weight loss, strength gains or sports performance. All that is required is to take an individual with a poor movement pattern and improve the quality of that movement pattern in one session.

What exercises did you pick? Did you pick both mobility and stability exercises? Do you have equal portions of each? Put those exercise in your corrective toolbox. Use the tools you have and add more as the need presents itself. You may have great strategies to improve single-leg stance stability, but have limited success changing shoulder mobility. In that example, invest in learning more about effective exercises for shoulder mobility.

Now pick all the exercises you would perform on someone with a great FMS score achieve new performance goals. Here you're looking at your conditioning toolbox. Many professionals are surprised how most of their exercises target conditioning and fall into the conditioning category, and this may be true for you also.

As you look at your corrective toolbox, don't assume anything. Always check the effectiveness of a well-executed corrective exercise technique against the movement pattern it is designed to improve. As you refine your corrective exercise skills, do not be surprised if you remove more exercises than you add.

Our FMS team members have a small core of exercises we use in a strategic manner to change movement in a single session. Obviously, these changes need to be reinforced and progressed effectively, but remember we also delete all possible counterproductive activity at the same time. If you pit one corrective exercise against poor conditioning practices and unhealthy lifestyle habits, the poor little exercise will lose every time.

Those who have mastered corrective exercise know how to set up the trick. They do mobility exercise when mobility is the limiting factor, not just when a patient or client claims of feeling tight. They do stability work when stability is the limiting factor, not when hearing someone feels weak. They work on movement pattern quality and continuously check the standard. Their feedback is instantaneous—this is how they got so good.

Each of us will have a slightly different playbook of preferred corrective exercises. However, that playbook will be ordered and organized with a specific framework. That framework is structured around motor learning and motor control and the natural orders of movement acquisition. It is refined to consider movement pattern quality as the foundation of functional movement.

Please see www.movementbook.com/chapter11 for more information, videos and updates.

BUILDING THE CORRECTIVE FRAMEWORK

THE SIX Ps OF CORRECTIVE EXERCISE

In my first book chapter on corrective exercise, I proposed four words starting with the letter *P* that help keep us on track when designing corrective exercises. These were published in *Musculoskeletal Interventions: Techniques in Therapeutic Exercise,* in Chapter 24, *Essentials of Functional Exercise: A Four-Step Clinical Model for Therapeutic Exercise Prescription.*

We now have two more words to add to list to refine and improve our corrective exercise choices and enforce principles of movement. These effective words can define and prioritize each point to consider when making corrective exercise decisions for a particular client or patient. Consider these as a professional development checklist—use this as your framework to keep you from missing something fundamental. In nearly every case, this little checklist will improve your outcome.

- **Pain**—is there pain with movement?

- **Purpose**—what movement pattern will be the target of the corrective exercise and what problems do we find within that pattern? Mobility, basic stability and dynamic motor control?

- **Posture**—which moderately challenging posture is the best starting point for the corrective exercise?

- **Position**—what positions demonstrate mobility or stability problems and compensation behaviors?

- **Pattern**—how is the movement pattern affected by the corrective exercise?

- **Plan**—how can you design a corrective exercise plan around the information collected from the screen or assessment and the initial corrective exercise session?

Use each *P* word as part of your checklist, and do not attempt to solve a movement pattern dysfunction without clear answers to each question.

Let's look at each point in more detail.

PAIN

Aristotle said *we cannot learn without pain,* which is very wise because pain is usually life's most powerful teacher. But pain is simply the brain's interpretation of a neurological signal normally associated with trauma, dysfunction, instant and continuing damage.

The Selective Functional Movement Assessment (SFMA®) has taught us that as clinicians we have historically mapped movement pain far more effectively and consistently than we have movement-pattern dysfunction. There are many dysfunctional movement patterns not associated with pain provocation that go unmanaged in conventional rehabilitation. By mapping both, we instantly develop greater perspective for treatments and corrective exercises.

Previous models focused on pain and isolated impairments, and when efficient rehabilitation solutions were not effective, we saw patients turn to medication to make painful movement tolerable in order to return to activity.

The point about pain and exercise should be clear, but it always seems to create questions and negotiations. It's simply a case of health versus fitness. If an individual seeking exercise training advice initially reports pain with movement, the FMS is not necessary. The most responsible advice you can provide is to instruct clients and athletes to get healthy before seeking fitness or conditioning for improved performance, but be prepared to explain yourself.

Many people unknowingly seek fitness and exercise as potential solutions for undiagnosed pain and musculoskeletal issues. Although lack of exercise and activity may have caused physical decline and painful issues, simply reversing the state of activity may not reverse the problems.

Many highly active and athletic individuals are told they had to live with a certain degree of pain, or are even provided with medication to cover the pain in order to stay fit, train or play. In some cases, the prognosis is true—certain situations will simply not be free of pain. However, in many cases this is simply a failed medical management model at work. Rehabilitation professionals who use this convenient explanation on a frequent basis will not fare well in the coming years. Consumers are getting smarter, and they know there are curative options, not just cover-ups.

We should agree that pain with movement resulting from orthopedic or musculoskeletal problems is a health problem. This will also likely increase risk factors associated with exercise. Painful episodes should either resolve in a reasonable period with rest and proper recovery methods or be evaluated by an expert.

Exercise and activity in the presence of pain present more risk than reward. If you note pain, you know what to do. If you discover pain while performing the Functional Movement Screen (FMS®), direct the client toward a healthcare professional who can provide an examination and diagnosis, preferably a professional who can perform the SFMA. The SFMA will provide the most complementary outline of painful and dysfunctional movement patterns and will provide insight into the FMS findings.

Remember, pain changes the rules of motor control and greatly reduces the effectiveness of your best corrective exercise choices and conditioning effects.

Even proper and responsible rehabilitation may not correct all painful problems associated with movement. Some will unfortunately continue to have pain with movement because of chronic damage or due to a structural problem. To manage risk, these individuals should exercise within specific guidelines created by collaboration between a rehabilitation professional and an exercise professional. Routine checkups with the SFMA may reveal improvement or decline over time that could modify the exercise program and improve the condition or prevent further complications.

PURPOSE

The FMS will identify dysfunctional movement patterns and the SFMA will identify dysfunctional patterns not associated with pain. Within those patterns, each of the systems has a hierarchy of priority. The purpose of corrective exercise is to address the dysfunctional movement pattern of greatest priority.

Once you have established the movement pattern priority, the FMS will follow corrective exercise progressions that first explore baseline fundamental mobility and symmetry. Once managed, the progression will move to stability and baseline the fundamentals of control and symmetry.

Next, the progression will introduce movement-pattern retraining. The SFMA pre-manages the progression through its selective breakout tests. The SFMA breakouts separate mobility and stability problems or suggest more specialized evaluation to address the priority dysfunction without exacerbating the pain.

POSTURE

Posture is an important consideration in corrective exercise. From supine to standing, each progressive posture imposes greater demands on motor control and balance. The most common postures used in corrective exerciser include—

- Supine and Prone
- Prone on Elbows
- Quadruped
- Sitting and Unstable Sitting
- Kneeling and Half-Kneeling
- Symmetrical and Asymmetrical Stance
- Single-Leg Stance

It is best to choose exercises in postures that allow moderate challenge, successful completion, successful breathing and absence of compensation. Prone on elbows, quadruped, sitting, kneeling and half-kneeling all require unique demands on the stabilizers. These also offer unique opportunities to observe progressive loads to motor control.

Imposing static loads and dynamic movements provide opportunities to observe asymmetry at each level in the progression.

The specific posture of the body is as important as the movement you introduce onto that posture. You might already know the movement pattern you want to train, but also consider the posture of the body as the fundamental neuromuscular platform when making a corrective exercise choice. The posture is the soil and the movement pattern is the seed. A chop pattern with the arms can be performed in supine, seated, half-kneeling, tall-kneeling and standing, and each posture will require different levels of stability and motor control.

When stability and motor control seem to be the primary problem, a posture must be selected to start the corrective exercise process. Let's think this through.

The single-leg stance pattern starts developing early, even before crawling, and, in fact, it actually starts with rolling patterns. To roll to the left or to the right, the body turns around an axis and that axis eventually becomes the foundation for single-leg stance.

If rolling from prone to supine does not present a problem, a more complex posture can performed, an obvious choice being quadruped. From the all-fours position, alternate arm and leg can be lifted to an extended and flexed position, respectively. They can also be tucked into a flexed and extended position by bringing the alternate knee to the alternate elbow. This causes a significant motor control load, moving from four points of stability to two.

The load becomes even greater as the movement of the extremities causes weight shifting that must be managed continuously. The concern here is not a perfectly flat spine; the concern is gross deficiency or loss of balance, particularly on one side compared with the other. If the movements are not compromised, the next progressive posture would be half-kneeling with a narrow base. If this narrow base, half-kneeling posture demonstrates asymmetry and dysfunction, this is the posture where corrective exercise will be developed. Slightly widening the base improves control, and as control is developed, the base can be narrowed to challenge motor control.

This little testing drill becomes the corrective exercise. You would not cue the client or patient to engage muscle or exert any particular force. The half-kneeling position requires core control, not strength. The posture forces reflex stabilization through the automatic attempts to not lose balance or fall. There is no set time or repetitions. You must allow the need for balance to be addressed with automatic righting reactions.

For the individual who does not have this particular functional problem, holding this position requires not effort at all. However, for the person who has a single-leg stance problem due to poor hip and core motor control, this presents a significant challenge. This person might even break into a sweat, not because of the physiological load, but because of the motor control load and uncommon perceptions.

Note shallow breathing or even breath holding during corrective exercises regardless of the exercise type—mobility, stability or movement pattern training. When breathing is compromised, the brain switches to survival mode. The client or patient is surviving the corrective exercise, not learning from it.

Keep the person relaxed; often it's the tension and stressful breathing that actually compromises movement. The middle ground between autonomic extremes of the sympathetic and parasympathetic systems is your goal—relaxed but not asleep or bored, engaged but not panicked or obsessive.

POSITION

Note specific positions where stability or mobility is compromised, and pay particular attention to alignment and end ranges. You may note poor hip extension, shoulder flexion or thoracic spine rotation. Address mobility problems with passive and active techniques—you should see some degree of improvement before prescribing a corrective exercise. The improved mobility will require improved motor control that should be facilitated and produced by the corrective exercise.

Pay attention to positions of body segments not directly involved with the posture or movement pattern. The anterior neck musculature is a key, as the neck musculature is often engaged as an attempt at extra stability. The shoulders will often be elevated and rounded in stressful situations, and the hands are also an indicator of non-constructive exercise stress.

You must focus on the area where the greatest compensation will occur. In the case of half-kneeling example above, proper maintenance of static hip position is the key.

Make sure the kneeling hip is not flexed or extended; neutral or zero degrees the goal. The same goes for the pelvis—look for neutral.

• When a person is pulled into a slight posterior pelvic tilt—not so much to engage the abdominal musculature but to fully extend the hip—it produces zero degrees at the hip. Another more automatic way to achieve neutral hip alignment is to push down through the person's shoulders. If natural postural musculature is active and functioning, the client or patient will be non-compressible.

If you feel sponginess, tell the person to stay aligned and not allow the compression to occur. Don't allow a shrug—make sure the shoulders remain relaxed. You want the integrity to come from reflex deep core stabilization. You are not looking for bracing here; you're looking for natural, sub-maximal reflex stabilization.

• Once you get the client or patient in the start position, ask for the arms to be raised overhead. Then request torso turns to the left and right. Have the person move a weighted ball around or try to touch your hand as you move from point to point in the air. As proficiency is gained, return to a narrower base and begin the challenge again.

• To avoid unmanageable compensation, avoid positions of compromised mobility and stability. If the person cannot even get into the start position, why would you expect anything to improve once movement starts? In computer language, they call this *garbage in, garbage out*.

Correct starting posture and alignment is the optimal starting point for healthy motor control. When the start position is compromised, the entire pattern is compromised.

Just make sure you don't just randomly test. You must follow steps to arrive correctly at the problem. Once you have found the correct problem, you need to establish the level of training difficulty. It can be difficult for an individual to maintain full hip extension and a tall spine during the hurdle step, but that same person may easily maintain both in half-kneeling because the ability to stabilize becomes better at that reduced postural demand.

Positions of compromised mobility and stability should not be reinforced in the corrective exercise—they should be managed so they do not reduce movement quality.

PATTERN

Once you have watched your client or patient perform a trial exercise, note the effect on the original dysfunctional movement pattern. Positive changes should confirm the exercise choice, or you'll be using replacement exercises that have limited influence on the dysfunctional pattern. Sometimes you need to replace these with more challenging corrective exercises and postures, and sometimes you need to replace them with less challenging options.

If you note progress, you'll compare this to the other side. If symmetry is noted, recheck the original test. This sounds like lots of work, yet it can all occur in less than a few minutes.

As you start your practice in this corrective model, use video to document the sessions. On later review, you may see unnecessary struggle or difficulty you did not note during the session. You may document an *ah-ha* moment that will give you the confidence to try new things. Multiple forms of feedback are great when used constructively.

PLAN

After checking the exercise against the original dysfunctional pattern, you will have real-time feedback. There are only three possibilities—better, worse or the same, and you'll learn something no matter what the outcome. The confirmation opportunity creates a user-friendly feedback loop. The mobility, stability or movement pattern retraining either improved the pattern or did not. The improvement does not need to be complete; it simply needs to measurable or appreciable. With this feedback, you should then be able to develop a corrective exercise plan or make another attempt at success.

The framework of six **Ps** combined with the FMS and SFMA hierarchies provide a systematic corrective path. All this framework discussion is critical for professional development. Presenting specific corrective exercises without clear definitions, systematic structure and movement-pattern hierarchy would not develop critical thinking or professional-grade problem -solving.

Now that the general framework has been presented, a few more considerations must be reviewed.

COMMAND OF THE STRATEGY

There are a number of corrective exercise options that can fit a particular movement problem. The number of options can be refined further, since equipment limitations and specific individual needs may negate some of the choices and options. With our options refined, we may see an obvious single corrective exercise that fits the situation.

We can all debate the single best option, but only the outcome will expose the truth. Don't debate or deliberate—let the system tell you if you followed the correct strategy and picked an effective corrective exercise. The client or patient will move better or will not, and either way you just learned something.

The two biggest errors with corrective exercise strategy are missing one of the **P** statements or neglecting the corrective hierarchy. The most important task is to make sure the corrective options fit the entire criteria within the corrective framework. You can remove all the other options and the choices can be implemented based on your professional preference, or each can be tested for overall effectiveness.

Learn to categorize, prioritize and plan effectively, because corrective exercises will evolve and equipment will change. Your professional skill must be based in a systematic approach. Just being great at a technique is not good enough. Technical aspects of exercise will change, but don't worry—this system is not based on exercise. It's based on human movement, not equipment, techniques or trends.

In the book *On Intelligence*, Jeff Hawkins, the creator of the Palm Pilot,® The Treo Smartphone® and other handheld devices, and his co-author, Sandra Blakeslee, discuss how computer scientists attempted to create artificial intelligence without first completely defining human intelligence. Many saw the human brain as a cool memory device with some computing power, and with that limited perspective, it's easy to assume a fast computer could be as good as a human brain. Sure, the brain has memory and it can compute, but it also learns, adapts and works efficiently even with imperfect information. It can deal with abstract concepts, and can pull information from patterns. It can predict things that could never be cleanly programmed. Jeff obviously understands the fundamental differences of the human brain and the modern computer and he criticizes contemporary technology for not learning more lessons from the greatest computer of all, the human brain.

It's easy to draw parallels with similar mistakes in movement science. We map anatomy and movement with our study of kinesiology and biomechanics, and we use that map to study and observe movement. No problem there—the problem starts when we use that map to design exercises to correct faulty movement patterns. Unfortunately, the map is not the territory. A brain already knows how to learn, but computers must be programmed, and in this instance we're treating the brain like a computer. We take our clean little single-plane, isolation resistance exercise and try to program a movement pattern, but that's attempting to program the brain by exercising a group of muscles.

We have long assumed that exercises based on a kinesiological and biomechanical model will program movement quality. The brain and body originally learned to move without the benefit of exercise, individualized muscle work, a kinesiology text or biomechanical analysis. Why don't we look at that model?

The six **Ps** follow that model. It's a simple model—avoid pain, build patterns, play around, get comfortable moving and provide a challenge. Then repeat the position extremes to check the baseline.

The model produces corrective exercise considering the sensory input as important as the motor output. The kinesiological and biomechanical models put nearly all the focus on motor output and assume it will refine itself. The best corrective exercise is a rich sensory experience. It has to be, because when there's dysfunction, something between the sensory and motor systems is unplugged.

The brain often rejects the diet of random movement information provided by conventional exercise. The diet is rich on movement and lean on sensory input. Sure, fat is lost and muscles get firm, but movement quality doesn't improve. The body must comply with activity, but the brain does not learn to move better because it doesn't learn to sense and perceive better. When alignment, posture, balance and awareness improve, movement improves.

Not all exercise interferes with learning; exercise has some of us moving pretty well, and others not so much. This is because we practice exercise randomly since we don't train quality movement patterns.

Poor quality seems to be compounded and not easily corrected with conventional exercise. Movement screening and assessment show us as many differences in the fit and in those considered fit as those considered non-fit. The main reason is that in many cases the brain is expected to unlearn a repeated and habitual activity, a behavior. Poor movement patterns are basically poor movement behavior. Behavior modification requires breaking an old pattern and introducing a new pattern in such a way that it can be effectively learned.

We step up the movement buffet and try a little of this and a dab of that. We don't train, we hope. If we train to fight, we expect every aspect of fighting should improve... offensive maneuvers, defensive maneuvers, reaction time, poise, stamina. When we train movement, we usually have a single goal and it is not even close to the primary problem or the weakest link that drives the movement system.

LEARNING MOVEMENT OR DOING MOVEMENT

When exercises are based on muscle maps, the brain is not provided with the natural opportunity to learn quality movement. There is an inherent difference between learning movement and doing movement. Both create a physiological load, but only through learning opportunities can the brain and body create more refined, efficient functional movement patterns. By sticking to this strategy, you'll remove the roadblocks to movement learning. You feed the brain a rich diet of sensory opportunity uncomplicated with reasons to compensate.

- By not exercising into pain, you don't insult motor control.

- By identifying purpose, you follow the natural hierarchies of mobility, stability and movement, and also create a fundamental base that moves from basic to more complex patterns.

- By identifying a challenging posture, you tap into reflexive and reactive situations that require sensory motor interaction. You honor the developmental language the brain prefers and used in its initial movement-pattern building.

- By paying close attention to joint position and postural alignment, you will increase the challenge by removing previous compensations.

- By rechecking the pattern that is the focus of the correction, you will instantly have valuable feedback that indicates a move forward or a move back.
- Your corrective plan will not be a handy, one-size-fits-all protocol you can tear from a three-ring binder. It will be unique and dynamic, and it will be effective.

The system will not only help the client or patient learn movement at an accelerated rate, it will also help you move easily from screening and assessment to a corrective exercise plan. Of course, the ease will not be there in the beginning. Your brain will be forced to answer more pre-exercise questions than ever before, and you might even hate all these rules. You might stress your brain at first, but a little constructive stress makes us stronger. Soon you will answer these questions automatically as you start to become intuitive with the system.

SPECIAL CONSIDERATIONS AND POPULATIONS

There are always special circumstances and situations. Each will be unique, but let's review how the basic framework can complement each situation.

REHABILITATION GOALS FOR THOSE UNDER MEDICAL CARE

These rehabilitation patients have been through the SFMA and have had their movement patterns categorized. They should not exercise or rehearse movements that cause pain or are dysfunctional and painful. Put them on a corrective strategy for patterns that are dysfunctional and non-painful, and use exercises in functional, non-painful patterns when not counterproductive to the rehabilitation goals.

Check for success with corrective exercise against the original patterns to note changes in functional movement and the provocation of pain. Along with appropriate manual therapies, modalities and other treatments, the corrective exercises should produce rapid changes in mobility or stability. The concept of rapid change suggests you should recheck the dysfunctional pattern or parts of it regularly. Only by rechecking will you have the appropriate feedback to progress rehabilitation, persist with the same treatments and exercises, or know to re-evaluate the situation.

Once pain is not the problem, use the FMS as part of the discharge assessment if the patient is returning to an active lifestyle. Map any potential problems and risks, and design a corrective plan. Another choice is to refer the patient to an exercise professional with the capability to perform the screen as well as corrective exercises.

TRANSITION TRAINING GOALS FOR THE SEVERELY DECONDITIONED

These are people transitioning from inactivity to activity, and are deconditioned. They might be recovering from a long infirmary or may have an extremely sedentary lifestyle and have decided to become active from a base of limited exercise capacity. You should focus all their training energy toward the weakest links of the FMS, the scores of one and any asymmetry. The corrective strategy can serve as both restorative exercise for the poorest movement patterns and the metabolic load to restart the energy expenditure and rebuild the recovery system.

Use intervals to build exercise capacity by interposing corrective strategy with active rest and constructive recovery. These could include light flexibility and constructive breathing practices. As movement patterns improve, you should observe a two-fold improvement in productivity—the reduction of dysfunctional movement patterns increases mechanical efficiency, and the work to attain the movement patterns improves metabolism.

These renewed enthusiasts will perceive increased capacity because of better mechanics and better metabolism. The basic rule in this situation is to conserve all energy for the corrective strategy, and let the corrections be the metabolic load. It is the most efficient path to increased activity, and presents the minimal risk.

METABOLISM AND WEIGHT LOSS GOALS

With a primary personal focus on weight loss, most of these people will not have optimal movement patterns. If the FMS reveals movement dysfunction, it implies poor efficiency, which can cause premature fatigue with exercise and activity. The vicious cycle of inefficient movement and quick fatigue produces a poor environment for improved metabolism.

This paradox is a great discussion to have with clients interested in weight loss. We need to educate this group so they understand increased activity alone may not produce the desired result. Generalized physical activity is not the goal and will rarely produce the best possible outcome. For these clients, select only cardiovascular exercise and functional movement patterns for resistance, functional movements scored as twos or threes on the FMS, without asymmetry.

Dysfunctional movement patterns will require corrective strategy. The best time for the corrective work is in movement preparation, warm-up and active rest between sets of resistance and cardiovascular exercise. In many cases, the overweight person is also severely deconditioned. If this is the case, the starting point for this client is to establish basic movement and conditioning to create the best base for weight loss with the minimal risk of musculoskeletal injury.

PHYSICAL CAPACITY AND ATHLETIC GOALS

Most goals for physical capacity and athletic achievement do not incorporate durability or resistance to injury in the performance programming. Athletes seeking higher levels of physical capacity should understand that performance programming does not guarantee durability. This faulty assumption is extremely prevalent, and we need to clear this up by using the screens to demonstrate that dysfunctional movements can undermine performance by increasing the injury risk. Dysfunctional and asymmetrical patterns can also reduce training effectiveness because they create an environment for inefficiency and compensation.

Law enforcement, fire service, first responders and the military all have levels of physical capacity they must achieve and maintain. Likewise, industrial workers routinely expend the same energy as athletes—they must be durable and perform. Injuries are unwelcome intruders in industry and in an athletic season or career.

We have tried to manage athletic and work-related injuries with preventive programs, but we have not effectively created systems to gauge risk before participation. When the FMS reveals scores of one or asymmetry, research demonstrates compromised durability: People with this movement profile have greater injury risks.

The goal here is not to get a perfect score on the FMS, but to get into a safety zone to limit the risk, as shown by a score of two or better on each

test and with no asymmetry. FMS goals should be paramount to performance goals when risk is present.

THE YOUNG AND OLD

Fitness and healthcare professionals often exclude younger athletes and active older adults from movement screening discussions because they assume these people will score poorly, and that the test is not worth the clients' time or frustration. Do not dismiss them—they deserve the benefits of pre-activity movement screening just like other populations.

Screening is foremost for risk, and the identification of pain in a movement screen is actually more important than finding dysfunction. Pain with movement means an acute or chronic problem or injury is present. It can also expose problems potentially complicated by exercise and activity, and of course, the FMS will identify dysfunction. Age can potentially present greater diversity when movement screening the young and old, but screen whenever possible even if modification is necessary.

As discussed earlier, these groups may have some predictable problems, implicating the need for group-wide modifications. It is true that older adults may score poorly on movement patterns requiring full range of motion, such as the deep squat and shoulder mobility. They may also have difficulty with the hurdle step due to balance issues. Likewise, the normal eight-year-old soccer player may have difficulty with the trunk stability pushup test, since core strength may not be fully developed.

Regardless of the score, the movement screen provides an extra measure of safety and it sets a baseline. This baseline will demonstrate the effectiveness of the exercise program and provide an extra measurement for improvement along with the usual parameters for performance. Let the screen point to the greatest limitation, and then manage and monitor it.

JUST DO IT

Ultimately, most people want to just get on with it when it comes to exercise. Many may even think screening is overkill. This may stem from the old Nike® slogan, still a popular sentiment regarding exercise—*Just do it.*® Nice, but... just do what? Sometimes the inference is to just move, and that can be good. However, the suggestion in the preface instructing *move well, then move often* is supported by all the references in this book. The *just do it* instruction can still apply, but we need to change *just do it* to infer moving well before we use *just do it* to infer moving often.

The cardiovascular, cognitive, emotional and psychosocial evidence is compelling regarding exercise. We all need to move for more than physical and musculoskeletal reasons. We need to get our clients and patients moving more often, but as Dr. Ed Thomas says, "Maybe if they could move well, they would move more often."

In reality, musculoskeletal issues and movement-pattern problems limit many individuals who have good intentions to exercise and become more active. They never tap into the natural pharmacy of good chemicals or the self-confidence that awaits the enduring and consistent exerciser. The perspective of movement presented in this book is designed to remove the bottleneck where poor movement, risk and injuries limit more active lifestyles and athletic longevity.

We want to move, we should move, but when we move, our options are limited by our abilities to move well. We bounce between a minimal training effect, nagging injuries and idle times when we become sidelined by our movement problems, tuned into injuries and flare-ups. Quality exercise requires a minimal investment of time and effort, but what if our bodies cannot handle the frequency, intensity or duration of the required minimal dosage? Easy answer: We turn to incomplete movement practices, like only training the legs, only doing machines, or only participating and practicing the same diet of incomplete movement patterns.

When our movement options are limited, we don't do technical or invigorating things; we do boring, safe routines like stationary bikes and treadmills that do not require movement competence and don't reinforce good technique. These things do move us, but we act like hamsters on a wheel for 20 minutes, and assume we are fit and physically balanced. The problem might be that many are forced into the limited movement experience by the confines of their movement patterns and a poorly managed injury history. Many will never make it back to authentic movement, but who knows how close some can get.

The cognitive and emotional benefits of exercise are presented best in a book called *Spark,* written by John Ratey, MD. This book should be on the shelf of every exercise and rehabilitation professional. It presents remarkably positive evidence for the exercise professional to develop exercise dosages for benefits in the human brain that go far beyond the physical and cardiovascular data you already know.

If you work with the young and old in particular, get the book when you finish reading this one. He discusses how exercise improves the brain in many measurable ways. He also suggests exercises that require more technical precision might even carry over into cognitive abilities. As you read it, think about the self-limiting exercises suggested earlier and you'll see these activities are a perfect complement to cardiovascular and weight training programs. They offer greater precision, progression and variety opportunities to exercise programming. Dr. Ed Thomas identifies these three things in particular when he discusses why some exercise programs fail and some flourish.

MOVEMENT PSYCHOLOGY THEORY

This book has described how survival instincts help us compensate when we perform exercises appropriate for our metabolic load capacity but not our movement capacity. This overload to our movement system causes movement compensation, substitution and poor technique. Modern conveniences give us forgiving exercise equipment for poor technique, braces where we should be stable, cushy shoes for our bad running strides, and anti-inflammatory medications so we don't need to listen to our bodies' instructions. All these advancements have allowed us to continue perusing activities that nature would limit or not allow.

Two compounding factors make this a hard problem to remedy. The survival instinct is strong, and there's pain memory from previous injuries. Pain memory actually creates a fear memory and these memories trump most other memories. These two factors compound each other and combine. If you have been injured, you probably have a kinetic perception of the incident and associated pain. This is where the limp comes from, and it sometimes continues even after all is well.

We address poor movement patterns with a firm understanding that they must be reconstructed as a behavior, and not just through biomechanics and supplementary mobility and stability drills. Without the evidence of a corrected movement pattern, mobility and stability exercises offer no guarantees—they only provide potential. Movement confidence does not come from supplementary exercises. It can only come from correct movement patterns repeated over time and across various situations.

Sometimes the biggest problem that must be overcome is body knowledge and trust in a movement associated with a previous bad experience. Sometimes the behavior is not driven by an injury, but by continuous survival strategies seen as compensation, and avoidance—the fear memory—even after the injury is resolved.

If you are trying to correct the squat movement pattern in an individual who has been performing loaded squats with poor mechanics for a long time, you will be confronting a pattern that has been designed to protect from further damage. The person hasn't been training the squat, more like surviving the squat and has the eroded movement patterns to prove it.

Likewise, as a result of pain memory, patients and clients with a history of back pain might unconsciously avoid the movement patterns that will allow them to fully recover. This is where the FMS and SFMA are vital. Once patterns are corrected, movements like the deadlift and modifications and variations of a loaded hip hinge will build the trust and confidence that long-standing back dysfunction can erode. This is why the FMS and SFMA structures suggest stopping certain activities so you can reconstruct them authentically. Using developmental movement patterns promotes

reconstruction because these don't elicit the fear memory or compensation that practicing higher-level functional movements might cause.

If a triathlete has learned to run, cycle and swim with unmanaged back issues, the techniques are not authentic. The athlete is using a combination of the most efficient metabolic strategy, coupled with the least movement damage available in each pattern. The person might run with increased hip flexor tone to block complete hip and lumbar extension and avoid a painful movement associated with acute inflammation. Even after the inflammation is gone, a survival memory has been constructed by the mileage and months of training. The overactive hip flexor is part of the package and the memory that constructed it is stronger than the technical memory that appreciates more complete hip extension.

You can't train this stuff away. You reconstruct faulty patterns from a developmental model while you cease repeating them for a time. Big problem here: You have to successfully explain how stopping running will make running better—good luck. Your professional confidence must be greater than the athlete's obsessive paranoia and need and to keep training. This may give you a headache, but you will do fine, and the person will eventually thank you.

SCREENING THE BREATH

Functional Movement Systems can account for a very interesting human phenomenon. Patterns using the same levels of metabolic and mechanical demand can have varying degrees of capacity and efficiency between individuals. This means mechanics and metabolism are not absolute—they are highly variable between individuals. Emotion, anxiety, breathing patterns, efficient timing, effective motor control and familiarity all play a role in the efficiency.

When we see a dysfunctional movement pattern, we should also note changes in breathing patterns. The best yoga practitioners argue that efficient and effective breathing should precede and then complement movement. As exercise and rehabilitation professionals, we can see it is nearly impossible for movement to be efficient if breathing is not efficient. We must also acknowledge that dysfunctional breathing, like dysfunctional movement, may be present in some selected patterns or in all patterns.

Initially, it is more important to identify this than to explain it. In the past, breathing dysfunction has been associated with disease or severe disability, but now clinicians and exercise professionals are starting to consider more subtle deviations from efficient breathing patterns. We are starting to understand what martial arts masters and yogis have always known—the breath is key to consistent, quality movement.

Breathing is important because it demonstrates metabolic efficiency, physiological capability and it can influence neuromuscular tension and tone. There are two basic ways for you to examine breathing when prescribing corrective exercise.

- **The first way** is to develop the skills and learning to appreciate the subtleties of correct and incorrect breathing. Professionals must understand and appreciate the basic difference between *apical*—upper chest breathing—and *diaphragmatic*—abdominal breathing, and learn concepts like over-breathing. This is an art, but some simple observation tools will help you identify basic dysfunction.

- **The second way** is more objective but requires mechanization—it's called *capnography*. This is a system that measures exhaled CO_2 with a sensitive measuring device and it can be used to accurately demonstrate deviations from normal breathing efficiency baselines. For more on breathing, see the appendix on breathing, where you'll also find an introduction to capnography.

It is especially interesting to note that in some instances poor breathing quality is present with all movement patterns in the screens or assessments. In other instances, breathing quality seems to be specific to some movement patterns and not others. In this case, some movement patterns will display a visible reduction in breathing quality and others will not. This is likely the autonomic nervous system responses to perceptions and behaviors associated with certain patterns. The brain will perceive varying levels of stress associated movement patterns. In advanced levels of SFMA clinical education, we discuss having patients cycle a breath at the end of each of the top-tier movement patterns. A pattern is not considered functional unless a full breath cycle can be performed at the acceptable end range of the movement pattern.

This breathing movement connection can pose a question: *Does poor movement-pattern quality cause the poor breathing pattern or do poor breathing patterns cause poor movement-pattern quality?*

Instead of debating the chicken or the egg, we should strive to return authentic breathing and movement quality by designing corrections that influence both. Proper corrective exercise choice is the key. By dosing the corrective exercise in such a way that proper breathing quality is reinforced and compromised breathing quality is avoided, we can use breathing as an indicator of difficulty and unnecessary stress.

Even though this book is about movement screening, movement assessment and corrective exercise, when screening and assessment identify dysfunctional movement, you should also be aware of dysfunctional breathing. Corrective exercise must consider breathing—it's fundamental.

Prediction: Objective functional breathing screens and screens for heart rate variability will follow movement screening as qualitative standards in holistic forms of fitness, conditioning and rehabilitation. You'll learn more about both in the appendices beginning on page 353.

Please see www.movementbook.com/chapter12 for more information, videos and updates.

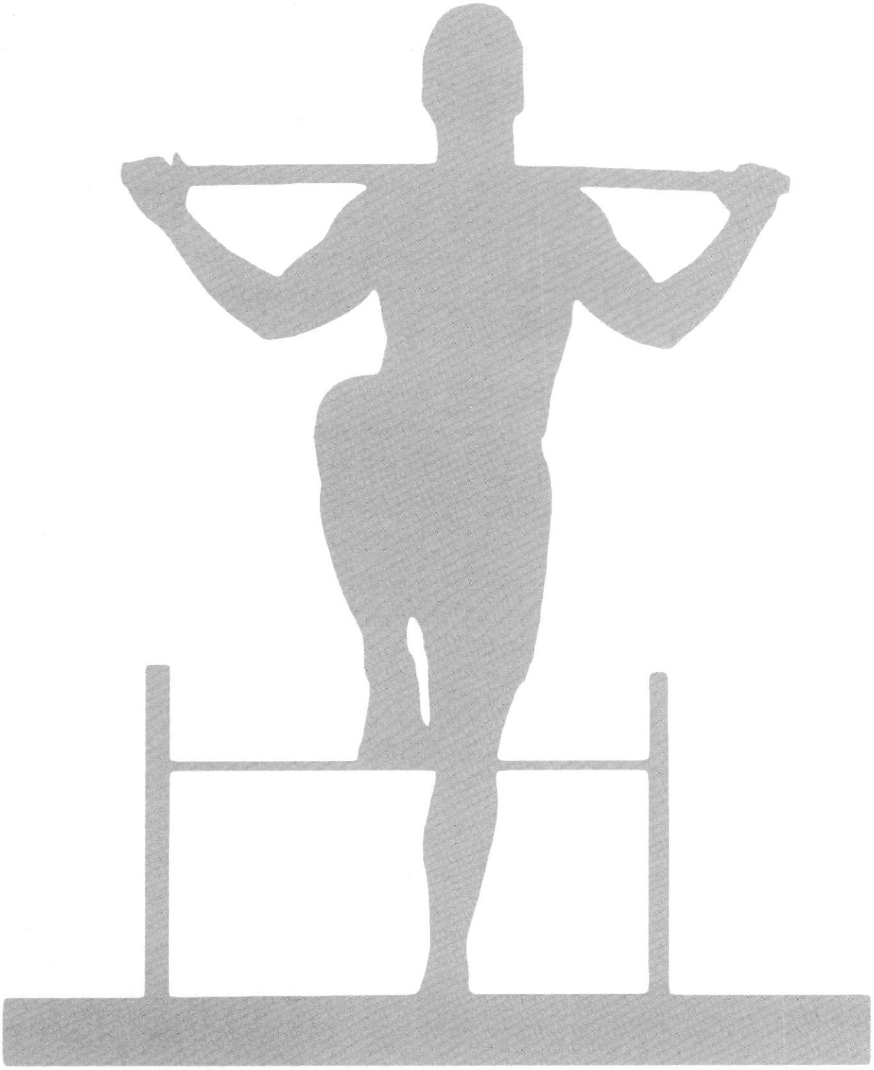

MOVEMENT PATTERN CORRECTIONS

CORRECTIONS

As we begin to talk about the technical categories of movement pattern correction, we'll start with a few cautionary words and guidelines, and then present the categories and examples of each technique to help provide clarity. You'll learn the specific way the brain prefers to learn movement, drawn from the natural examples are right under our noses. Decisions about which corrective category to use and when to progress from one level to another are a direct product of work using—

The FMS, the FMS hierarchy
and the 6 Ps checklist

The SFMA, the SFMA hierarchy
and the 6 Ps checklist

Each primary category is ordered in levels of progression identified by subcategories. The first subcategory will be the most fundamental level of correction, and the last will be the most advanced. At the end of the corrective section, we will summarize the learning brain in more detail and see how it learns movement.

The three primary categories of movement pattern correction are—

Basic Mobility Corrections
Basic Stability Corrections
Movement Pattern Retraining

GUIDANCE ON CORRECTIVE GOALS

The suggestions in this section are academic and should be practiced to the point of practical proficiency before you introduce them to clients and patients. The general concepts will be covered, but the art of practical application must be learned and practiced one technique at a time and one case at a time.

The exercise techniques discussed are used as examples that may encompass fundamental and common exercises, but they do not represent the infinite modifications and progressions that become possible once you've gained a command of the basic principles. After you have followed the corrective framework and applied techniques within each category, you will start to reinforce and accelerate your own learning and professional development.

BASIC MOBILITY CORRECTIONS

Your clients and patients must understand your rationale for mobility corrections, and for this, you'll need to learn to develop your dialog when discussing mobility. In physical terms, mobility is essentially freedom of movement. It represents tissues with acceptable extensibility and joints with actable ranges of motion, but it does not guarantee functional movement. Mobility is simply the first brick in the wall.

Together, healthy tissues and joints create the moving segments contained within the structural framework of the body. When mobility is limited in one segment, it causes systematic compromise to some degree, at some level, in some region. Compensation, substitution, asymmetry, reduced efficiency, poor alignment and faulty posture can all possibly be traced back to a mobility problem.

It is ironic that we are sometimes unaware of our tightest and stiffest regions, and our clients and patients are no different. They will more often complain of the stiff and sore back than the significant mobility restrictions in their hips. However, if we tested both the hips and spine for mobility, they might be surprised to find the actual mobility of the back is closer to optimal than are the hips.

In this all-too-common example, the back is the overworked victim, not the slacker causing the primary problem. The hips are further from optimal mobility than the back and therefore are a larger problem. The back must bend a little more, twist a little further, and actually give up some reflex stability to allow postural control and movement patterns. The back must compensate for asymmetry and move in ways inconsistent with its natural structure, movement patterns and general function. With all this going on behind the scenes, the back is the first to fatigue in almost every activity. Therefore, it must be weak, tight or dysfunctional… right?

In this particular situation, correcting hip mobility will have more potential positive effect on core stability and return of normal spine function than a spine stability program. That is not to say the individual's core cannot become more proficient in a controlled stability exercise by performing stability exercises. It simply implies that successful improvement with core stability will not positively affect function.

In this scenario, the client or patient can improve the exercise capacity of the core region and still display stability problems during functional testing. This is because every time the core stability meets the stiff hip, the hip will win. The newly improved core stability will concede to the stiff hip in the name of function—not quality function, but quantity function. Survival dictates it.

Our earlier discussions about survival demonstrate that compensation is a favorable temporary attribute. It can get us out of trouble when things are not perfect, and that is good for short-term survival. However, the long-term incorporation of compensation in place of an authentic movement pattern can compromise efficiency, cause micro-trauma and distort proprioception. The compensation interrupts the delicate sensory motor balance our brains and bodies develop as we grow, and that can be bad for long-term function.

The starting point to stability training begins with improvements in mobility. Every time mobility is improved, new opportunities for sensory input and motor adaptation are potentially available. Remember that trainers, coaches and therapists don't make stability; the brain makes stability. They might need to reintroduce the recipe, but the brain does all the work.

The recipe is actually simple—

Structural Integrity—pain-free structures without significant damage, deficiency, or deformity

Sensory Integrity—uncompromised reception and integration of sensory input

Motor Integrity—uncompromised activation and refinement of motor output

Freedom of Movement—mobility adequate to perform within functional ranges and achieve appropriate end ranges and structural alignment

If these fundamental elements are present, the brain will make stability. The brain will do this over trials and challenges provided at each progressive postural level. This will be discussed further in the stability section under the heading *Postures for Stability Corrective Exercise*, page 270.

If mobility is a problem, always attempt to improve it before attempting stability corrections. Following this directive, you can move to stability corrections in the same session to reinforce a mobility gain. You may also choose to spend a few sessions really focusing on mobility. If you need confirmation, a post–corrective-exercise test will tell you which recipe is best. Simply repeat the screen or assessment test that directed you to the mobility corrective to validate your corrective program design.

It's as simple as this: *If mobility is measurably improved, use it. If you gain hip extension, use it. If you gain shoulder flexion, use it.*

The stability work reinforces the new mobility, and the new mobility makes improved stabilization possible because new mobility provides new sensory information. New and improved sensory information is required for new and improved stability. The subcategories of stability corrective exercise will show you the best way to introduce the new sensory information, but you must agree with the concept or you will skip steps and compromise the outcome.

The three subcategories within basic mobility corrections are—

Passive Mobility Corrections
Active Mobility Corrections
Assistive Mobility Corrections

PASSIVE MOBILITY CORRECTIONS

Passive mobility corrections are directed at limitations affecting normal passive freedom of movement quality.

1. *Self-Passive Mobility Corrections* include static stretching, self-mobilization, roller stick work, foam rolling or any other self-administered maneuver that produces improved mobility or flexibility through lengthening and manipulation, and is not considered active exercise.

Note: When we stretch, compress, foam roll or bend the stiffest and tightest parts, we tend to breathe poorly. We tense, and our breathing becomes shallow and even faster in some cases. We show our stress in our breathing, and this actually increases our tension and makes the mobility work ineffective or even counterproductive. Be very aware of this. We know better and still make this mistake, so don't expect clients and patients to not slip into stress breathing—continuously observe and remind them of this.

Slow, steady breathing with a three-to-one ratio of exhale to inhale can help. As exercise demand increases, many people move to a one-to-one ratio, but corrective exercise and mobility work does not have a high metabolic demand.

Some will breathe too fast and some will hold their breath. The long exhale ratio is a guide to coach relaxation. Don't just teach a mobility technique and think you've done something—teach the client or patient how to control the situation and make it effective and reproducible. Some may debate the exact ratio, but the point is to linger with a steady exhale beyond the inhale time. Use what works for you, but always watch breathing.

2. *Manual Passive Mobility Corrections* are provided by professionals with manual skills and other forms of passive mobility work, which can fall under a broad category of manual techniques. These techniques are provided through some form of mechanical means with the hands or tools for the specific purpose of changing the state of tissue in a positive way. Some typical examples are general massage, deep soft tissue work, augmented soft tissue work with tools, mobilization, manipulation, chiropractic adjustment, acupuncture and other forms of soft tissue dry needling.

Once again, watch the breath—always!

ACTIVE MOBILITY CORRECTIONS

Active mobility corrections are directed at limitations affecting normal, active freedom of movement quality. These include any form of self-started movement with a focus on mobility and mobility gains. Examples of active mobility are dynamic stretching and work on the agonist and antagonist relationships.

A good rule of thumb: If a muscle appears tight, its antagonist usually appears weak. This is probably not as simple as tightness and weakness. It can be reciprocal inhibition where increased activity of the agonist reduces tone, activity and contractile quality of the antagonist. Any activity that exercises the antagonists and lengthens the agonist can be considered an active mobility technique.

A therapeutic technique known as proprioceptive neuromuscular facilitation (PNF) uses specific exercises like *contract and relax* and *hold and relax* to combine both passive and active techniques, but are considered active since the client must perform muscular contractions with some degree of active control at some point in the drill. Formal and informal study of PNF principles is highly recommended since most of our corrective exercise principles have roots firmly entwined in this respected and valued work.

We use passive mobility corrections before active mobility corrections to reduce mechanical resistance, improve sensory input, reduce guarding, to create familiarity with new positions and to reduce stress breathing.

ASSISTIVE MOBILITY CORRECTIONS

Assistive mobility corrections are directed at limitations affecting normal, active freedom of movement quality and quantity. Assistance is a combination of active and passive movements where each contributes to the correction or completion of a movement or movement pattern. If performed for correction and not conditioning, any form of spotting can be considered assistance.

- **Assistance for quality**—manual or mechanized help provided to improve alignment, support, balance, form or facilitate sequencing

- **Assistance for quantity**—manual or mechanized help provided to improve exercise volume or complete range of motion

Assistive mobility corrections are techniques that combine active mobility with assistance from you, or a resistance device essentially used backward. This means the load supports and aids a movement pattern or improves postural control. Assisted corrections are opportunities for an exerciser to perform a whole movement pattern with a load less than bodyweight. Aquatic movements can be considered assisted exercise, but resistance devices can also be used to provide assistance if positioned correctly. These are more practical, convenient and adjustable.

Assisted movement patterning allows the individual to perform more repetitions or explore greater range of motion without the full load of bodyweight. Assisted work also reduces energy expenditure and provides increased exercise volume to improve posture and movement learning opportunities through repetition.

Assisted work can be a transition between passive and active mobility corrections. Assistive mobility corrections are helpful when there's a large difference between active and passive movement, which in reality is a motor control or stability problem when there's no passive limitation. The problem is that coordinated control is not present toward the end range.

The passive assistance into greater range is not forced in the sense of a stretch. There is gentle guidance into the potential range, blending active movement with passive guidance and taking some share of the load.

Less help is always better than more help, because input is more important than output. The individual is not just doing this to finish a movement; you want the person to feel what it actually takes to make this movement happen actively.

Make sure the client or patient is breathing normally and not fighting the movement.

Assisted work can be used to add volume to newly acquired active mobility. This provides greater opportunity for sensory motor interaction, while avoiding fatigue that could reduce volume and learning opportunities.

SUMMARY OF MOBILITY CORRECTIONS

Since many poor movement patterns are associated with abnormalities in muscular tone and coordination, all of these methods play an important part in mobility efforts. We can't simply lengthen a tight muscle or move a stiff joint and think we have effectively changed a movement pattern, even though these very simple acts may be the fundamental starting point. We instead need to broaden the breadth and depth of our perspectives of poor mobility on movement patterns.

When we find increased muscular tone indicating facilitation, we will also find the reduced muscular tone of inhibition. This normally occurs simultaneously in different or opposing regions of the body. Effective normalization of mobility and muscle tone should precede efforts to reconstruct or improve basic motor control or fundamental movement patterns.

This is usually the most efficient and effective way to initiate a change, and it is often missed. Screening and assessment will keep us all where we need to be in this respect. The act of breaking a dysfunctional pattern down before attempting to replace it with a functional pattern is the starting point of correction. Remember, this is not a blind mandate or impractical rule. If no mobility problem is discovered or if mobility changes with minimal effort, the work is done and you should move on. You don't need to rehearse mobility if it is present. You need to gain control, to gain stability.

Restrictions, stiffness and inflexibility can cause a mobility problem that might be present for two fundamentally different reasons. These mobility problems can be produced by damage or dysfunction.

DAMAGE OR AN UNMANAGED PROBLEM

Pain, compensation and incomplete normalization after an injury can all produce a region of chronic stiffness or tightness, which can be considered a primary restriction or limitation. It is a physical limitation that restricts freedom of movement passively and actively. The damage could be traumatic or micro-traumatic. This is essentially an unresolved injury or problem that has not returned to a normal level of function.

In clear cases, a loss of mobility can be traced back to a single incident, yet other histories will be more complex. Scar tissue, fascial and connective tissue limitations, trigger points, capsular restrictions, post-surgical complications and degenerative problems can all measurably reduce mobility.

Increased muscle tone can be a major limiting factor, and this can be a result of local or segmental dysfunction. Local dysfunction would be dysfunction caused by poor contributions of agonistic muscles or dysfunctions in nearby joints. Segmental dysfunction involves the problems at the spinal segment associated with the key nerve roots for a muscle or group of muscles.

DYSFUNCTION

The compensation might be a crutch to deal with some level of dysfunction, a naturally evolved compensation created by a brain that can't locate a good, reproducible pattern for motor control in static or dynamic situations. In this case, stiffness and tightness is its only option. We often see this when global muscles are recruited to work in situations requiring local stabilization.

This case of stiffness and tightness is not a result of an injury. More likely, it is the result of a habit, faulty movement pattern or an activity performed repeatedly with poor form, alignment, posture and coordination. The body leans on this stiffness or tightness when unable to use fundamental or functional motor control and the stiffness becomes part of the person's movement.

Poor tissue extensibility, increased muscle tone, joint degeneration and general stiffness can be byproducts of poor authentic stabilization. These form naturally adapted restrictions that reduce the need for normal sensory motor interaction. The system creates mechanical integrity, but this has limitations. Stiffness is not stability. It is not authentic and it lacks refinement and situational adaptation. Therefore, the current mobility problem may very well be the result of an underlying stability problem that may reemerge once the mobility problem is corrected.

Just be ready—observe responses to mobility gains and stability corrections.

It might be impossible to discern the initial cause of a mobility problem. It is convenient, but regardless of the cause, you need to remove as much of the limited mobility as possible. Then you need to provide an environment for the individual to relearn the levels of stability. It could take a few days or a few months, but if you continually and correctly progress into stability training as you make systematic gains in mobility, you will reinforce the mobility and improve sensory input.

All conventional and common forms of passive and active mobility and flexibility work can be considered in the mobility-correction category. One powerful addition to the mobility corrective exercise toolbox is the use of reverse patterning. These techniques reinforce the increased mobility and provide alternative movement learning opportunities. You'll read about this in greater detail in chapter 14, under the advanced corrective exercise techniques called movement pattern retraining.

BASIC STABILITY OR MOTOR CONTROL CORRECTIONS

Once we have mobility, we need to control it. If your client or patient has just gained mobility, he or she needs to own it. The new mobility needs to be challenged, not exercised. Unfortunately, exercise often becomes mindless motor rehearsal with the assumption of lasting benefit. As you read this section, keep the first and second rules of stability corrections in mind.

1.   Establish that adequate mobility is present, and if it isn't, improve it in an appreciable way.

2.   Provide a rich sensory experience to stimulate sensory motor memory and reestablish postural control and movement patterns.

We'll look at three subcategories within the basic stability corrections. Facilitation techniques in each subcategory will also be presented. They are—

Fundamental Stability— Motor Control Corrections

Assisted Exercises
Active Exercises
Reactive Neuromuscular Training—
    Facilitation

Static Stability— Motor Control Corrections

Assisted Exercises
Active Exercises
Reactive Neuromuscular Training—
    Facilitation or Perturbation

Dynamic Stability— Motor Control Corrections

Assisted Exercises
Active Exercises
Reactive Neuromuscular Training—
    Facilitation or Perturbation

Your clients and patients must understand your rationale for these corrections, so again you must learn to develop your dialog when discussing stability. Stabilization is all about motor control, and motor control is not about motor output or simply practicing movements. It's about optimization and refinement of the interplay between the perception of the senses and the movement behaviors.

Each corrective technique presents a greater level of difficulty. Assistance techniques provide a gradient of increasing sensory information and safety. Active techniques provide time to learn, create memory and refine control. Active techniques also allow for increases in volume for learning fatigue management.

Reactive neuromuscular training (RNT) uses resistance in a unique and non-conventional approach. Resistance is not applied to foster strength or muscular hypertrophy; a small amount of resistance is applied to facilitate and refine movement patterns. When body segments do not maintain favorable alignment or contribute to the overall pattern in a complimentary manner, we can use resistance to *feed the mistake*. RNT will be explained in greater detail in Chapter 14 in the section on retraining movement patterns and again in the Jump Study appendix, beginning on page 359.

What follows are some simple statements about regaining command of stability. These represent steps toward progressive control and corrections for stability—they are not exercises in the classic sense—they are experiences. Experiences are opportunities to explore a challenging posture or movement pattern. This may seem remedial, but a true stability problem is a subconscious problem. This means the brain is actually causing or allowing poor stability before the conscious brain is aware of it.

This is not an annoying strength problem to overcome by forcing underactive muscle groups into submissive slave labor. True stability problems must be safely challenged with minimal opportunity to compensate. Stability corrections must be managed to maximize sensory motor interaction. Mistakes will be common, but compensation options will be removed. The experience should be designed to put the brain right on the edge of control.

The challenge is to the brain, not the body. The brain has developed and memorized and repeated faulty patterns. It would prefer to continue the same behavior as long as it perceives it successfully can. Corrective exercises can be designed to make the brain perceive challenges it cannot control without developing a new behavior. This will jump-start the learning process for two reasons.

• The brain is challenged.

• Normal methods of compensation have been removed, therefore a new solution must be created.

Reinforcing a positive experience can easily become an exercise, but first make sure the experience is positive, yet challenging. Remember, this does not mean mistake-free; it means compensation-free.

Stability experiences are opportunities to—

Coordinate body segments without postural load

Disassociate body segments without postural load

Not fall over with a postural load

Not lose balance in different postures

Maintain control as the base of support narrows

Shift weight without falling

Move some body segments without falling

Resist external force without falling

Manipulate external weight without falling

Transition between postures without falling

Move to a functional posture without falling

Perform movement patterns in reverse order

Perform functional movement patterns with alignment and coordination

Perform functional movement patterns with alignment and coordination unaltered by external weight or load

Perform functional movement patterns with challenges that exaggerate mistakes in alignment, posture and coordination

When appropriate, challenge movement by reducing sensory input

The list above is descriptive of your first two or three years of life. In those important years, you gained more motor control than any other time in your life. Your greatest movement achievement is behind you! Everything else you've ever done is just a variation or refinement of the patterns you learned, refined and developed without a movement coach, teacher, trainer or therapist.

Your brain likes to see, feel and hear things, and it needs your body to put it in the best possible places and positions to make that happen consistently. You like to perceive. Your movement is a result of your need for the sensory exploration of your environment. Your sensory experiences molded and refined your movement achievements. The more you moved, the more you compounded your sensory experience. Two key ingredients are necessary for a smart system like your brain and body to teach itself to move: a rich sensory experience and a safe environment to explore.

Let's develop that thought.

Your visual, vestibular and proprioceptive experiences all blend together in a totally silent and dark chamber inside your head. Your brain takes the flood of input and information and creates a space and time experience—your perception. The experiences are stored as input and output patterns resulting in a chain of experiences linking perception to behavior and behavior to perception: perception—behavior—perception—behavior. All of it becomes sensory motor memory.

You start to access sensory motor memory each time a pattern seems familiar. If everything works out, you consider the experiences related to older memory, and if things don't work out so well, you develop new sensory motor memories based on the differences of similar patterns.

An example would be stepping in a puddle versus stepping into a pool. The first part of each experience is the same—shiny water, one wet foot. The experiences are similar in the beginning, but completely different at the end. That constitutes a good reason to write a new sensory motor memory. In this way, the brain starts to learn.

Experiences produce sensory motor patterns that produce sensory motor memory. All new experiences are viewed against previous patterns in memory for similarities and differences. The brain either develops a new sensory motor experience pattern or uses an old pattern. Obviously, all situations differ in some way and the brain makes the necessary adjustments, but the underlying pattern that started the process is taken from a pattern memory.

POSTURES FOR STABILITY CORRECTIVE EXERCISE

The three categories of posture used for stability corrections are integral to developing the appropriate sensory experience and regaining functional stability. We have already reviewed techniques, but these are not enough. They are simply increasing portion sizes of sensory motor experience that must be introduced at each level of development.

The real essence of corrective exercise for stability can be found in the natural stages of growth and development. Many postures and positions are used to get from lying to standing. Each stage is a mile marker of stability, and each posture and position creates a fallback platform for the next. Every new level of difficulty requires sensory motor integration. It is defined by a new perception that stimulates a new behavior. The new behavior sets off another new perception, and the process continues until control is gained and efficiently reproduced.

The three levels of postural control and movement coordination used to regain stability are—

Fundamental

Fundamental postures are simple. Just lie down on your stomach or back, and then change between the two. That's as fundamental as you can get—rolling—yes, just rolling, and it's extremely powerful as a test and as an experience when used correctly.

Transitional

Transitional postures are all the postures between prone and supine and standing. They include prone on elbows, quadruped, sitting, kneeling and half-kneeling, as well as all the positions between them. These don't look like exercises any more than rolling does, but they are definitive when sensory motor control is compromised.

Functional

Functional postures are non-specific postural variations of standing. The three basic foot positions in standing are symmetrical stance, asymmetrical stance, and single-leg stance. More advanced functional postures can be explored for specific activities, but the best foundation for specific functional activities is competency with non-specific functional postures.

Fundamental stability corrections are performed when mobility restrictions are removed and stability is noted as dysfunctional at essential levels. This means adequate mobility is present, but stability is compromised at functional, transitional and fundamental levels. Functional stability problems are noted in functional positions like single-leg stance and squatting, and transitional stability problems are noted in postures like kneeling, half-kneeling and quadruped.

Therefore, fundamental stability corrections are needed when stability problems have been consistently observed in all positions requiring postural control. This leaves activities that require motor control without significant responsibilities for postural control.

Any activity above prone or supine will require postural control, making supine and prone positions the platform for all movement. Rolling is close to ground zero as far as movement patterns go. We use rolling to reset fundamental programming that may provide improved levels of motor control at higher levels of postural control and function.

Supplementary exercises can be performed in supine and prone, but that usually only represents a partial pattern. Bridging, leg raising, leg extensions and PNF patterns can be performed in prone, supine or side-lying, but these are mostly performed for supplemental reasons. This means they are normally used to train individual body parts or partial patterns prior to full movement-pattern work. If these are necessary prior to rolling work, they should be performed. If rolling difficulty is noted, these may prove to be temporary options to facilitate movement into the rolling patterns.

ROLLING

It is inappropriate to perform rolling movement patterns when mobility problems interfere with the relaxed prone and supine starting and ending positions. Unrestricted prone and supine positions are necessary to even consider rolling as a test or as a corrective strategy. Furthermore, full or near-full open-chain shoulder and hip mobility is required for rolling tests to be considered reliable. These tests are done to both observe and correct the most fundamental levels of motor control and sequencing of body segments.

In growth and development, the ability to sequence the head, neck, shoulders, thorax, pelvis and hips precedes activities involving loaded postural control. Rolling is the single corrective in this category and is often overlooked in fitness, conditioning and orthopedic rehabilitation. Rolling is commonly used when rehabilitating neurological problems, but for some reason rolling is not widely incorporated in conventional corrective exercise strategies that don't involve neurological rehabilitation.

Rolling is mostly performed as an active movement pattern. There is rarely a need to facilitate rolling with RNT exercises, but they have been developed and can occasionally be helpful.

When rolling presents too much difficulty, some form of assistance must be used. Manual assistance is an option, but a wedge or unilateral elevation is consistent and practical. This can be a half foam roll, a thick mat or a rolled-up blanket or beach towel extending from the glutes to the shoulders. Any small amount of elevation will create an advantage and make rolling easier. Imagine rolling downhill—the assistance makes coordination and sequencing possible as it helps the brain access the rolling memory. In most cases, you'll be able to remove the assistance within a single session.

Rolling can be repeated as an exercise to overcome a particularly difficult or faulty pattern. It may also be helpful to reinforce rolling for up to a week. For chronic problems, rolling can be used to check fundamental integrity before activity and even as movement preparation if it has been found helpful.

Once rolling is performed successfully and not limited or asymmetrical, it is advisable to advance to a transitional posture. There is no need to try to turn rolling into a repeated exercise or a conditioning circuit. Remember, rolling is fundamental. This means when you can, you can and when you can't, you can't. When you can't, you need to fix it, and when it's fixed, you need to progress it. There is no gold medal for rolling—fix it and move on.

Initially, we should avoid making the rolling experience too complicated. Below are some steps to reduce confusion and improve efficiency when appraising rolling and developing it as a corrective strategy. Think pass or fail.

Use the steps below to help eliminate confusion.

FMS ROLLING

In the FMS, rolling is used as a corrective strategy for dysfunction noted in the rotary stability test. Only one rolling pattern is used, a difficult cross-body flexion-based rolling pattern. The start position is lying flat with arms extended overhead. Your client will perform the pattern by bringing a flexed elbow to the opposing flexed hip and knee.

It is necessary maintain contact between the elbow and knee throughout the rolling pattern. The roll is always performed to the side of the non-flexed elbow. The neck should not be flexed, and the head should lie flat, in line with the spine. Head and neck movements into rotation initiate the rolling motion.

- Make sure neck stiffness or neck problems are not a limiting factor to rolling.

- Observe breathing, watch for breath holding and unnecessary strain—this should not be a struggle.

- Use assistance as needed. Assistance can be in the form of manual assistance, or in the form of chocking. A chock is simply a lift for one half of the body to help gain an advantage with rolling. The chock can be a pad, mat or half foam roll, placed under the shoulder and hip opposite the rolling direction.

- Another way to reduce difficulty is to change the rolling pattern from a crossed-body pattern to a unilateral pattern. The unilateral pattern allows the user to acclimate to the rolling movement with less difficulty. It is performed by bringing the same-side elbow and knee together, and rolling to the opposite side. This should not be considered a rolling exercise, since rotary stability correction ultimately needs the challenge of the cross-body pattern. It's simply a transitional phase.

- This is a difficult movement, but it is appropriate if you use the screen correctly and observe contraindications.

The FMS rolling correction is a cross-body roll pattern and it is very difficult. Some individuals will attempt to sample the exercise and will immediately become frustrated. They seem to forget that rolling may not be the correction they currently need—perhaps they have a fundamental mobility problem or asymmetry in the active straight-leg raise or shoulder mobility that compromises the rolling pattern. These little examples help reinforce the *mobility before stability* rule.

Sample any corrective exercise you like, but use your head. In most cases, you will feel the exercise is too easy or too difficult and you will be right. Randomly sampling corrective exercise is like randomly sampling medication with no consideration to diagnosis or dosage. Corrective exercise is highly specific to a particular movement problem and will not necessarily produce results when it is randomly sampled.

The entire appraisal performed correctly can easily be completed in less than 30 seconds, so don't make a big deal of it.

SFMA Rolling

In the SFMA, we use rolling as a breakout test when mobility is established and motor control is dysfunctional throughout all testing requiring weight bearing and postural control against gravity. Rolling shows up throughout the SFMA as a base test for sequential control.

- Unlike the FMS, the SFMA uses four quadrants of rolling to observe dysfunction. The upper quadrant is the upper extremity shoulder girdle, upper spine, head and neck. The lower quadrant is the lower extremity, pelvic girdle and lower spine. Each provides information about movement pattern, sequence, symmetry and direction information. By performing four movements from prone to supine and from supine to prone, the eight patterns of rolling create four opportunities for bilateral comparison.

- Prone-to-supine rolling patterns look at overall stabilization and sequencing with movement initiated primarily in the posterior chain musculature.

- Supine-to-prone rolling patterns look at overall stabilization and sequencing with movement initiated primarily in the anterior chain musculature.

Below is a list of the eight rolling patterns.

1.   Prone-to-supine rolling to the left—right upper-quadrant initiation

The right upper extremity and neck movements are performed to initiate and complete the rolling pattern. No part of the lower body is used, and the left upper extremity is not used.

2.   Prone-to-supine rolling to the right—left upper-quadrant initiation

The left upper extremity and neck movements are performed to initiate and complete the rolling pattern. No part of the lower body is used, and the right upper extremity is not used.

3.   Prone-to-supine rolling to the left—right lower-quadrant initiation

The right lower extremity and lower spine movements are performed to initiate and complete the rolling pattern. No part of the upper body is used and the left lower extremity is not used.

4.   Prone-to-supine rolling to the right—left lower-quadrant initiation

The left lower extremity and lower spine movements are performed to initiate and complete the rolling pattern. No part of the upper body is used and the right lower extremity is not used.

5.   Supine-to-prone rolling to the left—right upper-quadrant initiation

The right upper extremity and neck movements are performed to initiate and complete the rolling pattern. No part of the lower body is used and the left upper extremity is not used.

6.   Supine-to-prone rolling to the right—left upper-quadrant initiation

The left upper extremity and neck movements are performed to initiate and complete the rolling pattern. No part of the lower body is used and the right upper extremity is not used.

7. Supine-to-prone rolling to the left—right lower-quadrant initiation

The right lower extremity and lower spine movements are performed to initiate and complete the rolling pattern. No part of the upper body is used and the left lower extremity is not used.

8. Supine-to-prone rolling to the right—left lower-quadrant initiation

The left lower extremity and lower spine movements are performed to initiate and complete the rolling pattern. No part of upper body is used and the right lower extremity is not used.

Only compare a rolling pattern to its contra-lateral counterpart. This means to look at prone-to-supine upper-quadrant rolling for left-to-right symmetry. If no asymmetry is found between the four patterns, consider the pattern that produces the greatest overall symmetrical difficulty as a potential dysfunction. If all rolling patterns are intact, do not consider stability as a fundamental problem. Move to transitional postures.

Do not make the rolling experience too complicated. Below are some steps to reduce confusion and improve efficiency when appraising rolling and developing it as a corrective strategy. Again, think pass or fail. You are rating and ranking rolling, not measuring it.

- Make sure the starting position is possible and comfortable. Both supine and prone positions should be viewed with arms overhead and slightly abducted.

- Make sure all available mobility is present for rolling. This includes the cervical spine, since four of the eight patterns involve C-spine range of motion.

- Don't ponder or deliberate perfect rolling. Look for substitution in quadrants that are not involved in the pattern. If there's no substitution, look for struggle and difficulty with rolling.

- Observe breathing, watch for breath holding and unnecessary strain—this should not be a struggle.

- Use assistance as needed.

The entire appraisal performed correctly can easily be completed in less than two minutes. Once you identify a difficult or faulty quadrant, use assistance to make the pattern possible. Allow the struggle, but remind the person to breathe and relax.

STATIC AND DYNAMIC STABILIZATION CORRECTIONS

Static and dynamic stabilization corrections can be applied in both transitional and functional postures. Think of fundamental stabilization and rolling as attempts to make sure perception and behavior of movement systems are in working order. With rolling, you're simply checking for faulty circuits.

Static and dynamic stabilization corrections introduce progressive levels of integrity to the movement system. Some exercises require static stabilization at one body segment and movement at other segments; in these movement patterns, there's both a dynamic and a static component. Static stability corrections focus on challenges to the static component, whereas dynamic stability corrections focus on challenges to the dynamic component.

STATIC STABILITY CORRECTIONS

Static stability refers to a body segment that must remain stationary under either a consistent or a changing load. Corrections are performed when mobility restrictions are removed and stability is seen as dysfunctional but present at fundamental levels. This means adequate mobility is present, but stability is compromised at functional and some transitional levels. Functional stability problems are noted in functional positions like single-leg stance and squatting, and transitional stability problems are noted in postures such as kneeling, half-kneeling and quadruped.

Static stability corrections are indicated when stability problems have been consistently observed in some positions requiring postural control, but when rolling is not compromised, indicating fundamental motor control. Many corrective options are possible within transitional and functional postures, but the focus should always start with balance and postural control.

Static stabilization corrections are designed to challenge the individual to hold a posture or joint position against gravity, and eventually in the presence of force or perturbation. These techniques are performed in postures where some degree of balance and postural control is needed. Force or perturbation can be applied manually or an exercise device can be used.

TRANSITIONAL POSTURE— STATIC STABILITY EXAMPLES

Half-kneeling is a common posture to transition from quadruped to standing, therefore, each technique will be discussed in half-kneeling posture as an example. Half-kneeling connects quadruped postures to standing postures. First, let's address the use of half-kneeling as a corrective exercise posture.

Half-kneeling is kneeling on one knee with the other hip and knee flexed to 90 degrees, with the foot in front of the body for support. In contrast, tall-kneeling means both knees are down with both hips extended. By narrowing the base in half-kneeling to a central line in the sagittal plane, a challenging experience can be created. Be prepared to adjust the amount of cushioning under the down knee to create a level pelvis. Half-kneeling with a narrow base can be difficult for the person with a stability problem.

The transition from quadruped to half-kneeling requires a significant change in base of support. In quadruped, there are four points of stability and a large base of support, and the center of mass is well contained within that base. The transition to half-kneeling reduces the points of stability to two, one foot and knee and lower leg. The starting position should provide a base as wide as the shoulders. Most people can easily make this transition.

The true stability experience is delivered when the base is progressively narrowed to the point that both the foot of the front leg and the knee of the rear leg are on the same line. In this case, the base of support is narrow and the mass of the body extends beyond the base on the left and right sides. The majority of weight should be on the rear knee with the front leg used mostly for added balance and control.

The benefit of half-kneeling is the challenge to the weight-bearing hip. The hip should be in neutral position, with a neutral pelvis.

As most people get into half-kneeling, they initially make one of two mistakes, either restricted range or excessive range. A slightly flexed hip with an anterior-rotated pelvis represents the restricted-range position. For these people, instructions toward a posterior pelvic tilt will neutralize the pelvis and bring the hip into a neutral position. The excessive-range position turns the posture into a hip flexor stretch. Instead of using motor control for stability, the person uses a lazy posture and just hangs on the hip ligaments and hip flexor muscle tone.

- A great way to help the client or patient get into the correct position in half-kneeling is to push down through the shoulders—actually push and let up repeatedly. The person will either feel squishy or rigid and stable. If you feel squishiness, tell the person to become rigid or stiffen posture, but watch for shrugged shoulders. This is not the goal—it is the compensation. You want a tall spine on a neutral pelvis with the hip at zero degrees. Don't tell the person this; just keep pushing until the individual finds it. You want to feel a rigid connection between your force and the floor under the down knee.

- You will have situations where half-kneeling is not possible and you will need to skip the posture altogether or modify it in some way. Here the best modification is to allow the person to kneel on a soft pad placed on an elevated surface. Half-kneeling on the floor is performed with one hip at 90 degrees and one hip at zero degrees. With an elevated surface, the weight-bearing hip will still be zero degrees, but the flexed hip can be 45 degrees. Remember, it's all about the hip position. The most important thing is to get the hip to zero and see if the person can maintain the position.

- Half-kneeling offers a unique perspective since it effectively removes the foot, ankle and knee from the weight-bearing pattern. The mostly single hip, one side of the pelvis and spine take the load. This offers a unique perspective into core stability that is often overlooked.

- It also creates an interesting stabilization experience, because many individuals who have poor core stabilization can compensate with increased activity and compensation at the foot, ankle and knee. They can also compensate with poor hip, pelvic, spine and shoulder positions, as well as faulty alignment. In half-kneeling, all compensations are removed. You have a clean perspective of transitional motor control symmetry.

- Always look at both sides, even if you're convinced the problem is unilateral.

ASSISTED

The assisted situation is necessary for some people each time the base becomes narrower. You can provide the assistance by giving nudges to help maintain balance, or the person can steady him- or herself with one or even both hands using support. Start with as much assistance as needed and then remove it. Allow a little struggle so the brain can become acquainted with the limits of stability. This is the rich sensory environment that challenges the brain to access an old unused pattern of motor control.

The experience and exercise seem so simple, no fancy ratio of sets and reps, no cool equipment— no thought-provoking verbal cues or coaching tips. Just, *"Do this and try not to fall over."*

For the people who need this corrective, this is just enough to totally tax them. It has nothing to do with fitness or level of conditioning. Many people will try to muscle it out, using every muscle in the body. Let them struggle.

Remind them to breathe; remind them to relax. Remind them a three-year-old can do it with much less effort, which is the key. *Don't fall, and use minimal effort to do it.*

The problem is that those who fail are so busy acting that they cannot feel. To say it differently, they are so busy behaving, they cannot perceive. Remind them that if balance is not automatic, it's almost worthless. Tell them to control their breathing and relax the neck, shoulders and arms and just work it out.

ACTIVE

Once the client or patient no longer needs assistance to maintain a narrow base in half-kneeling and has no asymmetry, have the person move the arms into different positions. This movement will cause a weight shift and will require greater balance reactions and motor control. Next, request a turn of the torso, first with the arms by the sides and then with the arms overhead. Make sure the movements of the arms and torso do not alter the pelvic or hip position. Make the person reset and start over each time after losing balance or static posture in the pelvis or weight-bearing hip.

RNT

When the person can perform active movements with the arms and torso with control and symmetry, you're ready to try reactive neuromuscular training. This activity will use dynamic work in one part of the body to cause a perturbation through torque generated when external force is applied.

Two levels of RNT can be performed. The first is *don't let me move you out of position*, and the second is *do this and don't move out of position*. The first only requires a reaction; the second requires an action followed by a reaction or countermeasure.

- First, have the person try to maintain a static posture as you try to cause a shift. Do not confuse this with a strength test—force is not the issue. Handling a change in force direction is the real goal. Look for a delay in the ability to perceive a change in direction—the longer the delay, the bigger the problem. You can twist the client or patient at the shoulders, or have the individual's arms in front while you push the hands up or down along a diagonal going from the down knee to the opposite shoulder.

- Remember, this is a perception drill, not a force drill. Some people will try to make up a big perceptive delay with a large amount of force, so just keep switching directions. Remind them not to push you. Say, *Don't push me; just don't let me push you.* Say, *"ust tap the breaks so I can tell you have control—you don't need to slam on the breaks; just tap them.* This seems to be a good communication tool if used correctly.

- Second, have the individual perform a push or pull movement in the half-kneeling position, keeping the force in one plane. Oscillatory movements can also be performed, where the person moves from a higher resistance to a lower resistance in a rhythmical fashion. You might also request multi-planer movements like chopping and lifting.

- Lastly, the client or patient can perform impulse movements such as throwing and catching a medicine ball. This is often the most difficult, moving from a loaded situation to an unloaded situation in a rhythmical fashion.

FUNCTIONAL POSTURES— STATIC STABILITY EXAMPLES

Supported single-leg stance is an often-overlooked posture between half-kneeling and the single-leg posture. We'll cover each technique in a supported single-leg stance posture. First, here are a few thoughts about the use of supported single-leg stance as a corrective exercise position.

Supported single-leg stance is *almost* single-leg stance. It's not 100% single-leg weight-bearing; it's more like 80 or 90 percent. We achieve this with a small step or stool around six to ten inches tall. The step-supported leg helps with balance and mostly offsets its own weight. The position allows for perturbations without significant mistakes or compromised safety. Start at shoulder width and progressively narrow the base until reaching the midline.

ASSISTED

Apply the same model here as used in half-kneeling. Attempt to transition into single-leg stance by having the person lift the foot off the stool for short periods without losing control.

ACTIVE

Apply the same model here as used in half-kneeling. Transitions to single-leg stance can also be attempted.

RNT

Apply the same two levels of RNT used in half-kneeling. Remember, this is still static stability, so all reactions should successfully maintain posture, alignment and control.

DYNAMIC STABILITY CORRECTIONS

Dynamic stability refers to a body segment that must remain controlled in one direction of movement while moving in another direction under a consistent or changing load. Dynamic stabilization corrections are designed to challenge the individual to maintain lines and angles under movement and load. Therefore, dynamic stabilization corrections challenge the person to maintain movement pattern quality or hold correct alignment while maintaining balance in the presence of load, force and movement. Corrections are performed when static stability is present.

TRANSITIONAL POSTURES— DYNAMIC STABILITY EXAMPLES

Quadruped is a transitional posture. A common exercise called the quadruped diagonal is usually performed from the neutral position on all fours, moving into shoulder flexion and hip extension. The shoulders and hips work in opposition; if the right shoulder is flexed, the left hip is extended. In this maneuver, the static segments are the un-moving shoulder and hip, along with the spine.

When the exercise is performed with a focus on one pattern at a time, the static stability of the supporting side can be observed. It is not uncommon to see awkward or jerky movements on the flexing shoulder or extending hip and assume the problem is in the moving segments, but most likely the problem is in the transition from four points of stability to two alternate points of stability.

This exercise becomes dynamic when the patterns are alternated between one hip-and-shoulder pattern and the other, alternating diagonal patterns. Opposing hips and shoulders quickly transition between static support and dynamic movement.

ASSISTED

To assist alternating patterns between supporting and moving limbs, you can position the person over a partially deflated stability ball. The ball should not fully support the body—it should only assist. Just a little contact with the torso is enough in some cases.

ACTIVE

The active pattern could initially become more challenging by narrowing the base. This would have all the supporting limbs nearer to the midline of the body. Placing an object like a small towel roll over the upper or lower spine could challenge active alternating patterns. You could place the towel roll either parallel or perpendicular to the spine, depending on whether you wanted to challenge left-to-right excessive weight shifting or excessive lower spine and pelvis movement.

Movements could also be performed on a slide board without a lift by having the person slide opposing hands and knees away from and toward each other in a smooth, rhythmical fashion.

All three activities provide opportunities for observation and the experience of dynamic stabilization.

RNT

Again, two levels of RNT can be performed, the first being *don't let me move you out of position,* and the second, *do this and don't move out of position.*

- First, you can have the person try to maintain a static posture as you try to cause a shift. Do not confuse this with a strength test; force is not the issue—handling a change in force direction is the real goal. Look for a delay in the ability to perceive a change in direction.

Have the client or patient perform the quadruped diagonal pattern alternating from one pattern to the other. Push the hips or shoulders left to right to create a perturbation. You can also try to resist forward progress by pushing into the shoulders and instructing the person to try to crawl. Remember, this is a perception drill, not a force drill.

- Second, you can place light resistance across one flexing shoulder and the opposite extending hip. Lifting one set of extremities against a resistance load and one side without a load will create a unique dynamic experience. This activity will provide opportunities for observation and the experience of dynamic stabilization.

Pay special attention to asymmetry in both quality and volume. Quality may appear equal at first, but over a few repetitions, it may diminish significantly on one side.

FUNCTIONAL POSTURES— DYNAMIC STABILITY EXAMPLES

Single-leg deadlifting is a great example of dynamic stability in a functional posture. Static single-leg stance ability and the knowledge and ability of the deadlift movement are prerequisites for single-leg deadlifting. This should be obvious, but often it is not.

Dynamic stability training requires static stability and the ability to perform the movement pattern. The quality of the movement pattern can be questionable, but you need a raw pattern as the learning platform.

Single-leg deadlifts are performed with a flexed knee of around 20 degrees at the bottom of the movement. The knee is extended at the top of the movement, but this is in no way a squat or a partial squat. The tibia must remain vertical or near-vertical as the femur flexes backward on it. The object is to sit back as far as possible with the hips. The arms should hang in a vertical line just in front of the tibia.

ASSISTED

Assisted single-leg deadlifting is performed with one or both hands in contact with a support. This exercise can also be performed with a resistance device in reverse. The resistance is from above as you ask the person to pull into the movement. This helps pattern the hip-hinging movement, and can also be used to reinforce proper spine position.

ACTIVE

Active single-leg deadlifting is performed with no resistance or support. Always remember this is an opportunity for bilateral comparison. Active movements reinforce patterning and provide volume. Active movements should be used to maximize range in the movement pattern within the limits of good form.

RNT

Single-leg deadlifting is often used as a strength exercise, but it is also an excellent RNT corrective exercise. It serves as an appraisal and a drill for symmetry in single-leg dynamic stability.

Single-leg deadlifting can be performed two different ways. If performed with a weight in each hand, the load can be heavy but balanced. This strength drill is a perfect move for connecting the hip and core for functional activities. It also fits the definition of a self-limiting exercise.

The real RNT application of this exercise is found in the single-arm, single-leg deadlift. This is done with the arm opposite the stance leg performing the lift, which creates a cross-body load, and stresses the core muscles that control rotation. It loads the hip in three planes and forces the arch of the foot not to collapse. Knee valgus is not a problem when the foot and hip behave correctly.

Once again, the theme is not performance; it's comparison of symmetry along with form and co-ordination. Here we're less concerned with range than symmetry. Balancing the abilities of the hips and shoulder keeps the core balanced, and that is a fundamental goal. A rule of thumb is for the fist to make the midpoint of the tibia on each side. This provides enough mobility to exercise dynamic stability for perception and movement.

It is important to completely finish the single-leg deadlift. This means pull through and stand tall at the top. The weight should be released after each rep and the movement be cycled free of weight for every weighted rep, one set of ten actually looks like ten sets of one. The drill is about motor control, and this provides ample opportunity to both observe differences and correct them.

SUMMARY OF CORRECTIVE EXERCISE EXAMPLES

Here's we want to keep the focus on corrective exercise structure and not on specific exercises. When details were provided, it was to demonstrate that we try to control every aspect of a corrective session. Leave nothing to chance and never assume the exercise will do your work for you. The music does not play itself.

Throughout, it has not been our intention to discuss possible exercise options for each stage in corrective exercise progression. Exercise instruction is a multisensory experience and should be taught using as many forms of input as possible.

Map the presented framework. See where your exercise knowledge is sufficient and where it is lacking. Invest your time and resources in areas where you feel your influence is limited. If you feel lost when asked to produce multiple mobility options for a single area, invest in your education and become proficient in mobility work. If you feel lost with stability work, study in that direction. The education is only one step in four. The other three steps to expertise are practice, practice and practice.

If you're really good, you will make your practice deliberate, and screen and assess more often so you can have instantaneous feedback from your efforts. Practice and feedback turn knowledge into skill. All your peers have knowledge, but there are a select few with skill.

SUMMARY OF STABILITY CORRECTIONS

The senses provide all the information about time and space, but the information is often not pure or perfect. The problem is perception. This is the reason exercise and rehabilitation professionals often become frustrated with clients and patients, some of whom just don't seem to perceive movement.

Those of us working in fitness, conditioning, sports and orthopedic rehabilitation make assumptions about the neurological system. We roll out of bed and head to work armed with tools to fight tightness and weakness in the body without realizing the neurological system may have a perceptual or behavioral disadvantage. For some reason the sensory motor system may not perceive or behave optimally in certain movement patterns. As we push these patterns to improve through conventional exercise practices, we often reinforce the problem. We impose demands on output without improving input. *It's all about input.* Our clients and patients need time to remake their perceptions.

Strength coaches want to get on with the work. They want to lift, run, jump and build athletes. Personal trainers want to create energizing, entertaining, life-changing exercise experiences. Rehabilitation professionals want to solve complicated problems and create efficient outcomes. We all want to do what we were trained to do. The problem is, while we were studying our specialties, human movement patterns were eroding at the most accelerated rate in history. The result is we need to slow down and capture movement quality.

In manual medicine, we pride ourselves in our manipulation skills, but why do we stop at muscles, joints and fascia? We master the skills of creating mobility, but we often assume the system of motor control will reset or rebuild itself. When it does, we appreciate the miracle, but don't assume we will always get off that easy.

We need to master the sensory game and we can't master that by just studying exercises. Instead, we have to study nature's lessons about movement learning. Life provides obstacles that cause problems. These problems create perturbations. Perturbations are challenges and disturbances against posture, balance and movement patterns. Perturbations are ways to fool the system into reactivating a fundamental movement pattern and gaining stability in the process.

So why don't we also master the art of perturbations? Mastering perturbations is not as simple as introducing wobble boards and mushy pads to juice balance reactions. It is about dissecting the most fundamental posture to the point where balance and control can be challenged and still receive facilitation feedback to learn a consistent perception.

It's about finding the fundamental movement pattern that's awkward, poorly coordinated or asymmetrical, and introducing a simple sensory experience to improve it.

*Please see www.movementbook.com/chapter13
for more information, videos and updates.*

ADVANCED CORRECTIVE STRATEGIES

MOVEMENT LEARNING CONSIDERATIONS

In this chapter, we will review a case study as we look at the very important first corrective exercise session. This will be an involved description of something completely unimpressive to the outside observer, something designed for the recipient only. The outside observer will be looking for activity and will often judge the exercise by the output, but the focus of corrective exercise is not output. The focus of corrective work should start with optimal input. We are not trying to expend energy; we're trying to change movement. We are creating the most optimal movement learning opportunity based on the most appropriate information available—the current level of movement-pattern function and dysfunction.

You'll learn some of the concepts behind advanced corrective strategies and look at a few examples. Then we'll consolidate the rationale behind the framework for movement correction.

Here is where we stand: Movement screening and assessment assist in the identification, documentation and communication of problematic movement patterns. We know these issues are actually complex problems between perception and behavior. In the first part of this book, we discussed movement and presented it as a behavior. Later, the introduction of screening and assessment gave you the tools to specifically map movement patterns, and rate and rank them as functional and dysfunctional behaviors. Now, as we consider corrective strategies, the discussions of perception are intentionally becoming more frequent.

As professionals, we must realize that sensation and perception are not the same. The senses provide the brain with information, but that information is used at different levels of awareness.

Each individual can also interpret the same information in completely different ways. Experiences, memories, habits, previous injuries and lifestyle play a major role in perception. Ultimately, we know that perception drives behavior and behavior changes perception. From a truly intelligent standpoint, they are actually too connected to discuss separately.

The job of a teacher completely changes when a student starts to understand the subject matter, that point when the learner can actually connect the dots of why the action is performed. The same is true with perception. As soon as a client or patient perceives a movement pattern limitation or significant asymmetry in the form of a mobility or stability problem, the person can participate in the correction at a higher level.

LET YOUR CORRECTIONS DO THE TALKING

That heading doesn't mean what you think it means. At first glance, you might think the statement implies that we should let our results speak for themselves. Although this is important and absolutely true in a general sense, the idea really directly targets the individual perception of your client or patient.

Corrections are not simply solutions that we dispense. This statement is worth repeating. Corrections are not drugs doled out to mend faulty movement. They are first and foremost opportunities for our clients and patients to experience the actual predicaments that lie beneath the surface of their movement pattern problems. That is why we refer to corrective exercise as an *experience*.

First, we create an experience and then we use it to engage perception. Next, we repeat the experience so it can become the repeatable challenge. Discovering someone's movement dysfunction means nothing if the individual doesn't fundamentally understand what has been discovered. Be mindful that in many cases the person may not fully appreciate the movement dysfunction uncovered by your screening or assessment. The discovery is an experience, not an explanation. Explanations are helpful, but experience is fundamental to learning and correction.

Once the comprehensive testing and evaluation has indicated that corrective exercise is appropriate, it is best to identify the most fundamental limitation or asymmetry and move into basic corrections without lengthy explanations. Our discussions of a person's particular movement dysfunction, no matter how eloquent, are not as profound or enlightening as the multisensory experience of limited mobility or stability provided by a well-executed corrective drill.

When we move directly to the corrective for a fundamental problem, the individual will have a more focused experience with fewer distractions. Perception of the particular movement dysfunction is clearer, and that clarity creates a perceptive connection to the movement dysfunction. The perceptive connection sets a baseline of experience.

We often refer to this as *the ah-ha moment.* If you can learn to turn the information gathered in screening and assessment into that moment, you get it—you get it because you now know how to make the client or patient get it. This is why we discourage lengthy discussions of the *whys* and *hows* of movement during movement correction. Some people may want to have anatomical discussions of how and why, and if they are interested, that is appropriate, but not while they are juggling.

Think of this idea of *juggling,* because if you set up the corrective experience well, the person will be physically and mentally juggling to meet the goal. The individual will be engaged... and should be engaged. The engagement is with the senses, not through discussion or explanation. Save the dialogue for a rest break.

There is a zone where the person will naturally perceive and correct mistakes even when unable to verbalize the problem or the strategy to overcome it. The thoughts at this point may be *I just need to balance better* or *Wow, I can do it but I really need to concentrate.*

Don't interrupt during this time of juggling.

THE MANAGEABLE MISTAKE ZONE

The manageable mistake zone refers to a place in the corrective exercise stability progression where mistakes are frequent, obvious, perceivable and instantly correctable without lengthy instruction. This means the person is right at the outer edge of ability, with a clear picture of the goal of the drill. The key here is using mobility work and strategic position choices to remove compensations.

Of course, this setup guarantees that mistakes will be frequent and that's good, because constructive stress accelerates learning. Using a regression narrows the type of mistakes that will be perceived. In functional movement patterns like squatting and lunging, there are many possible types of mistakes. Movement mistakes can occur in lower body alignment, upper body posture, compensation, substitution, coordination or any combination of these. Therefore, the best corrective would reduce that squat into smaller segments.

SQUATTING EXAMPLE

If squatting is reduced to tall-kneeling and you instruct the person to find neutral positions for the hips and pelvis, perturbations can be focused on postural control and the limits of anterior and posterior stability.

Tall-kneeling reduces the mistakes seen in squatting to—

- **Loss of posture**—inability to maintain neutral hips and pelvis, either static or dynamic

- **Loss of balance**—anterior or posterior

In tall-kneeling rather than standing, there is less lateral control needed and more anterior and posterior control responsibility.

LUNGING EXAMPLE

If we reduce lunging to narrow-base half-kneeling and tell the client or patient to find a neutral position with the pelvis and loaded hip, perturbations can be focused on postural control at the limits of lateral stability.

Tall-kneeling reduces the mistakes seen during lunging to—

- **Loss of posture**—inability to maintain a neutral pelvis and loaded hip, either static or dynamic
- **Loss of balance**—lateral to the loaded side
- **Loss of balance**—lateral to the unloaded side

Using the half-kneeling posture offers less anterior and posterior control responsibility and more need for lateral control.

A CORRECTIVE EXPERIENCE CASE STUDY

Imagine performing a movement screen on a patient who no longer has pain and wishes to return to competitive triathlons. Alternatively, imagine a first-time client who wants your assistance to get in shape to attain the goal of competing in a triathlon. Whether you are rehabilitation professional or an exercise professional, participate in this thought experiment. Make notes for yourself and answer the questions asked before you read further.

The following is an example we can all identify with to some degree. Whether you work with patients, athletes or fitness clients, imagine yourself in this situation, able to identify with the responsibility of creating a corrective experience as well as making decisions about exercise progressions. The case study and thought experiment is designed to keep you from over-thinking movement. The design will also help keep you on the task of providing the best possible perceptive experience for setting up and initiating corrective exercises.

THE PROBLEM

A movement screen reveals lunging with the left foot forward and the right foot back is the most fundamental problem and greatest asymmetry. To reiterate, all scores on the movement screen are twos, with a one(L)/two(R) observed on the lunge test. At first glance, it appears the problem centers around the right hip region. During a left lunge—left foot forward—the right hip does not fully extend near the bottom of the lunge. The hip and body remain flexed at the bottom where they both should achieve a neutral position, and this seems to be the reason for the asymmetry. You don't need a doctoral degree in kinesiology to implicate the right hip extenders and stabilizers as the problem.

Thought experiment question: *Is this a glute problem or a left lunge-pattern problem?*

You may think calling this a lunge problem is a copout, an oversimplification of a problem that requires a much more specific answer. To be more specific, you could say something like, "We need to develop a glute-activation program for you." The average client or patient will probably just smile and say, "Okay, thanks," and may even think, *wow, you sound smart.*

He or she would be wrong.

The statement shows knowledge, but it's not that smart, and here's why. The glute you want to activate was totally active and symmetrical in all the other tests that required hip extension: the hurdle step, the squat, the pushup and the rotary stability tests. The only time hip extension was problematic was in the lunging pattern. Conversely, there was no indication of a hip flexor problem noted in the other screens. Lengthy discussions about the glute would actually be an oversimplification.

A thinking professional would deduce that the only thing unique about the lunge compared with all the other patterns is the movement pattern itself. The anatomical part—the glute—that was deemed dysfunctional and underactive was actually functional and active in four tests requiring hip extension. The deeper observation would be that the lunge movement pattern exposes or expresses movement dysfunction when the left foot is forward and the right foot is back, when the base is narrow, and when the hips are under load, a problem with loaded asymmetrical stance.

That's a mouthful, better to say there seems to be a left lunge-pattern problem. Logically it's not a glute problem, because that would have been noted in all patterns requiring glute activity.

- Movement-pattern talk often poses a problem because there does not seem to be an anatomical target or a specific location on which to work. It almost seems to be an intellectual letdown, perhaps a reluctance to identify the problem. This is only because we are stuck in a Kinesiology 101 mindset. It's more logical and defendable to discuss the problems within the behavioral aspects of the movement pattern that exposed the problem.

- Lunge patterns can improve with no appreciable change in glute strength, and glute strength can improve without automatically improving the lunge pattern. Even though it seems clean, concise and credible to implicate a single muscle, it is better to discuss the way the muscle is used within a particular pattern. If all movement patterns using hip extension were deficient, it would be logical to look at the nerves, muscles, joints and other tissues that could commonly present problems for hip extension regardless of movement pattern.

- Inconsistency has been observed in this scenario—hip extension is only dysfunctional and asymmetrical in this one movement pattern. This seems to implicate a motor control problem, which means something on the perception or behavior end of the left lunging pattern is problematic.

In contrast, a consistent problem would be present if all hip extension was deficient, regardless of movement pattern. Consistent movement problems point to a local problem involving the mobility, strength and integrity of a specific anatomical region. Inconsistent movement problems within a specific pattern point to a general problem not isolated to a specific anatomical region.

Deeper introspections would also imply that the problem does not seem to be right hip-extension mobility since that was reviewed in other patterns. The problem also does not look like a fundamental strength deficiency, because adequate muscular activity made the other movement patterns possible.

For argument's sake, you should be aware that this left lunge problem could easily be caused by the core, the right knee, rectus femoris, the left hip, knee, ankle, foot or the brain.

The main reason why the smart professional should not even discuss specific anatomy in this situation is that it will not improve the perceptive experience for the individual. The best possible approach for quick and clear perception is to double-check mobility just to be sure there is not a problem.

If mobility is acceptable, move right into the half-kneeling posture to view static stabilization abilities and see if an asymmetry is present.

Half-kneeling would seem to be the most appropriate choice of transitional postures when the rotary stability test did not demonstrate dysfunction or asymmetry in quadruped.

The lunge is dynamic and requires higher perceptive demands and motor control responsibilities—half-kneeling can reduce the movement to a static activity where the goal is simply balance.

Thought experiment question: *Move directly to the right-knee-down half-kneeling position since you're pretty sure that's the dysfunctional side, and that is where the correction should start. True or false?*

True? Don't go so fast. The correction is not as important as the findings and outcome of the initial experience. The answer to the above question is false.

A smart move would be to check half-kneeling with the left knee down first. This half-kneeling pattern represents the side where the lunge was not problematic. It would therefore be a logical deduction to expect less chance of dysfunction on the left-knee-down half-kneeling posture simply because the functional test was free of dysfunction.

By starting on the side with less potential dysfunction, you perform two helpful functions. First, you actually see if left half-kneeling is an additional problem. Second, if it is not a problem, you help set a perceptual baseline for the client or patient. This is the opportunity at a perceptual level for the person to appreciate the asymmetry you discovered.

Make no assumptions—there might be a problem on the "normal" side, but it's more likely the dysfunction in this scenario will be noted when the right knee is down. The experience starts when you move to the right-knee-down half-kneeling posture and instruct the person to slowly narrow the base.

First, the narrow base causes a perturbation. The loss of control and balance stimulates compensation behaviors. Increased tone in the neck muscles, shallow breathing, intense visual focus, shoulder shrugging, excessive anterior tilting of the pelvis, flexion and adduction of the hip and unnatural movements of the arms will show that primary control is not possible. The person needs to perceive this—this person *needs the struggle*, because moderate struggle can increase learning velocity. During this struggle for control, you need to keep the client or patient safe and relaxed.

Offer only small amounts of assistance. Give suggestions to help stay focused and in position, but don't instruct on how to do the move. Definitely don't tell the person to brace or contract anything. You are not trying to create an output—*you're trying to create an input*. Do what it takes to get to the correct start position and posture, but make the client or patient figure out how to hold it as the base and load are changed and advanced. Make suggestions to help stay between the over-thinking mode and panic mode, but the individual must find the stability; you cannot provide it.

Suggest that the person slow down and deepen the breath. Specifically for half-kneeling, the neck and shoulders should be relaxed; ask for a tilt in the pelvis in a posterior direction and get the hip totally vertical.

This is not to engage any particular muscle, but it is helpful and will accomplish two things. It will neutralize the poor pelvic alignment and put the weight-bearing hip in a neutral position instead of slightly flexed or hyper-extended. This might actually make the posture appear more difficult at first—this is because you took away a subtle compensation. Always remember that this cue was not made to get the person to fire a specific group of muscles; it was offered to create alignment. This neutral alignment can only be maintained with authentic stabilizers.

The position also provides the least amount of cheating or compensation options. In contrast, if you suggested that the person extend the hip, he or she would most likely extend the spine. Remember, the client or patient has lost this authentic pattern somehow, someway, and only he or she can find and remember it. You must force the person into the vulnerable position and let him or her work it out. Demand alignment and set it as a goal, but let the individual work out how to achieve it.

Thought experiment question: *From start to finish, the half-kneeling drill took about 10 minutes with breaks. From the outside, it didn't look like* much, but the narrow-base half-kneeling improved greatly in the short time. The next question you will get from the client or patient is how often it should be practiced. What do you say?*

Stop and try to formulate your own answer before you read the answer. Seriously, put this book down and write something in your notes. Try some corrective exercise program design on the fly based on the information you have.

The answer you might try:

"Don't practice this at all. Just retest it as often as you can.

"You see, after only a few minutes you improved your half-kneeling balance. Keep testing yourself and see how long it takes you to get your balance. The next time we work, I'll retest you, but my intentions are to move you up a level. Remember, this is a fundamental posture and you can do it fine on one side. It's not a new skill to learn; it's a lost posture and pattern you simply need to reacquire.

"When you can perform half-kneeling as good on one side as the other, I will advance your activity and keep you at the edge of your ability. This will accelerate your learning. Anytime the half-kneeling posture and activities present a problem, you simply need to keep testing and give your brain a chance to find the solution.

"Instead of testing for 10 minutes, try testing five times for two minutes. The more often you test, the quicker your brain will access the solution. Soon the solution will be automatic. I can't predict how long this will take, but just remember we are not waiting on weak or tight muscles to change—instead, this is learning. The stress of appropriate testing at the edge of your ability accelerates your learning."

Of course, this is actually attentive practice—encourage your clients and patients to test often. Remind them to test when they are fresh, not fatigued. Have them prepare for the test by doing anything that helps. They are welcome to stretch, perform a more basic drill, relax their breathing, whatever helps. Each time they test, tell them to expect improvement.

In the book *Naked Warrior*, Pavel Tsatsouline discusses *greasing the groove*. He directs the reader to practice strength movements without fatigue, with proper technique... and to do it often. To help the people you work with commit this to memory, use the saying *fresh, frequent and flawless*.

The repeated testing is corrective exercise in disguise. It feeds the curious and competitive mind. With this testing drill, the individual tests often, and only when fresh. It is unlikely the movements will be flawless, but your clients or patients need a clear image of the goal so they can shoot for flawless. The testing should be a chance to perceive mistakes and this is the point. For corrective exercise, we can change the rule to *fresh, frequent, with perceivable and correctable flaws*.

In the book *The Talent Code*, author Daniel Coyle presents the perfect case for this type of learning. He discusses stress, obstacles, learning velocity and all the best tricks to practice and accelerate learning. He also references research that reduced study time, increased testing frequency and produced better results in one group than extra study time and less testing did in other groups.

He contends it's not the practice; *it's the way* we practice. He refers to *deep practice* as a state where learning velocity is enhanced, and reminds us we're not to observe but to experience. His point is we're supposed to have obstacles and make mistakes, lots of mistakes. Without an answer, the brain is forced to find its own solution.

Think how often we teach exercises by demonstration. Oops—guilty as charged. We can't help ourselves sometimes—*Watch me do it, now you try.* We have all done this, but we can do better now that we know better. Remember, these are fundamental movements, not skills that often require observation and demonstration.

We just need to remember that perception is a highly personal experience. It's internal, individual and almost impossible to standardize. However, by creating standards for movement behavior, we are also capturing part of movement perception since the two are interdependent. We cannot make assumptions about perception, but we must always consider perception as we attempt to improve movement behavior.

It is often helpful to create a characterization for the complex problems we experience. It keeps things clear, light and even humorous.

We'll next represent the corrective framework in an unusual way.

THE PARANOID SYSTEM AND THE CLUELESS SYSTEM

We think of and remember information as patterns, and the following illustrations will help present the problems as patterns on a spectrum of movement perception and behavior. These characterizations are a blatant oversimplification, so please forgive the informality.

They are offered as an illustration of the contrasting problems within the movement systems that we must address and hope to correct. Remember, we are talking about systems here, not people or even personalities. Assigning personality characteristics to faulty movement systems turns them into outrageous cartoon characters to create contrast and to facilitate elemental understanding.

Screening or assessment identifies a problematic movement pattern. Next, our deductions take us to mobility or stability problems. For this characterization, consider mobility problems as the *paranoid movement system* and consider stability problems as the *clueless movement system*. Both are unflattering terms and labels, and hopefully no one will be offended by a bias in one direction or the other.

MOBILITY PROBLEMS

Mobility problems are identified when muscle tone or tissue stiffness restricts freedom of movement. The system is already behaving in a certain way, a restricted way. It has not waited on perception; it has already assumed and acted. For some reason, it has lost freedom of movement in one or more patterns. In the most basic terms, this system has actually lost some level of perception as a result of the reduction of movement. It's paranoid—let us say it's distrustful of perception and has decided to arm or brace itself against all things.

Obviously, we know this is not intentional, but the point here is that behavior seems to lead perception, or perception suggests that all situations should be handled with reduced mobility behaviors. A behavior of limited mobility has been thrust on all situations and will influence all other perceptions. Perception is disadvantaged, since it will always be influenced by limitations as the result of consistently limited mobility.

The perceptive system may even assume the limitation in mobility is actually normal. In biological terms, we call this homeostasis. The new normal is stiff and tight.

PROBLEM

To summarize for the purposes of this drill, let's just say that mobility problems are actually **paranoid movement behavior problems**. Previous injuries, movement habits, too much of one activity or too little of another have caused a reduction in mobility. Or, the body has chosen to deal with one or more of these problems by reducing mobility temporarily for the greater good—and that temporary act has turned into a chronic state.

SOLUTION

You must change mobility to change the behavior, breaking the behavior cycle. Once the cycle is broken, you have an opportunity to offer a new perception. As soon as mobility is improved, you'd move to a stability drill to use the new range.

Examples of the stability continuum—

Unloaded
Locally—active range of motion
Globally—appropriate rolling pattern

Static Loaded
Locally—isometric challenge
Globally—transitional or functional
posture challenge

Dynamic Loaded
Locally—PNF pattern
Globally—RNT drill

• *Locally* implies moving or engaging a body part or region, whereas *globally* implies moving or engaging the entire body. These progressions can occur in a single session or over multiple sessions.

• You can skip steps in the progression.

• The goal is to set up manageable mistakes to maximize the learning effect.

• Stay in the manageable mistake zone.

STABILITY PROBLEMS

Stability problems are identified when motor control does not create the most efficient and effective management of posture, alignment and coordination. Here's where we see sloppiness in movement appraisals and exercises and we can't seem to coach it away. In the past, we simply assumed the stabilizers were weak or lazy. We trained them to be strong, but they didn't become more reactive. We attempted to give the stabilizers endurance, but proper stabilization needs efficiency by responding faster and by improving the timing.

When we see poor movement patterns and can't find any underlying mobility problems, we need to question the efficiency of whole stabilization system—we can't just focus on the muscles. When we question efficient stabilization, we also need to consider perception. We've neglected to question perception altogether in our development of corrective exercises.

In this case, there is a lack of correct behavior in the system. The behavior is inappropriate, delayed or grossly incompetent—it's clueless. The entire system is bungling in its behavior choices. It is not efficiently and effectively providing static and or dynamic stability within one or more movement patterns. Once again, we know the act is not intentional or conscious, but the point is that perception is faulty or the corresponding behavior is faulty, or perhaps both.

PROBLEM

For the purposes of this drill, let's just assume for a minute that poor stability is a **clueless movement perception problem**. Previous injuries, movement habits, too much of one activity or too little of another have caused a reduction in stability. Alternatively, the body has chosen to deal with one or more of these problems by reducing stability for the temporary greater good, and that temporary act has turned into a chronic state.

SOLUTION

There is no behavior to break—here the goal is to enhance perception and to expose mistakes. There is actually behavior to make. You are looking for opportunities to introduce correctable mistakes. Thus, you'll move into stabilization challenges.

Examples of the stability continuum—

Unloaded
Locally—active range of motion
Globally—appropriate rolling pattern

Static Loaded
Locally—isometric challenge
Globally—transitional or functional
posture challenge

Dynamic Loaded
Locally—PNF pattern
Globally—RNT drill

Look familiar? It's the same continuum. The recipe is the same for both systems once mobility is not a limiting factor.

- These progressions can occur in a single session or over multiple sessions.

- You can skip steps in the progression.

- The goal is to set up manageable mistakes to maximize the learning affect.

- Stay in the manageable mistake zone.

Moving between mobility problems and stability problems can be challenging, but the rules are simple. You create appropriate mobility to perform a stability drill and then do the drill. If more mobility is needed to progress stability to the next level, go back and get more mobility before moving on to higher stabilization demands.

Never consider a stability correction if mobility is not appropriate or an appreciable improvement in mobility cannot be measured or documented. Corrective exercise for stabilization is all about being slightly vulnerable. Vulnerability only happens when you cannot lean on stiffness, poor alignment, acquired tightness, high threshold strategies or mechanical locking.

- Remove limitations to posture and alignment and then demand control of posture and alignment.

- Use postures, positions and movements at the edge of ability, but within the limits of possibility.

- Set up a situation for frequent mistakes that can be self-corrected.

- Use progressions to insure that challenge and mistakes stimulate learning.

People who have mobility problems need to use the new range, not lean on the edges of the old range. Stability corrections will expose mistakes, so use the progressions wisely and safely. The corrections will expose vulnerability, so don't let paranoid mobility issues return to save the day. It is so easy to tighten the neck, shoulders, lower back and hip flexors. Don't allow this—watch alignment and breathing.

Keep the person safe, relaxed but vulnerable to the mistakes to feel and correct. Let the client or patient know that the little mistakes are no big deal, and the more that are made, the more is learned.

Don't progress if you do not see a little success; progress when you do see a little success. Go back and forth between two levels of difficulty. Use one level to challenge and one to reinforce learning and build confidence.

Now that we have given a simple label to each problem let's see if the characterization will help us consistently manage each problem efficiently and effectively. Even though we know each problem is more complex than our simple labels, the labels may just force us all to apply the corrective framework more consistently. It will force us to challenge the system at its fundamental dysfunction.

CORRECTIVE SYSTEM OVERVIEW

MOBILITY FRAMEWORK

The next time you discover a mobility problem, you won't assume a movement pattern will improve until a change is created in the mobility behavior. You must locate the local or global mobility issue and make a primary step by producing some degree of change. For whatever reason, the system has decided to behave with limited mobility and there will be no need to attempt to improve a movement pattern until you remove the paranoid movement behavior.

All your efforts should be focused on changing mobility and documenting an improvement. From that point, you can stabilize until you run into another mobility problem or move onto another movement pattern. This all depends on the priority you identify through your screens and assessments.

STABILITY FRAMEWORK

The next time you discover a stability problem, you won't assume a movement pattern will improve until the stability problem is broken down to promote a change in the perception, demonstrated by a change in behavior. This breakdown will allow for mistake management as well as limit compensation. For whatever reason, the system has decided not to perceive posture, alignment and coordination correctly, and even if it does, it's not doing anything to fix it—it compensates.

Nothing can change until perception improves and drives more favorable stabilization behavior. All your efforts should be focused on facilitation, better perception and documenting an improvement. From that point, you can stabilize until you normalize the stability problem and then move onto another movement pattern. This all depends on the priority you identify through your screens and assessments.

The *paranoid or clueless* characterization idea has been helpful for many. It offers a focus on input strategy. If you identify a mobility problem, break the behavior somehow. If you identify a stability problem, find a way to make the problem perceptible at a correctable level.

Until now, a common approach to corrective exercise work resembled practice sessions repeating proper movements while hoping learning will occur. But most of the corrections needed aren't movement skills—*these should be fundamental*. In the past, we tried to make people memorize subtle adjustments within movements that need to work at a reflex level.

Fundamental movements don't require talent or skill, but the teaching methods we use to develop skill can produce accelerated improvements when applied to corrective exercise. Work to reduce verbalization and demonstration. Allow your clients and patients to *feel* as much as they can.

Try to create an experience. Then let the experience dictate your corrective exercise choices. This will also force you to stop leaning on instruction as your only input.

DEADLIFT: EXAMPLE OF A CORRECTIVE EXPERIENCE

Let's get some background and make sure we find common ground. If we have 50 exercise and rehabilitation professionals in a room, we will also have 50 different opinions about the deadlift. This makes this largely misunderstood fundamental exercise a great example for our discussion.

Many people familiar with weight training do not understand or use deadlifting as a primary lifting and movement exercise. They either perform it incorrectly, doing something between a squat and deadlift, or they have subscribed to misinformation that promotes the concept that deadlifting will injure the back, and so they avoid it altogether.

Instead, use the opportunity to present the deadlift as the most fundamental, natural and authentic way to move something heavy in a safe and effective way. Deadlifting should be the first exercise taught to anyone interested in weight training because it meets all the criteria of a great fundamental exercise. It can be modified; it promotes core stabilization; it demands good posture; it promotes shoulder stability and it forces the hips to be the main driving force.

Deadlifting also provides protection for those who have misused or injured their backs. Deadlifting is part of many back rehabilitation cases because of its both therapeutic and protective qualities. Obviously, we treat the underlying causes of the many back problems we encounter first; however when the condition is stable, we bring the exercise into the mix.

Most patients will never fully trust their backs once they have failed. If you have ever had debilitating low back pain, it's burned in your memory, so that's understandable. One solution to regaining confidence is to force the patient to constructively use the back in a strenuous way.

Whether they ever plan to weight train is beside the point—the deadlift is just a version of the *walk of fire*, a show of trust in their own physicality and recovery. Psychologically, our patients need to trust their backs, and perceptively they need to feel an appreciable load distributed across the tissues of the back, hips and legs. Most of all, they need to overcome the lifting tension with appropriate force, stability and alignment.

Basically, our clients and patients actually need to overcome the load and perform a deadlift. This act completes the learning for many patients who are pain-free, but still feeling vulnerable and protective about the strength of their spines. This is to reinforce confidence once corrective exercise has done its work.

The teaching process is the same for an exercise client, athlete or recovered patient, because screening has demonstrated appropriateness. If it had not, we would not be using this example.

THE DEADLIFTING EXPERIENCE

Teaching the deadlift is not about instruction; it's about education and learning. Most of the learning and education in the deadlift experience is not visual or verbal if you do it right. Don't show people how to deadlift or give them a set of lifting rules like *lift with your legs, keep your back straight, always stretch before you go to work.* People rarely apply these anyway.

Over time, begin to introduce a weight and let the client or patient feel where incorrect and correct leverage is in the start position. There is only really one good poison to start a deadlift. It is right in the middle of a bunch of bad positions, and it is obvious once you feel it. Note—*feel it*—not hear it or observe it.

Spend time on the start positioning—spend lots of time in isometric loading in the start position. If you get that right, the actual lift is a downhill walk from there; it's easy.

Do multiple reps of getting into position and developing tension without performing the lift.

If you do this, a good lift is the natural result of a great start. The goal is not lifting—it's perception of load, correct and incorrect loading and alignment. We want the person to feel how to pull the weight to the center, not just upward. We also want the weight pulled from the center of the instep, not from in front of the body.

Just getting the deadlift down will help develop a perception and behavior of fundamental lifting confidence. The load lifted and the tension produced is what made the perceptive and behavior memory, not our lifting rules.

Notice there's no mention if the weight was a loaded straight bar, a heavy medicine ball, sandbag, dumbbell or kettlebell. That does not matter—that's method; the principle is to set posture, produce tension and lift with the hips while maintaining stable shoulder and spinal positions. At the end of a deadlifting learning experience, your client or patient doesn't need to recite the rules of lifting well. You need to hear they get it. You're looking for, *Why would you try to lift any other way?* and *It just makes sense to sit back into your hips and let your big muscles do the work.*

You should apply basic fundamental principles to all corrective exercises. Our corrective exercises don't need to be a practice of the correct manner of doing things. True corrective exercises need to enhance perception of mistakes with a preset goal in mind. If the movement is too complex, it will have too many goals. If the movement is not challenging enough, it will not enhance perception or refine the behavior needed to reach the goal.

Don't talk with your clients and patients about lifting as a teacher or coach—they don't need that skill. They need movement ability, not exercise conversational ability. Sometimes we teach things the way we learned them without putting ourselves in the shoes of our unfortunate victims. Quit looking at methods and start applying fundamental learning principles to movement learning.

And guess what? It will help—

- Discover a more useful corrective framework

- Determine the most effective time to introduce a corrective exercise technique

- Indicate when to progress or regress an activity

- Refine and develop some of the corrective techniques presented in this book

If we discover a mobility problem, the number one goal is to create a positive change in mobility—all progress is dependent on that change. If mobility is clear and functional, but there's a stability problem, we find the posture and technique that demonstrates the most significant improvements in the shortest amount of time. The instant we hit a goal, we evaluate where we are and set a new goal.

Remember, if someone is paranoid and you want to change that, you must find a way to stop the paranoid cycle. Nothing will fundamentally improve until you do that.

If someone is clueless and you want to change that, you must find a way to help the person perceive things differently. Nothing will fundamentally improve until you do that.

This little characterization may not academically or eloquently capture the entire scope of mobility or stability problems, but it may keep you from drifting away from the fundamental problem and its systematic framework.

TRAINING WHEELS DON'T TEACH BALANCE

Training wheels don't teach balance; they just make learning to balance that first bike a little safer. These little devices simply prevent a mistake from turning into an accident. Training wheels are the tools, not the training, and in fact, they can actually hinder perception and slow the process of learning balance if used incorrectly.

Have you ever seen a child actually lean on one training wheel while making a turn? If you've witnessed this, you have watched the child learn a behavior that will actually need to be unlearned. This was not learning to appreciate balance; *it as learning to ignore it.*

If we put the training wheels on a child's first bike and took the time to demonstrate how the wheels work and challenged the child with a goal, learning would actually be accelerated. Saying "Honey, these little wheels are there to catch you, but try to never lean on them" would provide cleaner learning and would not set up a situation in which the child would need to unlearn something.

A great little perception and education tool would be to put some tape around the perimeter of each training wheel and have the little learner attempt to ride short distances without marking up the tape. You could let the child also just have fun and ride and play, but every now and then do the tape test. When you're not looking, the child would probably practice a little as kids always do. The subtle message you are sending is, here is the safety net, now try not to use it!

The beginning part of corrective exercise is no different. This is particularly important when creating a stability experience that could potentially become an exercise. Too much help or advice will not force the sensory motor system to perceive a problem and start searching for a behavior to recapture control.

For example, sometimes the narrow base in quadruped or half-kneeling is actually the exercise. People may work for minutes, struggling to find balance in a narrow-base half-kneeling posture drill. Some will get discouraged and say *wow, I'm sorry I can't even get in position for the exercise.*

You can counter that with, "You are doing great. Apparently, that is your exercise. It's right at the edge of your ability. Your brain and body are working it out. You're making connections and working out angles. You're learning to feel and do, and feel more. Thanks for helping me find your exercise."

Flight simulators are designed to allow a learning pilot to make mistakes without risking life and limb. The simulators still indicate and report mistakes, and that is where the learning comes from. Safe perception and correction of many movement mistakes will improve movement learning speed.

Once the training wheels come off, you can't verbalize how to ride a bike, can you? You can explain how it feels, but you can't say how you do it. Riding a bike is a perceptual and behavioral experience that does not lend itself to complete or compressive verbal description. The ability to perform a stable half-kneeling posture or a solid single-leg-stance posture is no different. Those who can do it cannot explain it as quickly as struggling can produce it in those who can't, provided the correctly dosed experience.

The point here is that correctives need to be challenging and mildly stressful, but they always need to be safe.

Modern conveniences are partially to blame for many of our movement problems. Adaptive movement behaviors are minimized every time a new group of modern conveniences is introduced. The loss of adaptability represents specialization, and overspecialization is the kiss of death with biological organisms. Modern conveniences are nice and we should keep them, but what happens when we actually start adapting exercise equipment to make that more convenient, too?

If exercise were a movement-quality stressor, movement patterns might be maintained. However, modern exercise equipment is usually designed to allow quantity regardless of the level of quality. This is the fundamental reason some self-limiting activities should be performed as part of a general exercise program.

Physical exercise is the last chance we have to maintain physical adaptability, and when we lose that adaptability, corrective exercises must fix it. Once corrected, better exercise choices should maintain it. Corrective exercise and general exercise must produce manageable stress in areas of both quality and quantity, and if it doesn't, it's not likely to produce appreciable authentic movement.

ADVANCED CORRECTIVE STRATEGIES

Advanced corrective strategies are drills used to incorporate and coordinate the attributes of mobility and stability into movement patterns. It's common for a person to have the mobility and stability required, but for some reason be unable to perform a movement pattern. This individual has the physical and mechanical ability, but just can't produce the behavior. We need to help connect the dots and turn ability into behavior.

Sometimes the person is thinking too much, obsessing too much, or concentrating on the wrong thing. Of course, the opposite can be true as well—the client or patient might be detached, under-sensitive, and out of sync with a posture or movement. That person is over-processing something natural, authentic and fundamental, or is not connected to it at all.

Advanced corrective exercises break these cycles four different ways. No single way is best; each one provides an option that you can employ to incorporate movement fundamentals into movement patterns. Each of these techniques can be called advanced, but these are really just movement pattern retraining.

Different problems and personalities will respond differently to each technique. Do not disregard the basic correctives in chapter 13. They create the foundation and might produce all the correction needed to change a screen or assessment.

The advanced corrective strategies are designed to work with fundamental mobility and stability already in place. Each drill or technique has a trick it plays on the perceptional and behavioral systems, but the trick can only be effective if the criterion is in place prior to the activity.

MOVEMENT PATTERN RETRAINING

REVERSE PATTERNING (RP)

The Trick: *Do something completely different*
The Criterion: *Basic mobility and stability must be present*

Some people have a hardwired way of doing things that simply makes it harder to break an old pattern. Pathways are set, and set deep. They might have all the necessary mobility and stability needed to perform a movement pattern, but they just can't make the pattern happen. They are too busy showing you the anti-pattern. They just can't unlearn the faulty way of doing things, because faulty is their default.

We need to stop trying to change the mistake at the conscious level—*just write over it.* By performing movements in reverse, the brain has no predetermined habit or preference. As far as the brain is concerned, it is a new activity.

All movement patterns have a circuit that runs a cycle. There is a starting position, an ending position and usually a return to the starting position. The ending position generally presents the big problem. If we can just get people into the proper ending position so they can perceive the position and remember the posture, they can develop or reset perspective.

That is the big problem—how to get our clients and patients into positions they cannot move into on their own. If they actually have the mobility and stability to get into the position and they still cannot, they're putting on the breaks somewhere, whether they know it or not. However, if we can trick them into getting there, they might just be more familiar with the ending position, which could improve the entire pattern. That is where reverse patterning comes in.

Many of the exercises in my book *Athletic Body in Balance* fit this description, and here are a few more examples.

EXAMPLE # 1
ACTIVE STRAIGHT-LEG RAISE

PROBLEM

Let's say you have a client or patient in whom the active straight-leg raise is limited on the left side. You have discovered this is the only asymmetry and it is the biggest problem. You've checked passive mobility and that does not account for the problem. Muscle testing does not indicate a deficiency in local or global stabilizers. It seems to be a coordination problem isolated to this movement pattern and nothing seems to help. You have worked on all the parts, but left leg raising remains difficult and asymmetrical. If you are new to screening and assessment, you might be wondering what is the problem with that left leg.

Big mistake, because with lots of screening or assessment under your belt, you will simply see a leg-separation-pattern problem when the left leg is up and the right leg is down. The problem could easily be the right leg, pelvic and core stability or the left leg, or all three.

Reverse patterning targets a pattern; trying to isolate a part does not recognize the totality of the movement-pattern problem. Don't focus on the left leg. The reverse patterning corrective will help you remove focus from the left leg and still change the pattern.

REVERSE PATTERNING SOLUTION

Get the person into supine and help lift both legs into an acceptable range, slightly better than the height of the best side. This is possible because lifting both legs simultaneously demonstrates greater range since the pelvis can tilt and compensate.

Hold both heels and get the individual to relax tension. Tell the person you will only be exercising the right leg, not the left. Even though you know the reverse patterning technique is designed to improve the entire pattern, it is best to keep the individual's focus on the right leg.

Let go of the right heel and ask client or patient to slowly lower the right heel to the floor. Emphasize slow controlled movement. As the right heel nears the floor, the person may feel tightness in the front of the right thigh or on the back of the left thigh and that should be expected.

Have the person cycle the right leg up to or past the left leg and down to the floor a few times. If it is difficult to get the leg down to the floor, use a mat or thick pad that allows the heel to stop a few inches above the floor. Once the movements of the right leg are going well, have the person press the heel into the floor or pad when in the extended position. This helps complete the full pattern.

Perform the drill a few times. As the movement becomes smoother, offer less assistance to the left heel with the goal of not supporting it at all. Ultimately, you want to transition the left leg from the passive support of you holding the heel to active isometric stability.

Don't rush the progression, but keep the person right at the edge of ability. After the drill has been performed a few times, retest the active straight-leg raise. Even though the left leg has been motionless in a flexed position during the drill, you will nearly always see a significant improvement on the FMS active straight-leg raise test on that same side.

EXPLANATION

Although the left leg was not moving, a subtle separation was occurring between the right and left hips. The person learned a new end range for the left leg raise by working on it in reverse. The left hip was placed in the new position and right-leg activity allowed time to perceive the new left hip position every time the right leg was lowered.

Some movement-pattern barrier or poor coordination was responsible for the limited left raise, and the reverse patterning is an excellent way to reset a problem such as this.

EXAMPLE # 2
SINGLE-LEG STANCE

PROBLEM

A client or patient has difficulty with right single-leg stance. It is evident in all single-leg exercises and with the hurdle step test in the screen. Hip mobility is adequate and half-kneeling stability is symmetrical, but the transition into single-leg stance seems to have a hidden barrier. You might expect a mobility or stability problem at the knee, ankle or foot, but nothing seems significant. No matter what you do, the person seems shaky in right single-leg stance. You can use reverse patterning to change the focus and overwrite the single-leg-stance movement-pattern problem.

REVERSE PATTERN SOLUTION

Using step box about one-half the height of the person's tibia, have the client or patient step up onto the box leading with the right foot. Then have the individual step back down, leading with the right foot. When the right foot is firmly on the floor, have the person move the left foot back to the floor.

Have this exercise performed very slowly. If appropriate, increase the height of the box as the exercise becomes successful. To look at quality, perform the movement leading with the left foot going up and down. If the movement patterns look similar, the faulty single-leg-stance pattern on the right is improving. To confirm, retest the single-leg stance or the hurdle step.

The slower the drill is performed, the more time the person trains single-leg stance, balance and control. The more time spent in single-leg stance, the more time the person has to perceive and behave at the edge of ability without recognize the right-leg balancing action.

EXPLANATION

By stepping onto a box or platform, the individual is distracted from the performance of single-leg stance, while nevertheless performing single-leg stance.

Here is why: Stepping up onto the box initially uses left single-leg stance, but putting the right foot on to the box initiates a weight shift to the right. The instant the left foot leaves the ground, the person is performing right single-leg stance. Of course, the right hip and knee are in flexion and the ankles are in plantar flexion, but the person is moving in the direction of neutral position.

The left foot reaching the box finishes the first rep of right single-leg stance. Moving the right foot back to the ground puts the right leg into a neutral position, and lifting the left foot begins a neutral, right single-leg stance. Moving slowly will perform right single-leg stance for a few seconds without the client or patient realizing it.

The person will mostly be focusing on the moving foot—the left foot—but right single-leg stance is responsible for the slow speed. Not only is this a distraction, but the individual has also gotten into right single-leg stance with a reverse pattern. Practicing this drill will help improve the right single-leg stance pattern.

As the person improves, reverse the lead foot in order to initiate right single-leg stance. Don't make a big deal of it, or even mention the right single-leg stance work. Just say, *Switch feet on the step up.*

SUMMARY OF REVERSE PATTERNING

Performing patterns in reverse is an efficient model to improve or restore a problematic movement pattern. Dysfunctional movement patterns are performed in reverse to circumvent preloaded, poor motor control and unproductive movement-pattern habits. One logical reason why reverse patterning is effective is the brain has no preconceived plan for executing the reverse maneuver.

More examples can be found in *Athletic Body in Balance,* and a classic example of reverse patterning to improve squatting is described in the section on analyzing the deep squat, page 199.

REACTIVE NEUROMUSCULAR TRAINING (RNT)

The Trick: *Don't do something*
The Criterion: *Basic mobility and stability must be present*

It is sometimes easier for people to stop doing something than to start. When we stop doing something, we often just react, whereas when we are asked to start doing something, we must program the task.

People often don't realize they are moving in a certain way. They might have all the necessary mobility and stability needed to perform a movement pattern, but they just seem sloppy and they can't perform the pattern correctly. They may not even realize how dysfunctional or incorrectly they are moving. They need to perceive the problem before we can expect them to improve the pattern.

Note the word *perceive*—not *think about, discuss or fix the problem.* Just perceive the problem. In this instance, it is better to ask them not to do something than to do something; the cognitive burden is actually less.

In this instance, we use the technique called reactive neuromuscular training (RNT) to exaggerate the mistake and bring the error to a clearer perceptual level. The exaggerated mistake forces natural balance reactions that serve to reset a faulty movement pattern. Consider RNT a preconceived perturbation or disturbance that you construct. It requires a fundamental righting reaction to maintain posture or balance.

Here is an example of RNT—

EXAMPLE # 1

PROBLEM

You note on a screen or assessment that a young female athlete can perform a full deep squat with unrestricted mobility at the ankle, knee and hip, but she tends to have excessive forward lean in the upper body. At the bottom of her squat, her lower body looks good, but her trunk is pitched forward and her arms are out in front instead of overhead.

Mobility and stability testing indicate that her upper body and shoulders should have the range and control to perform the squat correctly. She does not have kyphosis, lat tightness or weak lower traps and rhomboids, none of the things that sometimes get the blame for this dysfunction. You feel this is a movement-pattern problem and you are correct since you have done a good job of ruling out underlying mobility and stability issues. She is becoming frustrated that she cannot correct the squat pattern. Her conscious attempts at correction are actually making things look worse. You need to help her break the poor squatting circuit.

RNT SOLUTION

The problem in this example is the sequence and coordination of the squat movement pattern. Since the forward lean seems to be the obvious mistake and the biggest problem, we will use an RNT technique to attempt to increase perception and to challenge posture and balance.

Ask the athlete to stand erect with her feet slightly wider than shoulder width and her arms abducted above her head. Loop an end of a light elastic band behind each of her hands and tell her to keep her hands open.

Pull the band in the center with gentle oscillations to challenge her shoulder flexion, as well as the integrity of the entire posterior chain. The best instruction is simply *don't let me pull you forward.*

If the shoulders come out of the initial flexion position, you can move the bands to the level of the elbow or shoulder joint. The goal is not to challenge the shoulders; the goal is the challenge the trunk. The shoulders are just the best leverage point in this scenario.

The oscillations will usually produce a disconnect between the upper and lower body. This is observed by a jackknife or bending in the middle, or hip and spine flexion to maintain balance. The goal is to first produce the jackknife. Once you see that, instruct her to avoid letting that happen.

The entire drill is performed in standing until the oscillations cannot break posture. During the drill, have her routinely move her neck and head to show she is not trying to compensate with neck and shoulder musculature.

This drill can be extremely taxing, so offer rest breaks before fatigue is noted.

Eventually, stop the oscillations and provide a steady load with the band in a sharp, downward direction. If posture remains erect, instruct her to squat, encouraging her to go as deep as possible. Instruct her not to lean back, but to squat directly over her feet, using a spotter or having a wall close behind her in case she leans back.

The light band is not there as support or to hold her up as she leans back—it's not training wheels. The band is there to preset posture and core stability prior to squatting, and to keep an equal balance of anterior and posterior muscular stability throughout the squat movement pattern.

With improvement, provide lighter and lighter oscillations as she stands and lighter tension as she squats. The goal is to keep decreasing the load of the band until the individual can perceive how to squat without the oscillations or preload.

EXPLANATION

This person may have distorted her authentic squatting movement pattern with poor weighted back-squatting technique. She could also simply be very leg-dominant from athletics, which could put core stability at a disadvantage. Regardless of the cause of the problem, she goes into the squatting movement without a balance of forces between the anterior and posterior stabilizers.

The light elastic band provides an anterior weight shift in the same way that squatting facing downhill or with a heel lift provides an advantage. Most see a heel lift and assume it improves the squat because it negates ankle stiffness, but it also pretenses or preloads the posterior chain musculature. The oscillations in standing biases the posterior stabilizers and increases perception of the limited control even with light amounts of resistance. This is sometimes surprising to athletes when they have a hard time remaining erect with light band oscillations even thought they can squat with respectable weight.

This demonstrates the difference between strength and stability. Lifting weight is strength, whereas maintaining alignment, posture and moving with smooth quality is stability.

SUMMARY OF RNT

RNT drills work best by making the commands simple and letting the person fail, feel, and work it out. Instructions like *don't let me pull you off balance* or *don't let the band cave your knee inward* are usually enough. Then, the person moves in and out of patterns, struggling to maintain balance or alignment. When you see improvement, lighten the resistance so the individual can perceive and move correctly with less and less input until finally owning the movement.

RNT is a form of corrective exercise designed to stimulate reactions or reflexes that naturally enhance mobility and stability. Based on PNF, it provides an external force to increase the magnitude of a movement mistake. It is essentially a hands-off PNF approach.

The science of PNF is solid and has stood the test of time in exercise and rehabilitation. If you provide the force in the correct dosage, along the correct vector and to the correct body segment, the movement pattern should improve as a result of irradiation. This concept uses the stronger components of a movement pattern to facilitate the weaker components and refine the timing and coordination of each muscular contribution.

RNT brings this science into adaptable corrective exercise methods. You can affect quality, quantity and comfort with the correct RNT technique. You're not using resistance to produce a strengthening effect; it's used to produce the stabilizing or righting reaction that triggers improved movement pattern quality.

This concept, along with reverse patterning, is discussed more in the book *Musculoskeletal Interventions: Techniques in Therapeutic Exercise*, edited by Michael Voight, Barbara Hoogenboom and William Prentice, in Chapter 11, *Impaired Neuromuscular Control: Reactive Neuromuscular Training* by Michael Voight and Gray Cook. RNT is also covered in the appendix beginning page 359.

CONSCIOUS LOADING (CL)

The Trick: *Use loading to hit the reset button for sequence and timing*
The Criterion: *Basic mobility and stability must be present*

Some people do best with a little push. They might have all the necessary mobility and stability needed to perform a movement pattern, but they stop just shy of the performance of an acceptable pattern. These people need to perceive that what feels like a barrier or limitation is actually a sequencing problem. Sometimes a preload can provide this experience.

You can perform the loading with an active contraction or with some form of resistance or load. Here are three examples to establish how conscious loading can help improve sequencing for movement pattern quality.

EXAMPLE # 1
ACTIVE CONTRACTION—CL

PROBLEM

A 50-year-old man is ready to start doing upper body training. He scores symmetrical threes on shoulder mobility, and shows a score of two on the pushup. Shoulder mobility and stability seem appropriate for exercise and basic strengthening, but you note poor movement and control when the right arm is performing pulling and pressing movements overhead. You note scapular elevation when the right shoulder is flexed, instead of the natural depression and retraction you expect with good shoulder-girdle patterning. You easily observe the correct pattern when he presses weight on the left.

CL SOLUTION

Let's try conscious loading as a reciprocal action to improve stabilization. Reciprocal action refers to arm and shoulder patterns moving in opposition of each other. Running, walking and most swimming strokes are examples of reciprocal shoulder action where each arm moves as an opposing force and counterbalance to the other. We often forget that when we push or pull something with a single arm, we naturally employ some form of bracing with the arm not in use.

To perform this exercise, have the man lie on his back and perform a pillow press exercise. Supine with his left arm overhead—slightly abducted and fully flexed—and his right arm out from his side—slightly abducted and extended—ask him to press into the surface below with both hands. To set a baseline, lean over and try to pull each arm off the surface to check end-range strength. End-range strength is a great indicator of stability, perception and motor control.

Next, have him perform the same action in reverse. Check his right arm in flexion and his left arm in extension. Even though he can get into the position, perhaps he comments that it just doesn't feel the same. This is a great opportunity to explain that his problem is not his right shoulder. His problem is the entire movement pattern and posture associated with singular overhead work of his right shoulder. His left shoulder and posture do not complement his right-arm activities overhead. In contrast, his right shoulder and posture complement his left-shoulder overhead work.

Once again, you've found not a part, but a pattern that needs correction. When you test his stability, you might see his end-range stability is less than half of the previous reciprocal pattern. You're able to lift each arm a few inches from the surface before he can perceive or engage any usable form of control.

In this situation, you are not looking for how much force he can generate; you're looking for his arm to feel as if it's glued to the surface. You also don't want him pressing down into the surface when you test him manually. You only want him to match your pressure. When he can't effectively match your pressure in the second pattern, you know you have found the problem.

Now for the actual corrective exercise.

First, put pillows, pads, towels under both arms to lift them a few inches off the surface. Pillows make this a convenient home exercise because no one has the excuse of unavailable equipment.

By elevating both arms, you move the man out of the awkward end range and into a range where he feels more control. To start the exercise, first ask him to press the left arm down into extension. This is important because the left arm is the conscious load—the defining part of the entire exercise.

Second, ask him to press the right arm down into flexion. Using this sequence as an exercise, the person can start to slowly remove the pillow lifts and move his arms back onto the surface over one session or a few sessions. In extreme cases, use several pillows for the upper arm and suggest the person remove a pillow a week for following weeks.

Once this man displays equality between both patterns in supine, he can perform the same pattern in standing with two resistance bands or cables. It is helpful to stagger the stance to complement the arm action. This means when the right arm is in flexion, the left foot should be forward. If this is too difficult, consider half-kneeling as an option.

We must remind our clients and patients that the body does not get this way overnight, and sometimes corrections don't work overnight, but they do work in most cases.

EXPLANATION

When the conscious load is created in the extended left arm, the stability and posture generated complements flexion of the other shoulder. Thoracic spine extension is improved and retraction and depression of the flexing shoulder also become more authentic and automatic. You can also incorporate neck movements in this activity, depending on the issue you're addressing.

- Neck mobility problems can be addressed by having the exerciser turn the head in the direction of the flexing shoulder as the reciprocal action of the opposite shoulder presses into extension.

- Those with neck stability problems should just attempt to remain in a neutral position. In this instance, they engage the neck movers as secondary assistance to nearly any strenuous upper body movement. Having them maintain a relaxed neck and not compensate is actually stability work since the neck compensation is removed.

EXAMPLE # 2
FREE WEIGHT—CL

PROBLEM

A 20-year-old female yoga student has trouble at the bottom of her squat. Screening or assessment reveals that other movement patterns are functional and there are no fundamental mobility or stability problems. It simply seems to be a problem with the squat pattern.

CL SOLUTION

We'll use a squat ride-down to help her re-pattern. To be technical, we should call the exercise a *goblet squat ride-down, deadlift back-to-start combination.*

There are variations of this exercise using med balls and dumbbells, but the most popular version is now done with a kettlebell and it's most people's favorite. Dan John has discussed the benefits of the goblet squat, and it has quickly become commonplace as a warm-up and squat-quality drill when used correctly.

Here we add a little twist to really enhance the motor control aspect and the corrective effect. It can be performed with any weight, but a vertically stable dumbbell like a Powerblock or a kettlebell works best.

To begin, have your problem squatter curl or clean the weight to her chest. Have her get a good hand position on the kettlebell horns, Powerblock rails or under the top weight of a vertically held dumbbell. With the weight in position and with a totally erect starting posture, ask the person to first push the hips forward and pull the pelvis upward.

This assures that the hips are clear and unrestricted. It's common for squatters to start the squat with perpetual hip flexor activity. We want to engage the hip flexors and abductors on the way down, and we cannot engage a muscle if it is already on.

Once the hips are clear, have her pull herself down into the squat with an erect spine and abducted hips. This means the knees are pointed in a more lateral direction than the toes. This is key, because when most people hear the instruction to abduct the knees, they keep turning out the feet, which negates the effect. We want separation between the foot and the knee, with the knee being the more lateral of the two. Make sure she gets her knees outside of her feet to make room for the tips of the elbows to help nudge or hold the knees outward. At first, she may require help from the elbows, but as she progresses, encourage less and less assistance.

Once at the bottom, have the person curl the weight and move it left to right and front to back exploring the limits of stability. Here is where it gets different: Instead of having her return to the start position, have her set the weight down and try to not lose posture or balance or round her back. In a normal goblet squat, she would simply return to the start and repeat. In this movement retraining method, the person takes a conscious load and rides the weight down, but must release it at the bottom, thereby losing the training wheels.

The front-load weight helps to engage the posterior chain stabilizers as it creates an anterior weight shift. Both these things provide a mechanical and perceptual advantage to help stabilize and control the squat.

While still in the bottom position, we take away the help. Yes, you guessed it—there is also a little reverse patterning here because we get her into a position she could not get into on her own. We make her hold it for a few seconds to demonstrate static stability, and then get out on her own showing dynamic stability.

This is where the mistakes happen and this is where learning occurs. Every time she tries to set the weight down, she feels as if she is going to fall backward. Ask her to observe the other changes that occur as she sets the weight down, and after a few more attempts, she'll note that whenever she unloads the weight, her knees cave inward and she rounds her spine. Here you'll simply tell her to *stop doing that*.

One of these problems is primary, and one is secondary. You don't want to give her the answer and you don't want her working on too much at once, so tell her to only pick one problem and *stop doing it*. She tries both, and quickly finds the one that provides the most stability. The primary problem is always the same.

Did you guess the primary problem?

Valgus collapse or caving in with the knees is the primary problem. The rounding of the spine is an automatic countermeasure to maintain balance and not fall backward.

RATIONALE

By abducting the knees—actually the hips—the pelvis and buttocks are pulled forward into the stance. By keeping the posterior load closer to the center, the anterior load of the chest, shoulders and upper torso can remain more erect. You are welcome to argue the opposite point, but you better try this about thousand times first, and keep good records.

Alternatively, you could just think mechanically along muscular and fascial lines. Abducting the hips tightens the posterior chain and actually pulls the upper body into alignment as it centers the weight. Pulling the shoulders back and erect-

ing the spine does not have the same affect on the hips. It's mostly a leverage thing.

Back to the exercise—as you realized, we set a goal but we let her find the mistake. We tell the athlete not to come out of position or rush back to the top, but instead to set the weight down and count to 10 and then, if needed, grab the weight. We let her get comfortable setting down the weight.

Give the person a few rest breaks if needed. A rest break can mean getting completely out of the squat position or just grabbing the weight to reset balance. Then, finally, have her perform a ride-down to a deep squat position, have her set the weight down, gain balance and posture and then drive up to an erect standing position.

Once in the standing position, tell her to pick up the weight using the deadlifting technique and re-position for another ride-down. The deadlift is an important part once this initial experience becomes an exercise because it removes bad habits and reinforces posterior chain preloading and stabilization.

EXPLANATION

This corrective exercise is one of the best drills for the classic quad-dominant individual with squatting pattern problems. The drill forces hip and posterior-chain engagement. The load creates core stability and will exaggerate faulty posture, making the mistakes obvious. The load becomes the training wheels, which are taken away halfway through each rep. Setting the weight down internalizes the stabilization as the training wheels are removed.

From the depth of the squat, without extra help or support, your clients and patients are on their own. At the top, the weight is reloaded with a deadlift, which keeps the person in the hips and out of the thighs and shoulders. If you note a sticking point where the exerciser just can't seem to let go of the weight, offer a half-inch or one-inch heel lift to help reach success. As the person becomes successful, use the *halves rule* to remove the lift... moving from an inch, then to a half-inch and to a quarter-inch.

The knee abduction is the key in this drill, but don't tell the person to do it. Allow the knees cave in and then at the appropriate time, instruct the person not to let that happen. Pulling down into the squat and maintaining abducted knees at the bottom pulls the exerciser into the center balance point of the squat and greatly improves stability.

Pavel Tsatsouline has an insightful similar drill called *face the wall squat* that accomplishes the same effect.

EXAMPLE # 3
ELASTIC RESISTANCE—CL

PROBLEM

You screen an elderly man who is interested in getting back into shape after a long illness and successful recovery. He has always noted difficulty touching his toes and forward bending. The FMS reveals an asymmetrical score on the active straight-leg raise, in which you record a score of one on the right and two on the left. Interestingly, his passive range of movement and flexibility is almost equal for each hip and leg. He can also pull each knee to within six inches of his chest without pain or limitation.

You note no significant strength problems between the left and right side. Because of his age and activity level, he just assumed this was a flexibility problem, but now it looks like a coordination or patterning problem.

SOLUTION

Have him lie on a mat to attempt the active straight-leg raise on each side so he can appreciate the left-to-right difference and notice the extra difficulty when he lifts the right leg compared with the right. While still lying supine, have him hold both hands at eye level or 90 degrees of flexion. Ask him to not let you pull his hands backward or into greater flexion as you slowly increase the backward resistance.

Once he has engaged in the tension standoff, ask him to lift his left leg, then ask him to lift his right leg. He notes both are improved and the right is now equal to the left. He is amazed and wants to know what happened. You will explain in a minute, but you are not yet finished with the exercise.

Inform him that he does not need you in order to see this kind of instant improvement. Set up an elastic band attached to the wall with an end in each of his hands. Have him position himself far enough from the wall to allow moderate tension in the band. Tell him to pull the bands down to about six inches from his waist with his elbows in extension—straight arms. The tension is adjusted so he has to work but can perform multiple repetitions. Instruct him to lift his hands overhead and reduce tension on the bands.

Then tell him to pull the bands down again. Once the bands are pulled and the arms are in the tensed position, have him lift his left leg as high as possible and return it to the floor. Once on the floor, he should release the tension on the bands. He is only to lift his leg once he has pulled the bands into place and only to let off tension after the leg is returned to the floor.

Have him alternate the right and left leg for a number of repetitions. After a rest break, retest him to see that the asymmetry is nearly resolved. Remind him to repeat the drill to create change, but explain how he has introduced his brain to the lost coordination between the core and right hip.

EXPLANATION

This drill represents a normal sequence of events. Anticipatory core stability precedes most of our movement patterns. Somewhere in the very first part of every motor program is a trigger for general core stability. Once movement is underway, refined specific stabilization reactions help manage the particular demands of the unfolding movement patterns in each particular situation. Some of the stability may be the direct result of intention, while some is purely under reflexive control.

The conscious loading with the bands represents the anticipatory load. It accelerates the perception of stability and proves that the right leg-raise test is more than hip flexion against gravity with a straight leg. It is a coordinated effort to first stabilize the spine and pelvis, and then perform muscular contractions of hip flexors that anchor themselves onto the solid foundation created by the initial stabilization.

The drill is first rehearsed to demonstrate that normalization of the movement pattern is possible if the sequence is modified or facilitated. Then, the modification or facilitation is slowly removed, leaving a restored movement pattern in its wake.

SUMMARY OF CL

Conscious loading demonstrates accelerated control. The exercisers learn they are in control, and that the possibility of moving better is literally at their fingertips. They learn that simply following a few steps can boost timing and coordination. Once movement is possible with CL, the weaning process slowly removes the catalyst, while the movement pattern is practiced alone or as part of a more complex task.

Conscious control is practiced with the breath, posture, pressurization, intense concentration and *conscious loading* is no different. It uses a muscle contraction generated intentionally or as a response to an external load to improve movement quality and to create efficient timing and coordination.

RESISTED EXERCISE (RE)

The Trick: *Stress in the form of resistance for accelerated learning*
The Criterion: *Basic mobility and stability must be present*

Resistance exercise is not so much corrective as constructive, but it warrants recognition as an advanced corrective strategy when used correctly because it fortifies learning. Used properly immediately following corrective success, it has the ability to reinforce new patterns that can ignite the senses with greater magnitude. Tissue tension and balance responsibilities are amplified. Even though incorrect application of resistance exercises is one of the problems that got us into this movement mess in the first place, we must remember that the resistance did not cause the poor movement patterns we see. Poor movement-pattern quality under loads is actually the problem.

Incorrect applications of resistance exercise can produce some of the movement problems we need to correct. Even though resistance can reinforce a bad pattern, the same forces that go to work building bad patterns over time can be put to use to build good movement patterns. The first and most important step before resistance is that the corrective exercise session must produce some degree of improved movement-pattern quality. This improvement indicates a window of opportunity to keep pushing the movement limits and expanding the outer edge of ability. Resistance can be the push that produces constructive stress that can accelerate learning.

In a single session, corrective exercise can improve the active straight-leg raise movement pattern, which can normalize forward bending. Once forward bending is practiced a few times and cleared, you have a choice to keep performing corrective exercises to reinforce the new pattern or stress it to make it more stable and durable.

The deadlift is an excellent stress to reinforce the forward bend or toe-touching pattern. Don't get confused by the differences in the toe touch and deadlift. One is a movement pattern and one is a loaded lift. In the movement pattern, it is acceptable to round and relax the spine. In the lift, the spine must remain in a stable and safe position. The similarity between the two activities is the initial hip hinge and a posterior weight shift.

The load will reinforce fundamental movement. When the toe touch is compromised or limited, it is nearly impossible to perform correct deadlifting without constant coaching. Therefore, we often require acceptable leg raising and toe touching prior to deadlifting. When the leg raise and toe touch are compromised by limited mobility, significant and lasting gains are rarely obtained by mobility work alone. When improvement is noted, it often requires regular mobility work to maintain.

A favorable solution is resisted work in the form of deadlifting. This addresses the problem, not the symptom. Hamstring, back and hip tightness is rarely the primary problem. The tightness can often be acquired tightness and stiffness as a natural adaptation for adequate stabilization in a compromised forward-bending movement pattern. Simply stretching out the tightness or working on the stiffness will only produce a temporary change. That limited mobility is being used as a secondary support system.

If we introduce the deadlift within the temporary window of improved mobility, the brain will be required to produce another option without the familiar tightness backup. Success is made possible by attention to detail and correct technique. This will force the authentic stabilizers to work and the efficient movers to generate force.

Dosage is important. Overloading and over-stressing might cause a return or default to tightness. Under-loading may not force authentic motor control. Use a load that makes three to five clean repetitions possible. This will indicate a good choice in load.

Nearly every movement pattern has a complementary lift that will expose it to stress. Pick your lifts in advance and be ready to load movement patterns when you see the quality improve.

EXAMPLES

Here are a few examples of resistance exercises that can reinforce and strengthen movement quality. Try these and add to the list as you develop your skills.

To reinforce posture and improved motor control in torso and shoulder girdle try adding—

- Half-kneeling chops and lifts, kettlebell or dumbbell presses or kettlebell halos

- Tall-kneeling chops and lifts, kettlebell or dumbbell presses or kettlebell halos

- Half get-ups

To reinforce hip extension and core stabilization try adding—

- Regular deadlifts or kettlebell or dumbbell single-arm deadlifts

To reinforce single-leg stance—

- Single-leg opposite-arm deadlifts

Overall balance and connection—

- Bottom-up kettlebell cleans, squats and presses

- Full get-ups

SUMMARY

We use resisted exercise as a final way to reinforce a movement already considered functional and correct. Loading forces extra quality when lifts are performed correctly. Isn't it funny that many of the resisted exercises on the above list also appear on the list of self-limiting exercises in Chapter 10?

SELF-LIMITING EXERCISES REVISITED

As you recall that list of self-limiting exercises on page 233, you may have noticed something new that may not have been apparent a few chapters ago. The common thread between all of the self-limiting exercises is the rich sensory experience and continuous high degree of feedback. The goals of each lift and exercise are clear, and mistakes are obvious. You know when you do it wrong and you know quickly.

The self-limiting lifts and exercises have an uncanny movement-quality maintenance effect when used correctly as part of regular conditioning.

MOVEMENT LEARNING

A basic understanding of the brain will help explain why applications of kinesiology and biomechanical-based exercise may not provide the best platform for replacing a dysfunctional movement pattern with a functional pattern. The brain stores and retrieves the patterns of movement. The brain demonstrates plasticity, which means it can be molded and changed if the correct learning opportunities are provided.

As you identify dysfunctional movement patterns, you should understand that for some reason or another these patterns have been learned and reinforced, or they would not be present. Furthermore, they must have some purpose or practicality for that individual. This means the brain you are getting ready to interact with has assigned value to a movement pattern with less-than-acceptable functional quality. A dysfunctional movement pattern is being used and repeated alongside other functional movement patterns.

Your screens and assessments have removed a major obstacle to movement pattern correction. They have identified pain and removed it from the equation, meaning the movement pattern you are getting ready to correct is not a painful pattern; it's a dysfunctional pattern. It is also the most fundamental dysfunctional movement pattern because you used a hierarchal system to identify the dysfunction.

Now you stand ready to change the brain. You must first convince it to stop doing one movement pattern and then convince it to start using another in its place. As you have learned, stretching a few muscles and strengthening a few others will not consistently give you the best possible results when your goal is movement-pattern correction.

Movement is based on patterns stored in memory that is part of consistently repeated behavior, and a good part of this behavior is below the level of conscious control. If we endeavor to change a movement pattern, we must set up environments that allow for positive modifications of conscious as well as subconscious sensory motor interaction. As you apply the categories of corrective exercise, be particularly aware that each category and subcategory is designed to address both the sensory and motor systems.

In many cases, the sensory input is more important than perfect motor output. In fact, that is the essence of corrective exercise.

Sometimes we over-coach movement as we strive for technical correctness and perfect output with each exercise we teach. By over-coaching, we are forcing output that is not driven by natural input. Toddlers push the limits of their motor control on a daily basis and they don't look smooth, controlled or even correct, but the sensory motor system is working out patterns, making memories and then refining them. Each success refines a pattern and improves the memory. Each day, memory is retrieved and patterns are reinforced. Coaching a toddler to focus on output would interrupt the sensory motor interaction.

The best way to assist the situation is to provide more constructive sensory opportunities and clearly define goals. Outside of this, we should be very cautious in the way we try to control or direct fundamental movement patterns. As professionals, we must accept the fact that we cannot make movement patterns, but we can help them grow.

Most of the fundamental patterns we hope to improve are somewhere in the memory, but have been altered for some reason. Sometimes we can discover the reason and sometimes we cannot. The important thing is to see if it can be improved. The first improvement will be a simple, positive response in a single session, which should be a short-term goal.

Progress over multiple sessions resulting in adaptation is the long-term goal.

PERSONAL PERSPECTIVES

You now have many pages of steps in the corrective exercise process. You have clean academic statements, definitions and observations, and a step-by-step framework to guide you.

SEQUENCE VERSUS THE SHORTCUT

Now ask me what I do. Do you really want to know? I skip steps all the time. The better I get, the more I skip, always staying aware of how many steps were skipped. I see a flicker of stability and move right to an advanced drill. When static stability plateaus, I don't just automatically move to dynamic stability work. Sometimes it's good to jump back to mobility work to capture a little more mobility. More mobility provides greater opportunities to change perception—changed perception is the blueprint for changed behavior.

I've also learned to recognize my mistakes quickly. This is because every corrective exercise session starts with a specific intended outcome. I know where the patient is and where he or she needs to be. Knowing where the potential improvement in a single session is, I go for it. I move from one specific selected correction to another, searching for the one that provides the envisioned outcome.

Enjoy using the corrective framework. Let it guide you. It is not designed to make you rigid. It's designed to make you systematic.

Please see www.movementbook.com/chapter14 for more information, videos and updates.

15

IN CONCLUSION

A PICTURE
IS WORTH 10,000 WORDS

The cover of this book could have been highly technical, with lines and angles showing the complicated intricacies necessary to mechanically describe functional movement patterns. In fact, it almost was, but my writing, editing, reflection and a very wise publisher helped me understand the actual mission. Movement is not about passing screens and assessments. Those are just tools and methods we use to communicate deviation from accepted minimums.

Movement is about breaking into a run on the beach without thinking to engage the abs or glutes. It's not about worrying if your back can handle the stress of a load, and it's not about who is watching your form. It's not about looking at your watch to gauge pace or heart rate, or deciding whether to run when you forgot your fancy new shoes. I'd like to think the runner in the picture is no more concerned with technique than the nearby seagull, wings outstretched. Both runner and seagull are just doing what they do—authentically.

Use this information to help those in your care recapture and maintain all the authentic movement that they can possibly acquire. Most likely, your clients and patients will not even be able to verbalize the process; they will just move better and will have you to thank. They will do what you say and simply move better as a result.

Actually, that's what we signed up to do in the first place, isn't it?

A CHANCE
AT AUTHENTIC MOVEMENT

Books are judged on what has been read, not what was written. By only reading parts of a book, or reading a book out of sequence, the reader may finish confused. If your curiosity has caused you to read this material out of sequence and you arrive at Chapter 15 feeling uncertain, please review the material in the order it was written. Human nature and curiosity make us skip steps and nose around. Still, we are ultimately responsible for our professional practices, so take time to build your foundation, however long it takes.

Learn in whatever method you can and practice what you learn under the direction of those who will share their wisdom with you as you practice. Practice what you've learned under the supervision of an expert if possible. This habit will be reinforced when you realize how much you may have missed on the first pass and how much you still can't apply even with a second pass. Reading informs, doing teaches.

By introducing movement screening and assessment to the rehabilitation and exercise world, this text is an effort to reorganize exercise economy. The words *exercise* and *economy* are not often used together, but this coupling may explain many of our recent exercise failures. These range from ineffective weight-loss programs, to the all-time highs observed in scholastic sports injuries. As Steven Levitt and Stephen Dubner explain in their book, *Freakonomics: A Rogue Economist Explores the Hidden Side of Everything*, economics is above all a science of measurement. Economics is more a set of tools than a subject matter, tools designed to expose misconceptions and truths when discussing the causes and effects of things. That is what we are trying to do here.

Movement screening and assessment offer an additional tool, a method, in the appraisal of the effects of exercise and rehabilitation practices on fundamental movement patterns. Our industry can do better than we are currently doing, regardless of the endeavor. From conditioning our military to reducing the ACL injury epidemic in young female athletes, and from managing low back pain to addressing escalating obesity even in the young.

Exercise science has for too long maintained a bias toward the study of metabolism and physiology in contrast to movement perceptions and behaviors. We can't blame the physiologists because they are making their contributions within their specialty, but we need to balance the research and education scales and tip them back to the center. If not, we will blindly continue our attempts to strengthen painful backs when lack of strength may not be the problem. We will design exercise around caloric expenditure and focus on exercise quantities without a minimum movement-quality standard. We must professionally demonstrate a balance in our exercise economy through our methods and practices.

We often use our current science to bully fitness, to try to force it to produce on poor perceptive and behavioral soil. The qualities of good fundamental movement patterns provide the fertile soil for fitness and physical capacity. Movement screening has demonstrated that we can enhance fitness, metabolism and physiology without improving movement quality, but not without a loss, a silent deficiency only noted when we slow down and look at the movements behind the movements that interest us. By ignoring fundamental patterns and focusing only on metabolism, physical capacity and specialization, we have missed the natural progression of authentic fitness.

Our durability is compromised when we skip ahead to conditioning without first establishing fundamental movement capability. The consideration of movement shows we should not so much attempt to make fitness, as we should grow it. We practice our performance and skills, but fundamental and functional movements come early and support everything we do. They should be fundamental across the active lifespan. The ability to balance on one foot, to deep squat, bend backward, touch the toes, reach across the body or overhead and around the back are innately human unless an injury or disability has left a temporary or permanent limitation.

It is comical to listen as a person focused on high-end performance goals argues the unimportance of these basic abilities. These same people would rush their child to the family physician if he or she couldn't squat, balance on one foot or reach most parts of the body equally with each hand, but somehow they will not recognize their own limitations as fundamental problems or deficiencies. We as professionals also failed to notice these things, and in so doing we have missed a critical opportunity in fitness and rehabilitation evolution.

Movement quality seems to have acceptable and unacceptable levels.[58-59] People who initially perform well on movement screening and movement assessment don't specifically practice fundamental movement patterns—they never cease to use them. Once these patterns are lost, a degree of practice is necessary to regain control, but regaining a movement pattern and never losing one in the first place are two different ideas.

It sounds crazy to suggest that we consider movement in all aspects of exercise. To the exercise outsider, it must seem ludicrous to think exercise and rehabilitation professionals would need a resource for movement correctness, but we do.

We need a standard operating procedure. Movement is at the heart of the exercise and rehabilitation professions, and yet we have not mined the information within movement patterns and behaviors to its fullest potential. It does not currently drive our exercise and rehabilitation decisions in the same way anatomy, kinesiology, physiology, specialized activity and sports influence them, but that is changing.

If we use these tools to identify and monitor movement quality, we will better understand how to grow fitness, rather than to sprinkle on random activity and hope it sticks. You won't need your clients and patients to constantly practice the movement patterns of the functional movement screen and selective functional movement assessments. Correct exercise practices will produce acceptable screen results. Perfection isn't the goal anyway—only a select few would pass the screen with perfect scores. We don't use these tools to look for perfection; they're intended to identify risk, severe deficiency and a minimum level of movement-pattern quality.

Modern exercise and rehabilitation business is essentially the tuning, toning and retraining of the physical body, and we use it to advance, maintain or restore physical capability. This book has shown how conditioning and corrective exercise practices each have their place as we develop and redevelop performance and durability on populations that have drifted a long way from the authentic ideal.

THE BUSINESS OF EXERCISE AND REHABILITATION

Exercise and rehabilitation are part of a big business. They are, in a sense, big machines, and it's difficult to change directions. These machines can produce ideas with some degree of merit that can grow out of control if unchecked by logic and reason. These massive machines need to be governed by a balanced perspective—they need to be governed by principles.

As discussed earlier, a good example of a mechanical idea unchecked by a balanced perspective is the general assumptions around the modern running shoe. The modern running shoe is designed to provide cushioning and motion control, and creates the illusion of running on a perfect surface. These benefits offer comfort to anyone with tired, sore feet, joint pain or muscle stiffness. That's good right? Not really.

With the help of marketers, many incorrectly assume this increased comfort provides a degree of safety and performance enhancement, but injury rates have not declined. As more and more shoe companies adopted a marketing strategy of high tech cushioning and motion control, the public was blind to the fact that we have a much greater legacy of runners who used primitive footwear... or none at all. Obviously some cushioning and motion control are necessary for some people, but we are internally wired and beautifully constructed to control motion and to absorb shock naturally and authentically.

By protecting our feet to an absurd extreme, we deprive them of the sensory experience of running. The barefoot sensory experience will usually be uncomfortable without the purest of mechanical correctness and movement pattern competency, and that is precisely the point. Barefoot running is a speed bump. All self-limiting exercise is full of speed bumps and people hate them. When you're in a rush, you hate each annoying speed bump, and

that is precisely when you need it the most. The speed bump declares *slow down and pay attention, stupid.* And we hate that.

When we slow down and pay attention to movement and exercise, we are rewarded with an authentic experience, rather than the synthetic shiny chrome substitute.

Once the shoe industry produced the technology for cushioning and motion control, the endeavors became self-perpetuating. The race for shoe innovation became more important than the millennia of sensory experience that refined our movement without shoes.

Modern running shoe technology allowed people with poor running mechanics to run more often and for greater distances. They got the endorphins and that's good, but they got them dishonestly and that's bad. Fitness surpassed function. As we appreciate the balance of safety and protection with the authentic experience of developing sound running mechanics, let us heed an important lesson: As we endeavor to cover up one problem, another will appear.

A wonderful storyteller, *Born to Run* author Christopher McDougall reveals the story of how we once ran, then got smart and nearly screwed the whole the thing up, and then started to fix what we broke by returning to our authentic roots. Technology did not do runners a favor—it simply gave them a temporary solution, a little piece of foot candy in the form of a cushy shoe that caused cavities in authentic running strides. It made a dysfunctional activity comfortable when nature did not wish it to be. Nature was perfectly positioned and ever available to help each of us learn to run efficiently and gracefully at a pace and progression naturally adjusted to our individual limitations, but we were too impatient.

We found a path around the speed bumps—aren't we smart? Technology compensated for our bad mechanics and we laughed at nature's simple limitations. Less-forgiving footwear would have been a better teacher by enforcing favorable movement patterns while also introducing limitations on volume and distance. We would eventually reach our distances and gain the endorphins, but the experience would be the total authentic package.

There's a parallel between the running shoe analogy and the big business of exercise and rehabilitation. Exercise and rehabilitation rely heavily on the results produced. Most of the results revolve around physical capacity, physical production, performance or aesthetics. The focus is directed toward quantities because we lack an equal amount of qualitative standards based on the fundamentals of human movement. Modern running shoes made it possible to run farther and more often with poor mechanics, but running injuries soon balanced the equation. This is always the lesson when a pursuit of quantity does not parallel quality development. Nature always has the last laugh.

Likewise, modern advances in exercises and rehabilitation make it possible to develop fitness and elevated physical capacity in the presence of existing movement dysfunction. We must be diligent to keep the big business machines from creating assumptions and imbalanced perspectives in our field of expertise. Work to create a balanced perspective as your clients develop authentic movement that complements physical capacity, and the physical capacity to reinforce authentic movement.

The way to do this is to base our professional decisions on principles. Once our principles are in place, we can shop for the best methods that support those principles. The methods will change and that's okay—we are much more than our methods.

The principles are based on the highly refined and time-tested human movement systems. The methods are based on technical advances to improve efficiency and effectiveness as we address the principles.

PRINCIPLES VERSUS METHODS

The purpose of this work is to promote closer adherence to fundamental movement principles by providing methods that redirect or broaden our considerations when we seek to—

1. Recognize, rate and rank movement pattern dysfunction to standardize communication and promote systematic management of the movement behaviors produced by exercise and rehabilitation

2. Predict movement behaviors associated with increased level of injury in active populations

3. Develop more authentic exercise programs that naturally apply checks and balances to the development of physical capacity alongside fundamental movement-pattern competency

4. Develop corrective exercise strategies to target, manage and remedy movement-pattern dysfunction

5. Develop practices that clearly separate movement patterns that display dysfunction from those associated with pain provocation

6. Improve the functional diagnostic practices that identify and organize movement-pattern problems and their relationship to movement-impairment problems

The addition of movement-pattern screening as a routine standard operating procedure within organized exercise practices will address points one through four.

The addition of movement-pattern assessment as a routine standard operating procedure within organized rehabilitation practices will address points four through six.

Finally, the addition of movement pattern screening at the end of the rehabilitation process will reinforce points one through four.

This creates a theoretical net that captures poor movement patterns at every turn. By doing this, our practices fall in line with natural principles of movement.

Principle—
fundamental rule or law, usually unaffected by time or technology

Method—
how to do or make something, usually improved over time or with technical advances

In the first three chapters of this book, you read of how intense focus and development of exercise and rehabilitation methods has cast a shadow over some of our fundamental principles. In our rush to advance technical aspects of measurement and output, we forgot reductionism, and generalization in the form protocols has caused us to overvalue or undervalue certain aspects of human movement. We have learned if we step back and look at movement behaviors in the form of movement patterns, we add an important variable to measure the value of all the methods we debate and defend. We should defend principles and let the methods justify themselves. We hope this book will serve to redirect focus on the fundamental principles that preside over the way we learn to move.

On the surface, this book actually looks like a text dedicated to methods because it presents methods of applying screening and assessment to human movement patterns. However, if you look deeper, these methods only serve to keep us closer to principles of movement that have been devalued in recent exercise and rehabilitation. That is not to say the individuals who provide us with physical exercise and rehabilitation don't appreciate or promote fundamental principles. It simply suggests that we have not developed objective systems to manage human movement patterns with the same level of appreciation and organization that we approach physical performance and isolated measurements in biomechanics and physiology.

IF WE BUILD IT THEY WILL COME

We can expect professional migration toward organized, objective, practical systems and methods of movement management. This is because most exercise and rehabilitation professionals embrace a clearer understanding of movement principles than the current standard methods support. Current methods apply exercise and rehabilitation principles without complete consideration of natural movement principles. This is how we have been able to successfully but incorrectly put fitness on movement dysfunction.

Classic methods of movement training have more completely embraced movement quality and quantity, which allowed movement patterns and physical capacity to actually support and complement each other. Classic forms of movement learned by trial and error apply a balanced approach to movement endeavors. The balance produced in ancient systems like the martial arts and yoga negated the need for movement screen-

ing because incorrect movement was not practiced, and therefore was not developed or reinforced. Quality produced quantity.

In this book, you read about methods to reveal movement principles complementary to systems of exercise, rehabilitation and corrective exercise. Screening and assessment alert us to natural tendencies of movement perception and behavior, and help us refine our approaches to grow physical capacity alongside movement-pattern competency. Our methods will evolve and improve in their ability to represent the principles, and that is what methods should do.

Embarking on this current book project, we had to question the current practices we have all been taught. That hits some too close to home and people must defend the status quo.

When clinicians, researchers and educators question the approach, they don't criticize the principles or results; they criticize the methods and that's expected. They have invested time and effort into their methods, and they want to debate methods to reduce the threat of being incorrect. Our feeling is, base your standards on sound principles and let the methods justify their existence. Why fight over a method? Let it validate itself or die. It's just a thing, a tool, a recipe... not a person.

Our message has always been simple and clear. We need to add greater perspective to the way we deal with movement, because the model we have is not working. That's a problem that goes way beyond a battle over methods.

What we often use are random acts of exercise and rehabilitation without regard for natural principles of movement learning. Movement has always validated things. From manual treatments to corrective exercise choices, always look to movement to validate or refute your intervention. Journal articles and blogs just don't cut it. They might inform, but these are just opinions until they have utility on the floor. If you are unable to produce measurable results, it won't feel honest and you'll move in another direction.

You can trust movement because it rarely lies. It doesn't always make us look smart or feel confident, but it is always honest. Sometimes it rubs our faces in a mistake and hurts our pride, but we'll recover smarter with our pride in check.

When we teach workshops to practicing professionals and as they perform screens and assessments for the first time, they are amazed. The thing that impresses them most is a recurring theme: *Look what we would have missed.*

Or, to say it differently to those who are skeptics, it's not about what you find—it's about what you are comfortable overlooking.

FUNCTIONAL MOVEMENT SYSTEM PRINCIPLES

Five principles are introduced in the beginning of this book. Now let's expand those to incorporate movement-learning principles.

PRINCIPLE # 1

Separate painful movement patterns from dysfunctional movement patterns whenever possible to create clarity and perspective.

Pain produces inconsistent movement perception and behavior. We should not exercise around or into pain hoping it will get better without first attempting to manage it systematically. The movement screen at its core is designed to capture pain and identify situations that should be properly evaluated prior to consideration for exercises, activities and conditioning programs. The movement assessment improves clinical perspectives by separating pain and dysfunction, and placing equal focus on movement dysfunction to manage regional interdependence.

PRINCIPLE # 2

The starting point for movement learning is a reproducible movement baseline.

Professionals working in physical rehabilitation, exercise and athletics must adopt systematic approaches that transcend professional specialization and activity specificity. Movement professions need movement-pattern standards. This book develops two systems that logically rate and rank using movement-pattern fundamentals.

PRINCIPLE # 3

Biomechanical and physiological evaluation does not provide a complete risk screening or diagnostic assessment tool for comprehensive understanding of movement-pattern behaviors.

This text presents the case that we have investigated physical capacities and movement specializations in greater detail than we have the fundamental movement patterns that support and make them possible. Our application of knowledge regarding exercise physiology and biomechanics surpasses our application of what we know about the sensory and motor development of fundamental human movement patterns.

As professionals, we have tried to solve physical capacity problems with solutions exclusively targeting physical capacity. We have tried to enhance movement-specific skills by detailed maps of skill that are often practiced at the very edges of physical capability. These practices are valuable if they identify the weakest link in the movement chain. However, if they simply identify physical capacity and skill problems caused by some fundamental movement problem, focus on these areas actually overshadows a crack in the entire foundation. The roof isn't leaking, the basement is.

PRINCIPLE # 4

Movement learning and relearning has hierarchies fundamental to the development of perception and behavior.

The natural movement learning progression starts with mobility. This means unrestricted movement is necessary for clear perception and behavior through motor control. It may be unrealistic to expect a full return of mobility in some clients and patients, but some improvement is necessary to change perception and enhance input.

Active movements demonstrate basic control and are followed by static stabilization under load. This is followed by dynamic stabilization under load. From this framework, our freedom of movement and controlled movement patterns are developed for transitions in posture and position, maintenance of posture, locomotion and the manipulation of objects.

Principle # 5

Corrective exercise should not be a rehearsal of outputs. Instead, it should represent challenging opportunities to manage mistakes on a functional level near the edge of ability.

Technological advancements in movement and exercise science that neglect functional movement-pattern baselines ignore the natural laws that govern the sensory motor learning system that produces our perceptions and behaviors. This is the process that initially produces these patterns. Some conventional practices rehearse proper movement outcomes without establishing proper sensory inputs. They attempt to manage behavior without addressing perception.

It's common to see movement scientists identify the best technique for an exercise or an athletic movement. To create an acceptable standard, they map the sequence of movements that consistently produce great performance. Coaches and trainers come along and try to mimic those movements, and these become drills and exercises. The drills and exercises get recycled and modified. They're applied on top of dysfunction and they become protocols. After a few years, no one questions the logic.

This is not to discredit the high-end skill drills. It just points out that drills are applied whenever deficiency is noted without considering other aspects of movement or performance. The ironic part of the story is that the elite individuals who produced the near-perfect movement sequence that become the standard did not actually practice or use the drills.

To state it a different way, the analysis of the superior techniques produced exercises that did not produce the technique in the first place. How could they? The best arrive at excellence without access to drills because the drills are built on observations of their athletic output, but not their input.

Fancy drills are often developed by watching the end result of a movement, performance or skill, and not the fundamentals and deep practice that produce the superior outcome in the first place. We must be cautious at each level of movement learning not to practice rehearsals of outcomes. This might produce very fine imitation, but not authentic movement behaviors.

Principle # 6

Perception drives movement behavior and movement behavior modulates perception.

The question is, how does movement develop naturally and how do all these great performances come about? Could the same forces produce both a toddler's first step and the authentic running stride? They are both driven by inputs that influence perception. We get stuck in the practice of outputs and assume our input is the same as those we want to emulate. We perform step-by-step exercise and assume our brains will find value and therefore commitment it to movement-pattern memory.

We should know better, but we all expect that practicing outcomes will create favorable movement patterns. The fact is we should try to emulate all the sensory inputs that produce favorable general and specific movement patterns, rather than practice the motor outputs. This will put our focus on perception, and when we hit the correct perception dosage, movement behavior will provide the feedback.

Actors mimic the outputs of the characters they play and often give us convincing performances, but these are scripted. The actor is not the character, but for a brief time, they behave like the character. We treat exercise and rehabilitation in the same way. We coach movements in a controlled environment and assume we have changed behavior across other situations or even other activities. We forget that when the actor leaves the stage, he or she returns to daily life eventually forgetting the character life. Our clients and patients often do the same thing. The way they move will tell the story of what they have learned and what they have forgotten.

Principle # 7

We should not put fitness on movement dysfunction.

It is possible for fit people to move poorly and unfit people to move well. We measure basic fitness quantity and basic movement quality with different tools. We forget this and assume that fitness is the fundamental baseline, but it is not.

Fitness and physical performance or capacity is the second step in a three-step process. As you discuss the information in this book with peers, other professionals, clients or patients, keep it simple at first. Make sure you establish agreement on the fundamentals. If there is a problem understanding the basic logic of functional movement systems, you will have little chance creating weight and appreciation for the corrective parts of the model. People must understand the basics of the pyramid approach.

PRINCIPLE # 8

We must develop performance and skill considering each tier in the natural progression of movement development and specialization.

Try to keep it simple even when using the pyramid model. First direct the conversation away from perfection and exemplary performance and redirect the focus to minimums using blood pressure as an example. When we screen a group for blood pressure ranges, we're not looking for a perfect blood pressure number—we're looking for red flags. Without much thought, we will probably separate the group into high risk, borderline and low risk.

Why can't we just start our movement conversations the same way? Throw out three terms when discussing the topics of rehabilitation, exercise or training: Are we talking about competency, capacity or specialization? This usually gets a confused look, but it's a great way to start. It forces perspective. It forces a consideration of principles.

Each of these levels of movement must be cleared for minimum competency, and in a progressive order.

Competency

Capacity

Specialization

Competency
This we test with movement screening. If screening reveals pain or dysfunction in the form of limitation or asymmetry, there is a movement-competency problem. Alternatively,

there is a basic movement-aptitude problem—pick your term, but make the point. Adequate competency suggests acceptable fundamental-movement quality.

Capacity
Capacity is measured using standardized tests for physical capacity against normative data specific to a particular population or category of activity. Football players are compared with football players and golfers are compared with golfers. If movement competency is present and if testing reveals limitations in basic strength, power or endurance, there is a fundamental physical capacity problem. Adequate capacity simply suggests acceptable fundamental movement quantities.

Specialization
Coaches and experts grade skill with the use of observation, special tests, skill drills and by previous statistics when available. If capacity is present and if testing and statistics reveal limitations in the performance of specific skills, there is a specialization problem. Adequate specialization simply suggests acceptable specialized movement abilities.

This is a way to discuss the performance pyramid without a diagram. It's also a great way to see if someone has an appreciation of the natural developmental continuum that produces human movement.

A few words of caution: We cannot become movement pattern snobs demanding total perfection on screens. Practice balance and look for deficiencies at each level of movement. Our ultimate goal should be to identify the weakest link, because sometimes the problem is not movement quality. It is a deficiency within physical capacity or a shortage of skill or specialization that is causing problems.

PRINCIPLE # 9

Our corrective exercise dosage recipe suggests we work close to the baseline, at the edge of ability, with a clear goal. This should produce a rich sensory experience filled with manageable mistakes.

Our actual goal is silent knowledge—no words, just better movement perception and behavior. In

The Voice of Knowledge, former physician Miguel Ruiz discusses the silent knowledge of the body with eloquence and clarity. He states, "Your liver does not need to go to medical school to know what to do."

We can expand that brilliant and simple statement across the movement systems as well. These systems naturally use their perceptions to create their behaviors, and their behaviors to refine perceptions. Your abdominals, diaphragm and pelvic floor know what to do and how to work together if you let them. This is why we don't need to do core work with toddlers. Their curiosity drives exploration and their lack of control demands movement coordination if they are to explore. The exploration requires movement, and they work at movement to achieve exploration.

When your clients and patients arrive on the scene with movement dysfunction, you can't leave it to Mother Nature, because for a long time they have been working against her. To help them, you might need to break a behavior and reset an experience. From the experience, you will have to develop a corrective strategy.

PRINCIPLE # 10

The routine practice of self-limiting exercises can maintain the quality of our movement perceptions and behaviors, and preserve our unique adaptability that modern conveniences erode.

When corrections have done their jobs and it's time to get back to exercise, this is your opportunity to prevent future problems. The addition of self-limiting exercises to the exercise program or as preparation or cool down can keep authentic patterns maintained. Since self-limiting exercises offer greater challenges, you can create situations to use these as a form of play or self-competition.

TECHNICAL CONFIDENCE VERSUS SYSTEMATIC AUTHORITY

As you apply the functional movement systems to your work, don't let yourself be overwhelmed. Think about the systems in reverse. Don't think about them as dictating you actions; think of them as removing options unfavorable or inconsistent with movement principles.

Practice the screens or assessments whenever you can. Don't over-think things during the actual process—just execute the screen or assessment. Record your findings and then review them. Let your eyes scan the information and learn to lock onto the weakest link.

Consider this: Each process is scripted for you. In the early stages, it's best to function at the level of a technician. Collect your data and get it recorded properly. Move on to any other testing if warranted. Do not even feel obligated to correct anything. Don't burden yourself initially with the concerns of corrections.

Screen or assess friends, family, coworkers, whomever. Just run the process whenever the opportunity arises. Once you're alone, look at your notes and rate and rank your data and review the hierarchy of corrective importance. If movement correction is necessary, you should have a clear starting point demonstrated by your data.

You need to practice getting to the starting point before trying to master corrections and influence outcomes. It is best to practice one process at a time. Once your screens or assessments are smooth and you demonstrate confidence and authority, move into the corrective experiences.

As you embark on a corrective experience, plan your level of challenge and review a less challenging and a more challenging option so you can move easily forward or backward depending on the level of success.

SOME THINGS CANNOT BE FIXED

Unfortunately, you will experience things that will not change. Some limitations will be structural. Total joint replacements, fusions, significant degenerative changes will all hit a point where further functional improvement is limited by something that will not change.

Likewise, some movement dysfunction is so ingrained and so fixed within the central nervous system that improvement is not possible. Some cases are extreme, but just because you cannot resolve a dysfunction completely doesn't mean you cannot make small changes.

For people with severe movement problems, small changes may greatly improve the quality of their lives. In these cases, a corrective strategy may become a perpetual activity and part of the actual program, not because it advances function, but because it reduces further loss of function or degradation of structures.

THE AMAZING MOVEMENT-LEARNING BRAIN

Species endowed with a larger frontal cortex seem to play more when they're young. This play is not just the random result of a big new brain with nothing to do. More likely, play has an important developmental role. All this fooling around that looks so useless, random and unstructured serves to fire circuits across our vast cortical map. We have a very large and complex frontal cortex by biological standards, and we play a lot, for a long time. Our brains are two percent of our weight and use 20 percent of our energy—sounds like a pretty big muscle!

In the most basic terms, we are born, and our large circuit board is introduced into an unknown environment... and it must adapt. To be so big and complex, in the biological hierarchy of brains, our brains seem to have a very basic operating system. However, that is its fundamental and functional brilliance. It seems to be prewired with only three objectives: to stay safe, satisfied, and explore everything possible. If all three of these things are satisfied, the brain takes over and starts to program itself from that point on.

Nature's wisdom knows that too much pre-loaded software may not adequately serve every situation. Preloaded programs would reduce our adaptability to the particular clans or environments in which we pop up, because it's not known exactly where or to whom we will be born.

Exploration is vital to our learning and adapting. The safety element keeps the mistakes we must learn from causing permanently damage or from killing us. The satisfaction keeps our bellies full and our bodies warm so we keep growing. All the exploration takes on the appearance of play as we drift from activity to object, engaging all the senses and making sensory-rich, wonderful mistakes.

Our brains are so large we are born helpless. If we were born with our brains fully developed, our heads would be so large we could not fit through the birth canal, and if we tried our mothers would never forgive us. Instead, we arrive with a brain that continues to grow aggressively for two years after birth. We don't start off playing, but we get to it as soon as we possibly can.

As we play, the circuits we fire the most don't so much get bigger, but they get faster. The things we do often cease to be a combination of a few components and start to become a single program—thus the pattern is born.

A pattern is born to our perceptive brain as well as our behavioral brain, and recognizable situations and responses are linked. The more we observe and use the linked patterns, the more the chain of circuits fire and the more padding we dedicate to insulating the frequently used circuits. Speed develops as the insulation of the cables connecting our favored circuits gets better and thicker. This insulation is called myelin, and we make it and break it down according to the perceptions and behaviors we practice most or least. To quote Daniel Coyle in *The Talent Code,* when we put down more myelin, *we go from dial-up to broadband.* Movement-pattern learning and development seems to be a case of the same old computer, with much better connections.

Fundamental movement patterns seem to be largely developed by play, but consider the practice hours a normal toddler puts into walking. Obviously, higher skills require more specific play or practice and maturity, but the learning is the same.

Learning is about turning our most frequent movement perceptions and behaviors into memory patterns we can quickly access and execute. Some even make it to the automatic level; some remain at the conscious level and some linger between, and leave us options allowing us to modify the pattern one way or another. These memory-pattern circuits get insolated and become fast and efficient and—*voila!*—all is well and good in our world.

There is only one problem. What if we are somehow deprived of a full sensory experience as we develop? What if we are injured or disabled during a key learning period? What if the environment is hostile or unsafe? What if we don't get proper nutrition? What if we are emotionally deprived or distressed?

What if all goes well and we develop wonderfully, but then later in life, we choose to only move one way? What if we mold ourselves into jobs that force us to sit, stand, twist or bend in unnatural ways? What if we choose to perform activities so specialized that some of our fundamental movement patterns erode? Is it possible that our brilliant and miraculous automatic learning brains will make and memorize dysfunctional patterns for us? Absolutely.

THE DYSFUNCTIONAL MOVEMENT-LEARNING BRAIN

The same brain that learns function can just as easily learn dysfunction. In fact, after all this discussion about movement, you may ask the question, "How can some people keep their fundamental movement patterns without practice?"

That is a great question, best to keep the explanation simple. The elements probably go something like this—

- In the best case scenario, those who care for us allow us to develop naturally, and we install our fundamental patterns with an acceptable level of quality.

- Then we enjoy a diet rich with a variety of movement experiences and activities that we engage in frequently.

- If we become injured, we seek full recovery of movement quality and not simply pain relief.

- Although the fundamental movement patterns are not part of daily practice, there is no reason for them to erode because they are actually subcomponents of larger circuits and patterns.

- Since these patterns function close to an authentic standard without compromise or compensation, fundamental patterns are maintained as well.

- Lastly, some of us picked the right parents and got dealt a better set of functional movement-pattern genes than others.

This last point was intentionally not listed first, because it would inevitably be used as an excuse not to do all the constructive work associated with quality movement experiences, and instead blame poor movement on bad genes. It is true that some will need to work harder to maintain movement pattern minimums, but that is life. Some battle weight gain; some battle weight loss. Some put on muscle by looking at resistance, and some cannot seem to create hypertrophy no matter how hard they work.

In contrast to the components above, some people continuously work on flexibility, strength and endurance, and have success with one or two aspects of fitness, but have difficulty improving fundamental movement. Somewhere in all their activity is compromise, unbalance and compensation, which they practice every day. This practice actually myelinates patterns that compromise movement quality. They and their workouts are the unconscious destroyers of their own fundamental patterns.

The problem we often find associated with fit individuals and poor screens is over-specialization. These people cannot grasp the concept of balance. They pick and then practice a single activity, assuming sheer levels of superior fitness will generate holistic benefits.

Others mistakenly rush into activities with poor preparation, and compensation is their only option—they get great at all the compensations. Meanwhile, still others are not fully rehabilitated from an injury or ailment and too soon return to full activity. Confronting denial and slapping some basic logic on impatience can cure this.

Lastly, pharmaceuticals have numbed us to the sensations of pain as we persist in activities, ignorant of warning signs and signals of slow degeneration and continued damage. The drugs can cover the pain, but they can't hide the dysfunctional movements for long, and this is why screening offers an effective defense against this drug-induced pseudo-recovery syndrome. This type of individual can also be known as cosmetic anti-inflammatory junkie. These people would not have inflammation at all if they would stop the insulting exercise practices and correct the underlying problems.

The word *cosmetic* here does not refer to physical appearance; it refers more to the superficial ego that makes us exercise when we know we should seek medical attention and take the time necessary to fix the problem instead of covering it up. Admit it, we all hope stuff will just go away, but when it does not, we get to use our brains not only to figure it out, but to actually re-pattern the movement.

THE WISDOM OF OUR ANCESTORS

Our ancestors had it tough; they did not have the luxury of movement specialization. They had to toil, labor, fight and flee. They had to walk, run and carry on a daily basis. Those who had consistent food, shelter and safety could probably move well by modern screening standards. Civilization, specialization and modernization have reduced our need to stay adaptable and functional against even minimal movement standards. For a few thousand years, we have sought to maintain our bodies' movement capacities in times of peace and prosperity. The ancient ones adopted daily rituals to age gracefully, find peace and harmony and stay strong in the event peace was not possible. Some approaches have been brilliantly holistic and some have been comically shortsighted, stupid and fortunately short-lived, although they keep popping up like weeds in each generation.

Nevertheless, our ancestors realized that our quest for convenience caused a subtle decline in our movement competency, and endeavored to construct movement development and maintenance plans after realizing movement is important. Our ancestors devised games, competitions, rituals and rites of passage revolving around movements and physical prowess. Most of these devices represent the physical ideals that stress a balance of our physical and mental abilities and make them valuable to our tribes, to our families and to us.

Scientists think there is no common ancestor to the 40 species of flightless birds we know today. Each species lost flight simply by not using it. I hope we are not the first generation of the squatless humans, but we can't fix what we don't check.

Every now and then, we get off track, and more recently we have gotten way off track. Movement screening and assessment are the rulers we can use to measure our departure from our fundamental authentic movements. This book has prepared the foundation for you to use this in your work.

YOUR LEARNING

It might also be helpful to apply the same principles to your learning that you provide for your clients and patients. You will be expected to give them a favorable environment in which to learn. You need to provide lots of appropriate sensory input with some room to work things out or even better, to learn things out.

The same goes for you. The framework and all the rules are here to help you. They are here to reduce confusion, narrow your choices and allow you to retrace your steps when you succeed or fail.

Your corrective outcomes are less important than your corrective inputs. Your outcomes will represent the quality of your screening, assessment and application of the strategy. The goal is not to memorize the strategy. The goal is to apply the strategy frequently so your brain can start to learn the logic. The screen and assessments provide the starting point—your sensory input, your new perception. The framework provides the strategy, and rescreening and reassessing provide the feedback. Your brain requires all three steps to learn how to smoothly apply the system.

All three steps must be repeated in a number of different situations. Soon you will forget about rules, hierarchies and even flowcharts. You will practice systematically with the latitude and flexibility to fit each unique situation.

CLOSING THOUGHTS
AND THE GRAND SOLUTION

Earlier, you read that babies only have three preloaded programs. They want to be safe, be satisfied and to play. Given the opportunity, their brains and bodies develop beautifully and wonderfully. When we practice these basics in exercise, rehabilitation and in our own lives, we must heed the warnings of balance.

If we are too protective, our mistakes will not teach. If we get too hung up with total satisfaction, we become professional comfort-seekers; we never benefit from constructive stress. If we play incorrectly or explore only specialties and extremes, we may not maintain our authentic patterns and may compromise our durability.

The responsibility to screen, assess and correct movement is one we can all shoulder together. We have provided some science mixed with some commonsense to help you in the practice of your profession. You must develop the art.

Outside of that, develop the best methods you can, methods designed to keep you close your movement principles. You will do just fine.

Now go play.

*Please see www.movementbook.com/chapter15
for more information, videos and updates.*

APPENDIX 1

THE JOINT-BY-JOINT CONCEPT

Let's elaborate on the joint-by-joint approach to training to discover what's behind the concept of which joints need stability and which need mobility. First, for those who aren't familiar with the idea, we'll begin with an overview written by Michael Boyle. Following his thoughts, I'll expand with more detailed commentary.

THE JOINT-BY-JOINT APPROACH

Excerpted with permission from Michael Boyle's Advances in Functional Training

If you are not yet familiar with the joint-by-joint theory, be prepared to take a quantum leap in thought process. My good friend physical therapist Gray Cook has a gift for simplifying complex topics. In a conversation about the effect of training on the body, Gray produced one of the most lucid ideas I have ever heard.

We were discussing the findings of his Functional Movement Screen (FMS), the needs of the different joints of the body and how the function of the joints relate to training. One beauty of the FMS is it allows us to distinguish between issues of stability and those of mobility; Gray's thoughts led me to realize the future of training may be a joint-by-joint approach, rather than a movement-based approach.

His analysis of the body is a straightforward one. In his mind, the body is a just a stack of joints. Each joint or series of joints has a specific function and is prone to predictable levels of dysfunction. As a result, each joint has particular training needs.

This joint-by-joint idea has really taken on a life of its own, one I certainly didn't envision. It seems like everyone's familiar with it; it's become so common knowledge people fail to reference Gray Cook or me as the developers of the idea.

The table in the next column looks at the body on a joint-by-joint basis from the bottom up.

The first thing you should notice is the joints alternate between mobility and stability. The ankle needs increased mobility, and the knee needs increased stability. As we move up the body, it becomes apparent the hip needs mobility. And so the process goes up the chain—a basic, alternating series of joints.

Joint—Primary Need

Ankle	Mobility (sagittal)
Knee	Stability
Hip	Mobility (multi-planar)
Lumbar Spine	Stability
Thoracic Spine	Mobility
Scapula	Stability
Gleno-humeral	Mobility

Over the past 20 years, we have progressed from the approach of training by body part to a more intelligent approach of training by movement pattern. In fact, the phrase *movements, not muscles* has almost become an overused one, and frankly, that's progress. Most good coaches and trainers have given up on the old chest-shoulder-triceps method and moved to push-pull, hip-extend, knee-extend programs.

Still, the movement-not-muscles philosophy probably should have gone a step further. Injuries relate closely to proper joint function, or more appropriately, to joint *dysfunction*. Problems at one joint usually show up as pain in the joint above or below.

The primary illustration is in the lower back. It's clear we need core stability, and it's also obvious many people suffer from back pain. The intriguing part lies in the theory behind low back pain—the new theory of the cause: *loss of hip mobility.*

Loss of function in the joint below—in the case of the lumbar spine, it's the hips—seems to affect the joint or joints above. In other words, if the hips can't move, the lumbar spine will. The problem is the hips are designed for mobility, and the lumbar spine for stability. When the intended mobile joint becomes immobile, the stable joint is forced to move as compensation, becoming less stable and subsequently painful.

THE PROCESS IS SIMPLE

- Lose ankle mobility, get knee pain

- Lose hip mobility, get low back pain

- Lose thoracic mobility, get neck and shoulder pain, or low back pain

Looking at the body on a joint-by-joint basis beginning with the ankle, this makes sense.

The ankle is a joint that should be mobile and when it becomes immobile, the knee, a joint that should be stable, becomes unstable; the hip is a joint that should be mobile and it becomes immobile, and this works its way up the body. The lumbar spine should be stable; it becomes mobile, and so on, right on up through the chain.

Now take this idea a step further. What's the primary loss with an injury or with lack of use? Ankles lose mobility; knees lose stability; hips lose mobility. You have to teach your clients and patient these joints have a specific mobility or stability need, and when they're not using them much or are using them improperly, that immobility is more than likely going to cause a problem elsewhere in the body.

If somebody comes to you with a hip mobility issue—if he or she has lost hip mobility—the complaint will generally be one of low back pain. The person won't come to you complaining of a hip problem. This is why we suggest looking at the joints above and looking at the joints below, and the fix is usually increasing the mobility of the nearby joint.

These are the results of joint dysfunction: Poor ankle mobility equals knee pain; poor hip mobility equals low back pain; poor t-spine mobility, cervical pain.

An immobile ankle causes the stress of landing to be transferred to the joint above, the knee. In fact, there is a direct connection between the stiffness of the basketball shoe and the amount of taping and bracing that correlates with the high incidence of patella-femoral syndrome in basketball players. Our desire to protect the unstable ankle came with a high cost. We have found many of our athletes with knee pain have corresponding ankle mobility issues. Many times this follows an ankle sprain and subsequent bracing and taping.

The exception to the rule seems to be at the hip. The hip can be both immobile and unstable, resulting in knee pain from the instability—a weak hip will allow internal rotation and adduction of the femur—or back pain from the immobility.

How a joint can be both immobile and unstable is an interesting question.

Weakness of the hip in either flexion or extension causes compensatory action at the lumbar spine, while weakness in abduction, or, more accurately, prevention of adduction, causes stress at the knee.

Poor psoas and iliacus strength or activation will cause patterns of lumbar flexion as a substitute for hip flexion. Poor strength or low activation of the glutes will cause a compensatory extension pattern of the lumbar spine to replace the motion of hip extension.

This fuels a vicious cycle. As the spine moves to compensate for the lack of strength and mobility of the hip, the hip loses more mobility. Lack of strength at the hip leads to immobility, and immobility in turn leads to compensatory motion at the spine. The end result is a kind of conundrum, a joint that needs both strength and mobility in multiple planes.

Your athletes, clients and patients must learn to move from the hips, not from the lumbar spine. Most people with lower back pain or hamstring strains have poor hip or lumbo-pelvic mechanics and as a result must extend or flex the lumbar spine to make up for movement unavailable through the hip.

The lumbar spine is even more interesting. This is clearly a series of joints in need of stability, as evidenced by all the research in the area of core stability. The biggest mistake we have made in training over the last 10 years is an active attempt to increase the static and active range of motion of an area that requires stability.

Most, if not all, of the many rotary exercises done for the lumbar spine were misdirected. Physical therapist Shirley Sahrmann in *Diagnosis and Treatment of Movement Impairment Syndromes* and James Porterfield and Carl DeRosa in *Mechanical Low Back Pain: Perspectives in Functional Anatomy* all indicate attempting to increase lumbar spine range of motion is not recommended and is potentially dangerous. Our lack of understanding of thoracic mobility caused us to try to gain lumbar rotary range of motion, and this was a huge mistake.

The thoracic spine is the area about which we know the least. Many physical therapists recommend increasing thoracic mobility, though few have exercises designed specifically for it. The approach seems to be "We know you need it, but we're not sure how to get it." Over the next few years, we will see an increase in exercises designed to increase thoracic mobility. A leader in the field, Sahrmann was early to advocate the development of thoracic mobility and the limitation of lumbar mobility.

The gleno-humeral joint is similar to the hip. The gleno-humeral joint is designed for mobility and therefore needs to be trained for stability. The need for stability in the gleno-humeral joint presents a great case for exercises like stability ball and BOSU pushups, as well as unilateral dumbbell work.

In the book *Ultra-Prevention*, a nutrition book, authors Mark Hyman and Mark Liponis describe our current method of reaction to injury perfectly. Their analogy is simple: Our response to injury is like hearing the smoke detector go off and running to pull out the battery. The pain, like the sound, is a warning of some other problem. Icing a sore knee without examining the ankle or hip is like pulling the battery out of the smoke detector. The relief is short-lived.

Michael Boyle, 2010

Excerpted with permission from
Advances in Functional Training, by Michael Boyle.

APPENDIX 2

EXPANDING ON THE JOINT-BY-JOINT APPROACH

The original conversation Mike Boyle and I had about the joint-by-joint approach to training was more about the thought process than about physiological facts and absolutes. This has been the topic of lots of discussion, but here is the pearl: Our modern bodies have started developing tendencies. Those of us who are sedentary, as well as those of us who are active, seem to migrate to a group of similar mobility and stability problems. Of course you will find exceptions, but the more you work in exercise and rehabilitation, the more you will see these common tendencies, patterns and problems.

A quick summary looks goes like this—

1.   The foot has a tendency toward sloppiness and therefore could benefit from greater amounts of stability and motor control. We can blame poor footwear, weak feet and exercises that neglect the foot, but the point is that the majority of our feet could be more stable.

2.   The ankle has a tendency toward stiffness and therefore could benefit from greater amounts of mobility and flexibility. This is particularly evident in the common tendency toward dorsiflexion limitation.

3.   The knee has a tendency toward sloppiness and therefore could benefit from greater amounts of stability and motor control. This tendency usually predates knee injuries and degeneration that actually make it become stiff.

4.   The hip has a tendency toward stiffness and therefore could benefit from greater amounts of mobility and flexibility. This is particularly evident on range-of-motion testing for extension, medial and lateral rotation.

5.   The lumbar and sacral region has a tendency toward sloppiness and therefore could benefit from greater amounts of stability and motor control. This region sits at the crossroads of mechanical stress, and lack of motor control is often replaced with generalized stiffness as a survival strategy.

6.   The thoracic region has a tendency toward stiffness and therefore could benefit from greater amounts of mobility and flexibility. The architecture of this region is designed for support, but poor postural habits can promote stiffness.

7.   The middle and lower cervical regions have a tendency toward sloppiness and therefore could benefit from greater amounts of stability and motor control.

8.   The upper cervical region has a tendency toward stiffness and therefore could benefit from greater amounts of mobility and flexibility.

9.   The shoulder scapular region has a tendency toward sloppiness and therefore could benefit from greater amounts of stability and motor control. Scapular substitution represents this problem and is a common theme in shoulder rehabilitation.

10.   The shoulder joint has a tendency toward stiffness and therefore could benefit from greater amounts of mobility and flexibility.

Note how stiffness and sloppiness alternate. Of course, trauma and structural problems can break the cycle, but it is a present and observable phenomenon producing many common movement pattern problems. It also represents the rule in orthopedics evaluation of always assessing joints above and below a problem region. It would be illogical to expect to improve knee stability in the presence of ankle and hip mobility restrictions. Likewise, it would be impractical to assume that a recent improvement in hip mobility would not return to stiffness if improved stability were not also created in the lumbar and knee regions. Chronic sloppiness would always be more convenient to use than new mobility.

When Mike and I first discussed this layering of opposites, he did a great job of developing the topic to discuss a more comprehensive approach to exercise program design.

The point in the joint-by-joint approach is not so much the 10 Commandments of Mobility and Stability: *Make the ankle mobile. Make the knee stable. Make the hip mobile. Make the low back stable.* We'll find a person every now and then whose ankle has too much mobility or who's sloppy in the hip. We use the words mobility or stability to implicate a segment of the body that should be moving better or have more control. The whole point is to practice with a systemic approach to clear the joints above and below the one with the problem.

I was interviewed on this topic after it became popular, and many of my comments regarding the joint-by-joint discussion have been transcribed for you here.

When we talk about the ankle, we're talking about the ankle joint, the inverters, the everters, the dorsiflexors, the plantar flexors and all of the other stabilizers that control that ankle. We're not just talking about a joint—we're talking about a complex. Same thing with the knee; same thing with the hip; same thing with the back, the T-spine, and so on up the chain.

When you're about to do knee stability training or lumbar stabilization and you take the classic kinesiology approach of training all the muscles around the knee or all the muscles around the core, you're going to make a mistake nine times out of ten. You're assuming when you train the knee that the ankle and hip are contributing like they should, as much as they should. That's hardly ever the case.

It's the same is true with lumbar stability. Some of the people producing lumbar stability research today are very well intentioned about the muscles they want us to fire and the muscles on which they want us to focus our exercises. I don't have a problem with stability research or stability suggestions. All I ask is that the authors use a qualifying statement in front of their core stabilization talks: *These statements about stability have been made assuming that you know how to clear the hips and clear the T-spine and other regions where mobility will actually compromise stability. These regions should be considered as potential reasons for loss of stability and compensation behavior.*

Logically we must make sure these areas are mobile, because if the hips and T-spine aren't mobile, the lumbar stability we create is synthetic. It is not real. We develop enough stability and strength to do a side plank, but we don't authentically stabilize in natural environments. The central point of the joint-by-joint discussion is to assure we're working on what we think we're working on. Most of us make the mistake by assuming sloppy knee, stiff ankle, stiff T-spine without considering the potential problems above and below.

What would be a reason for the T-spine to become stiff? Probably there's a lack of stability somewhere else. Often if you don't have the necessary core stability, the T-spine will get stiff and this also works in reverse. If the T-spine is too stiff, the core stability will be compromised. It can work either way. It's not about finding what came first, the chicken or the egg—you have to catch both or you can't manage either.

The takeaway from a joint-by-joint discussion is this: Instead of trying to memorize how everything is supposed to be in a perfect world, ask yourself these questions—

- I'm getting ready to train mobility or stability in this segment.

- I either want this segment to move better or I want this segment to be more stable.

- Have I truly cleared the joints above and below that can compound the problem?

REVIEWING THE JOINTS

I often start at the discussion at the foot, where I defer to Todd Wright and Gary Gray. They have great perspective and discussion with respect to the foot. People have always tried to pull me into a top-down or bottom-up argument, but I'm not committed either way. Problems can come from either place and be corrected by either approach. The real question is what do you see.

Here is an example.

Let's say we do the movement screen and we learn that the active straight-leg raise, shoulder mobility, pushup and rotary stability patterns are great, but in standing, the squat, hurdle steps and lunges are bad. You need to consider the foot. This is because everything was going great until you asked the foot to contribute. It does not imply a foot problem; it simply suggests that perceptions and behaviors are compromised when the foot hits the ground.

Here's what I want people to know: The brain and its information pathways work two ways. We're not just sending information down the spinal cord out to the hands and feet. We're also uptaking information through the hands and feet.

If the feet are sloppy and the grip is off, not only will the person not activate the right muscles, but he or she is not even up taking the right sensory information. Let me say that again. If there are any mobility or stability compromises between the foot and the brain, it's like standing on two garden hoses wondering where all the water is. The information pathway is broken two ways... up and down.

The foot is no longer a sensory organ because any information that foot could collect in its normal alignment has to be compromised. The foot has to pronate even more because of a stiff ankle, or maybe the foot has to fire too much throughout the plantar flexors because of a sloppy knee.

The other reason we've got to clean up these issue is it's not just motor pathway down; it's sensory pathway up. The foot will keep flattening out to grab as much sensation as possible because the brain knows there is a problem. It's hoping more information will help. If you've got bad shoulder positioning in a push or pull movement, you're going to do things with your grip that aren't as authentic as they could be.

Let's look back at the foot. The foot needs to be dynamically stable, but it's inherently set up to be mobile. Look how many bones, how many joints are in the foot. There's movement all over the place unless there's arthritis. The muscular role in that foot should be that of stability, and that's why we have all those intrinsic muscles. These are muscles that dwell within the foot, within the arch of the foot.

Then we get to the ankle. It's a boney, stable joint. You're never going to see many people over-dorsiflex or over-plantar flex. But since people know of inversion or eversion sprains or strain, they think the ankle must be trained for stability.

Most of the time, the patient with the rolled ankle will also have restricted dorsiflexion, unless the person stepped on a foot or had a contact injury. There's a huge prevalence of restricted dorsiflexion in people who present with knee problems, whether MCL or ACL.

When a client can squat to parallel, we often leave that last 10 degrees of dorsiflexion on the table, thinking it's no big deal. We want the foot to be stable, but that doesn't mean the foot has to be stiff. We want a mobile foot to be instantaneously stable at contact and push-off, but also to be relaxed enough to accommodate great range of motion.

The foot has to be adaptable, but it also has to be instantaneously stable. The ankle has to have freedom of movement. You can't have ankle restrictions. The ankle also has to be stable, but one of the major problems we see is lack of dorsiflexion. Is it our footwear? Is it the way we train? It's all that. The muscles attaching around the ankle have great leverage and strength, but the mobility provides the best overall function to utilize the potential strength and power in the ankle.

We need that inherent reflex stability in the foot. We need to have a clear ankle when it comes to plantar flexion and dorsiflexion.

Knees are simple hinge joints. They're supposed to flex and extend, and when they rotate too much or move valgus or varus too much, we start seeing problems with the knee. Does the knee need to be mobile? Yes, but once it's mobile, it needs to be stable enough to stay inside the proper plane of movement where its functional attributes are possible and practical.

The rotating joints are the ankle and hip. The ankle doesn't just hinge, and the hip doesn't just move in one plane. The knee is more of a hinge joint. What we want to see at the knee is once we have the mobility, we need stability.

What are the common problems we see at the hip? Can we see a sloppy hip? Can we see a dislocating hip? Absolutely. But in general, we see a lot more hips that don't have the full authentic mobility.

- Common problems in the foot: People give up their stability.

- Common problems in the ankle: People give up their mobility.

- Common problems in the knee: People give up their stability.

- Common problems in the hip: People give up their mobility.

- Now we're at the low back: People give up their stability.

So once again, these aren't the 10 Commandments, but they're common tendencies when injury, poor training, unilateral dominance, one-dimensional training, a lack of training or an excess of training occur. These are common defaults the body will go to; they're not absolutes.

Ribs, vertebrae and lots of muscle and fascia crisscrossing the front and back of the thorax cause thoracic stiffness. We don't inherently have a lot of mobility there, but we need all we can get. However, stiffness isn't just something we need to get rid of. Stiffness is there for a reason. Biological mechanisms that move very well in childhood will develop stiffness following an injury or following repetitive bad mechanics over time. If the body doesn't stabilize correctly, it will figure out another way to get stability: *it's called stiffness.*

If you find tight hamstrings or a tight T-spine and you just hit the foam roller, you may change mobility, but you will see the stiffness return the following day. Mobility efforts without reinstalling stability somewhere else simply don't last. Those hamstrings were tight for a reason. That T-spine is stiff for a reason.

If you don't also backfill some of that new motion with reflex muscular integrity and motor control, you're going to have a problem. Usually we see tight hamstrings on people who don't extend their hips well. They don't use their glutes well, and so the poor hamstrings get double-time. The hamstrings get too much use, and they fatigue—a fatigued muscle and a tight muscle look very much the same. It's all just protection.

Most T-spine mobility problems occur in people who also don't have full range-of-motion

core stability and strength. We may see a tight T-spine on a person who can side plank or front plank for an hour, but who don't have great core stability through a full shoulder turn in the golf swing. This may be a stiffness developed as a protection. As we get up in the thoracic spine, we'd like to have mobility.

In the scapulothoracic complex, there is only one boney connection of the scapula to the entire axial skeleton (rib cage or vertebra) and that's at the sternoclavicular (SC) joint. This is where the top end of the collar bone and sternum meet. The acromioclavicular (AC) joint and the SC joint are at each end of the collarbone connecting the shoulder girdle to the rest of the body. But that poor scapula is floating on the rib cage, held in place mostly by muscles and by two joints that aren't much bigger than the joints in the index finger.

That scapulothoracic area needs stability. Does that mean we don't have to get rid of some trigger points in the upper trap first? No. But often that scapula is stuck in the wrong position. We think it's stable, but instead it's just not mobile. It doesn't mean it's stable where it ought to be. Sometimes we loosen that scapula up to make it more stable. We foam roll the upper back, do a little bit of stretching of the teres major, stick a little ball in the armpit, stretch that out, and reset the scapula. Then we train it for authentic stability, but only when mobility is acceptable.

Once again, we see tight traps, and we think the last thing we need to do to those shoulders is add stability, thinking instead we need to do mobility work. Maybe you get the scapula back where it belongs, but if you want to see if it's stable, watch the person deadlift and see if the exact same scapular position can be maintained throughout a deadlift. No? Then the individual has no stability. The deadlift represents distraction, and plank and pushups represent compression. The stable shoulder must be able to manage both situations.

At the glenohumeral joint we look for mobility. But certainly you can think of a person who dislocated a shoulder. Once you see the dislocation, you may think everybody needs to stabilize their glenohumeral joint, but if you actually go around and measure glenohumeral range of motion, you might start to feel different.

In past shoulder training, we'd work on the rotator cuff and try to strengthen it. Then we got better and realized the shoulder needed a stable base. That base was the scapula.

How can you make the scapula stable if the T-spine is stiff? The scapula may be moving around incorrectly or too much when the shoulders don't turn right. I've seen many golfers try this. They don't have T-spine mobility for rotation, so to get a good shoulder turn on a golf swing, they protract one shoulder, retract the other, and it looks like they're turning their spines. They're not. They're just destabilizing both shoulders and in doing so, they're really losing a lot of good contact and connection with the club.

We can take this a few steps further. Past the glenohumeral joint, we were back on the T-spine, we go up into the mid-neck, the vertebrae from maybe seven up to two. Most people need more stability there. They need their curve back, and they need good stability.

Most people in the computer age, in the driving age, are stiff in their suboccipital region, the joints between the base of the skull and C-2. That's why so many people with their teeth together can barely touch the chin to the chest or do 45 degrees of rotation without using the rest of the neck. They're very tight in the suboccipital region from many bad posture habits and from tension. They overuse the middle components of the neck, which are usually where we see degenerative changes.

Where do we see degenerative changes in the spine most? In the mid-neck and in the low back, areas that need to be more stable. Once these areas are degenerated, they become stiff, Many people don't understand that the stiffness is the body's attempt to stop the sloppiness.

We usually see quite a bit of degeneration in the knees. That doesn't mean we don't have it in the hips and ankles, but in the knees it just seems to be compounded. These are areas that could probably use better stability, and better alignment, better everything.

We can follow this out into the elbow and hands, but it gets complicated there because we've got injuries to consider. The elbow is more than just one joint, too; there are a lot of things going on there. When we get into the hands and all the manipulative things people do, one of the first things I always do is look for full wrist extension and flexion. Without that, the other mechanics all the way up the chain are compromised: elbow, shoulder, scapula, T-spine and neck.

In our *Secrets of the Shoulder* DVD, Brett Jones and I discussed all the neurons in the brain dedicated to the hand. These exceed all the neurons dedicated to the entire arm, scapula, and even the same-side leg.

There is a large amount of brain area dedicated to the effective management of the hand. When there are restrictions, compensations and problems in the hand, a person will nearly contort the whole body to accommodate it.

Because sensory information is so important, because foot information is so important, because hand information is so important, a person will sacrifice other parts of the body. This is to make sure to get a good perspective with grip, with stride and step, and the way the foot connects, and with the way vision interacts.

The whole purpose of the joint-by-joint concept is to realize generalities. It's a mobility stacked on a stability, stacked on mobility. The examples are there to make you think above and below the area you're working on and in the things you're asking for. That's why, in a strange sense, the joint-by-joint is simply another way to make people appreciate whole movement patterns outside of the movement screen.

Once you get it, if you decide to go on through the rest of your life without using movement screen, it won't bother me a bit. *It's simply a tool.* Once you get the perspective, that's fine. What happens, though, is this tool sets a great baseline and sometimes protects us from our subjectivity. A doc can get really good at calling fractures, but we still appreciate him shooting the X-ray.

It's very easy without an X-ray to get about 85-percent accuracy on a fracture, and anyone who's done sports medicine for a long time gets a sense of a sprain or a fracture in a joint. But, you'd always want to have that X-ray.

I have a pretty good perspective on how a person moves, but I want to revisit the baseline because if I improve the movement in some way, I don't just want my subjective information to say that. I want to know I followed a joint-by-joint perspective, and have something to show for it.

We often see somebody focus on core stability. They hammer the side plank, they hammer another core exercise. The core stability is better, and I won't argue that. But now you've jacked up the upper trapezius, threw the neck out of alignment, and the hip basically doesn't move any better than it did before the side plank. The side plank fired the core, didn't fix the hip, and jacked up the shoulder and the neck.

That's what? One step forward, two steps back? That's the problem we get into with the Kinesiology 101 approach. We find a movement error and we want to fix it. We map the major movers in that area. We exercise them concentrically, and think we did something. We didn't.

Honestly, we leave so much on the table in rehab, we can't throw stones at anyone in strength conditioning. The number one risk factor for a future injury is a previous injury. That pretty much means there are a lot of chiropractors, physical therapists and athletic trainers discharging people, or giving them a clean bill of health when patients say they feel fine. That's great, I am glad they feel fine.

If the doctor releases an NFL player to play, the strength coach might agree that the medical problem is resolved. However, being well and being ready to play in the NFL are two different things. The movement screen and other functional testing demonstrate risk factors, and the best strength coaches watch these risk factors constantly. The guy might have an asymmetrical lunge. He's pain-free; nobody's arguing that. But we as clinicians in the musculoskeletal fields discharge people feeling fine, but who are still moving poorly. We send them back to their personal trainers, back to their strength coaches, back to their yoga instructors.

Now we've got an entire fitness industry trying to deal with issues that should have either been cleaned up in the rehab situation or at least forecasted, meaning clinicians need to be ready to have another conversation.

"Insurance isn't going to pay me to treat you anymore, you've got no back pain and you feel fine, but you don't squat well. When you lunge on the left side, it looks great. When you lunge on the right side is very unstable. I want to get you hooked up with a trainer who gets it, but here's the deal.

"You've got to get your lunge patterns symmetrical and get your squat pattern back. I know you want to lose weight and get back in the gym but you need to move well before you move more. I know you want to get fit again. I know you want to play golf in the spring. These are the fastest ways to get you there."

That's what I talk about in our movement training workshops when to get people working together. The top risk factor for an injury is a previous injury. That is an insult to anybody who's treating injuries, because it means we leave risk factors on the table. It does not mean we need to fix all these problems, but we can use our professional network to give our patients options.

When we peel the onion, guess what we find these risk factors are? It isn't strength. It isn't even flexibility. It's left-right asymmetries. Not mobility asymmetries or stability asymmetries—*movement asymmetries*.

Break these down. Figure out what's causing them: dorsiflexion restriction, poor spine mechanics, whatever. Fix it, but recheck the movement pattern. If the movement pattern didn't change, you think you fixed it, but you didn't. Keep working, keep tweaking it. When the movement pattern changes, you've done your job.

MOTOR CONTROL

Motor control is the ability to balance and move through space and range of motion. People call it stability; we're going to call it motor control. It's not strength. It's just *can you balance on one foot? Can you control a deep squat? Can you lunge narrow without losing your balance?*

Asymmetries and motor control are the two underlying things that aren't addressed in rehabilitation. I want the entire fitness and conditioning community to learn from the mistakes we make. Just because a person feels fine doesn't mean he or she is not at risk for an injury, and it doesn't mean the person is not going to butcher the great exercise program you designed. It's not because it's a bad exercise program. Your clients are going to try to move around things because they can't move through the things.

Joint-by-joint is an excellent template to get you past that entry-level thinking that Kinesiology 101 is going to save the day. It makes you consider joints above and below, but if you really want another way to check yourself, look at the whole patterns of movement.

Movement, once we get through the mechanics, is still a behavioral entity that largely goes unaddressed. Really, when we train people and we're working on functional training, we're working on conditioning, training or changing movement behavior. To take joint-by-joint a little bit deeper, don't only focus on the segment in which you think you found a problem.

Realize this: Until you clear everything above or below, it cannot be a singular problem.

APPENDIX 3 SFMA SCORE SHEETS AND FLOWCHARTS

We have devised a color system to help guide you through the SFMA. It starts as the same colors found on a traffic light—red, yellow and green. These work well for the top-tier tests. To help you navigate the breakout tests, we added blue and orange, which are described below. Remember, the colors are guides. The hierarchy and severity of DNs ultimately dictate your corrective exercise path.

THE SCORE SHEETS

The score sheets use shapes to indicate direction.

THE TOP-TIER SCORE SHEET

The top-tier score sheet uses a hexagon to indicate red or stop, a inverted triangle to indicate yellow or proceed with caution, and a circle to indicate green—move forward with a breakout.

THE BREAKOUT SCORE SHEETS

The breakout shore sheets provide shapes to indicate findings for documentation. Refer to the flow charts and carefully consider the SFMA hierarchy to guide your breakout decisions.

THE FLOWCHARTS

THE TOP-TIER

A red bar means STOP—you do not have to continue with a breakout. These patterns will be functional and non-painful. Breaking these down will only expose imperfections and not major limitations.

A yellow bar means proceed with caution—you must break out these patterns, but there is pain involved, so be careful. Use these breakout findings as indicators that your treatments are working, and re-test them frequently.

A green bar means go—you need to break these patterns out to their termination and use corrective exercise and treatments appropriately.

THE BREAKOUTS

A red bar means you can STOP the breakout. Make note of the painful pattern and only continue that breakout if the flowchart indicates further action. All red box findings should be treated with medical modalities, not exercise progressions.

A yellow bar means proceed with the breakout—you must continue the testing since you need more information before you can treat.

A green bar means GO. You have your answer as to what is causing the dysfunction and should start appropriate treatments and exercise progressions if applicable.

An orange bar is a significant finding, similar to a green bar, only in this case you can't stop the breakout process. There may be more dysfunctions, so you should note the problem and continue with the breakout. Treat these with corrections as you would treat those with a green bar.

A blue bar can indicate a normal finding and it will direct you to another flowchart or breakout. It can also be dependent on a previous finding. If there is a dysfunction involved, treat this as you would an orange or green bar.

SFMA

SFMA TOP-TIER ASSESSMENTS

Cervical Pattern One

⬡	▽	▽	◯
FN	FP	DP	DN

Cervical Pattern Two

⬡	▽	▽	◯
FN	FP	DP	DN

Cervical Pattern Three

L	⬡	▽	▽	◯
R	⬡	▽	▽	◯
	FN	FP	DP	DN

Upper Extremity Pattern One

L	⬡	▽	▽	◯
R	⬡	▽	▽	◯
	FN	FP	DP	DN

Upper Extremity Pattern Two

L	⬡	▽	▽	◯
R	⬡	▽	▽	◯
	FN	FP	DP	DN

Multi-Segmental Flexion

⬡	▽	▽	◯
FN	FP	DP	DN

Multi-Segmental Extension

⬡	▽	▽	◯
FN	FP	DP	DN

Multi-Segmental Rotation

L	⬡	▽	▽	◯
R	⬡	▽	▽	◯
	FN	FP	DP	DN

Single-Leg Stance

L	⬡	▽	▽	◯
R	⬡	▽	▽	◯
	FN	FP	DP	DN

Overhead Squat

⬡	▽	▽	◯
FN	FP	DP	DN

PROVOCATION ASSESSMENTS

Pattern One

L	⬡	▽	▽	◯
R	⬡	▽	▽	◯
	FN	FP	DP	DN

Pattern Two

L	⬡	▽	▽	◯
R	⬡	▽	▽	◯
	FN	FP	DP	DN

SFMA

CERVICAL SPINE BREAKOUT

Active Supine Cervical Flexion (Chin to Chest)

⬡ FN   ▽ D &/or P

Passive Supine Cervical Flexion

⬡ FN   ▽ D &/or P

Supine OA Cervical Flexion Test (20°)

L ⬡   ◯   ▽
R ⬡   ◯   ▽
FN   DN   FP / DP

Active Supine Cervical Rotation (80°)

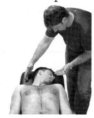

L ⬡   ▽
R ⬡   ▽
FN   D &/or P

Passive Cervical Rotation

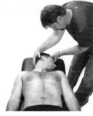

L ⬡   ▽
R ⬡   ▽
FN   D &/or P

C1-C2 Cervical Rotation Test

L ⬡   ◯   ▽
R ⬡   ◯   ▽
FN   DN   FP / DP

Supine Cervical Extension

L ⬡   ◯   ▽
R ⬡   ◯   ▽
FN   DN   FP / DP

UPPER EXTREMITY PATTERN BREAKOUT

Active Prone Upper Extremity Pattern

L ⬡   ▽
R ⬡   ▽
FN   D &/or P

Passive Prone Upper Extremity Patterns

L ⬡   ◯   ▽
R ⬡   ◯   ▽
FN   DN   FP / DP

Supine Reciprocal Upper Extremity Pattern

L ⬡   ◯   ▽
R ⬡   ◯   ▽
FN   DN   FP / DP

SFMA

MULTI-SEGMENTAL FLEXION BREAKOUT

Single-Leg Forward Bend

| L R | Bilat FN | Bilat D/P | Unilat D/P |

Long-Sitting Toe Touch

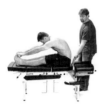

| FN Toe Touch | DP Touch NSA | Touches Ltd SA | DP Touches Ltd SA |

Rolling—FN _____ DN _____ DP_____ FP _____

Active Straight-Leg Raise

| Bilat FN | D (<70) or P | L R |

Passive Straight-Leg Raise

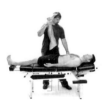

| L R | FN | 10 > ASLR | FP, DP, DN |

Rolling—FN _____ DN _____ DP_____ FP _____

Supine Knee-to-Chest Holding Thighs

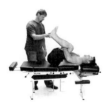

| L R | FN | DN | FP or DP |

Prone Rocking

| FN | DN | FP or DP |

MULTI-SEGMENTAL EXTENSION BREAKOUT

Backward Bend without Upper Extremity

| FN | D &/or P |

Single-Leg Backward Bend

| Bilat FN | Bilat D/P | Unilat D/P | L R |

Prone Press-up

| FN | D &/or P |

Lumbar Locked Unilateral Ext. (IR) 50⁰

| L R | FN | FP/DP |

Lumbar Locked Passive Uni. Ext. (IR) 50⁰

| L R | FN | Bil DN | Uni DN | FP/DP |

Prone-on-Elbow Extension (IR) 140⁰

| L R | FN | Bil DN | Uni DN | FP/DP |

SFMA

MULTI-SEGMENTAL EXTENSION BREAKOUT

Single-Leg Hip Extension

L	○	▽	▽
R			▽
	Bilat >10	Bilat D/P	Unilat D/P

Prone Active Hip Extension (10⁰)

L	⬡		▽
R	⬡		▽
	FN		FP, DP, DN

Prone Passive Hip Extension

L	▽		▽
R	▽		▽
	25% > Active		D &/or P

Rolling—FN _____ DN _____ DP_____ FP _____

FABER

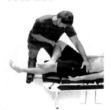

L	⬡	○	▽
R	⬡	○	▽
	FN	DN	FP or DP

Modified Thomas Test

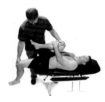

L	○	○	○	○	⬡	▽
R	○	○	○	○	⬡	▽
	Knee Strght Touch	Hip Abd Touch	Never Touch	Abd & Strght Touch	FN	DP/FP

Unilateral Shoulder Backward Bend

L		⬡	▽
R			▽
		FN	D &/or P

Supine Lat Hips Flexed Test

L		⬡	▽
R			▽
		FN	D &/or P

Rolling—FN _____ DN _____ DP_____ FP _____

MULTI-SEGMENTAL EXTENSION BREAKOUT

Supine Lat Hips Extended

L	⬡	○	▽
R		○	▽
	FN	Improves	No Change

Lumbar Locked Unilateral Ext. (ER) 120⁰

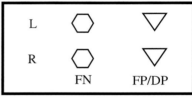

	⬡	▽	▽	L
			▽	R
	Bilat FN	Bilat D/P	Unilat D/P	

Lumbar Locked Unilateral Ext. (IR) 50⁰

L	⬡		▽
R	⬡		▽
	FN		FP/DP

Lumbar Locked Passive Unilateral Ext. (IR) 50⁰

L	⬡	○	○	▽
R	⬡	○	○	
	FN	Bilat DN	Uni DN	FP/DP

MULTI-SEGMENTAL ROTATION BREAKOUT

Seated Rotation

L	▽	▽
R	▽	▽
	> 45 Bilateral	D &/or P

Lumbar Locked Unilateral Rotation

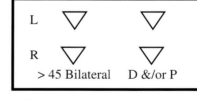

L	▽	▽	⬡
R	▽	▽	⬡
	Switched	DN, DP, FP	FN

Rolling—FN _____ DN _____ DP_____ FP _____

Lumbar Locked Passive Unilateral Ext. (IR) 50⁰

L	⬡	○	○	▽
R	⬡	○	○	
	FN	Bi. DN	Uni. DN	FP/DP

MULTI-SEGMENTAL ROTATION BREAKOUT

Prone-on-Elbow Rotation (30)

Asymm Bilat DN FN FP/DP

Rolling—FN _____ DN _____ DP_____ FP _____

Seated Active External Hip Rotation 40⁰

FN D &/or P

Seated Passive External Hip Rotation

FN DP/FP DN

Prone Active External Hip Rotation 40⁰

FN D &/or P

Prone Passive External Hip Rotation

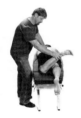

DP/FP DN FN

Rolling—FN _____ DN _____ DP_____ FP _____

MULTI-SEGMENTAL ROTATION BREAKOUT

Seated Active Internal Hip Rotation 30⁰

FN D &/or P

Seated Passive Internal Hip Rotation

FN DP/FP DN

Prone Active Internal Hip Rotation 30⁰

FN D &/or P

Prone Passive Internal Hip Rotation

DP/FP DN FN

Rolling—FN _____ DN _____ DP_____ FP _____

Seated Active External Tibial Rotation 20⁰

FN D &/or P

Seated Passive External Tibial Rotation

FN DP/FP DN

MULTI-SEGMENTAL ROTATION BREAKOUT

Seated Active Internal Tibial Rotation 20⁰

L ⬡ ▽
R ⬡ ▽
FN   D &/or P

Seated Passive Internal Tibial Rotation

L ⬡ ▽ ◯
R ⬡ ▽ ◯
FN DP/FP DN

SINGLE-LEG STANCE BREAKOUT

Vestibular Shake Test

L ⬡ ▽
R ⬡ ▽
FN   D &/or P

Half-Kneeling Narrow Base

L ⬡ ▽
R ⬡ ▽
FN   DN, DP, FP

Rolling—FN _____ DN _____ DP _____ FP _____

Quadruped Diagonals

L ⬡ ▽ ◯
R ⬡ ▽ ◯
FN DP or FP DN

Heel Walks

L ⬡ ▽
R ⬡ ▽
FN   D &/or P

Prone Passive Dorsifl—FN ___ DN ___ DP/FP ___

Toe Walks

L ⬡ ▽
R ⬡ ▽
FN   D &/or P

Prone Passive Plantar—FN ___ DN ___ DP/FP ___

SINGLE-LEG STANCE BREAKOUT

Seated Ankle Inversion/Eversion

L ◯ ◯ ▽ ⬡ ◯
R ◯ ◯ ▽ ⬡ ◯
Can't Evrt | Can't Invrt | DP/FP | FN | Both DN

OVERHEAD DEEP SQUAT BREAKOUT

Fingers Interlocked Behind Head

⬡ ▽
FN   D &/or P

Assisted Deep Squat

⬡ ▽
FN   D &/or P

Half Kneeling Dorsiflexion

L ⬡ ▽
R ⬡ ▽
FN   D &/or P

Supine Knee to Chest Holding Shins

L ⬡ ▽
R ⬡ ▽
FN   D &/or P

Supine Knee to Chest Holding Thighs

L ⬡ ◯ ▽
R ⬡ ◯ ▽
FN DN FP/DP

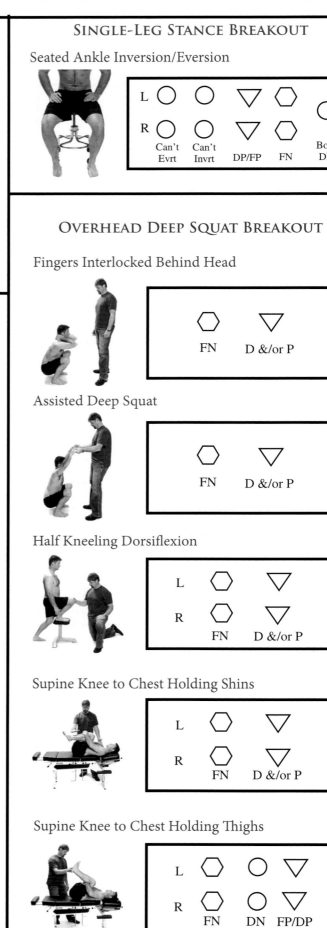

SFMA

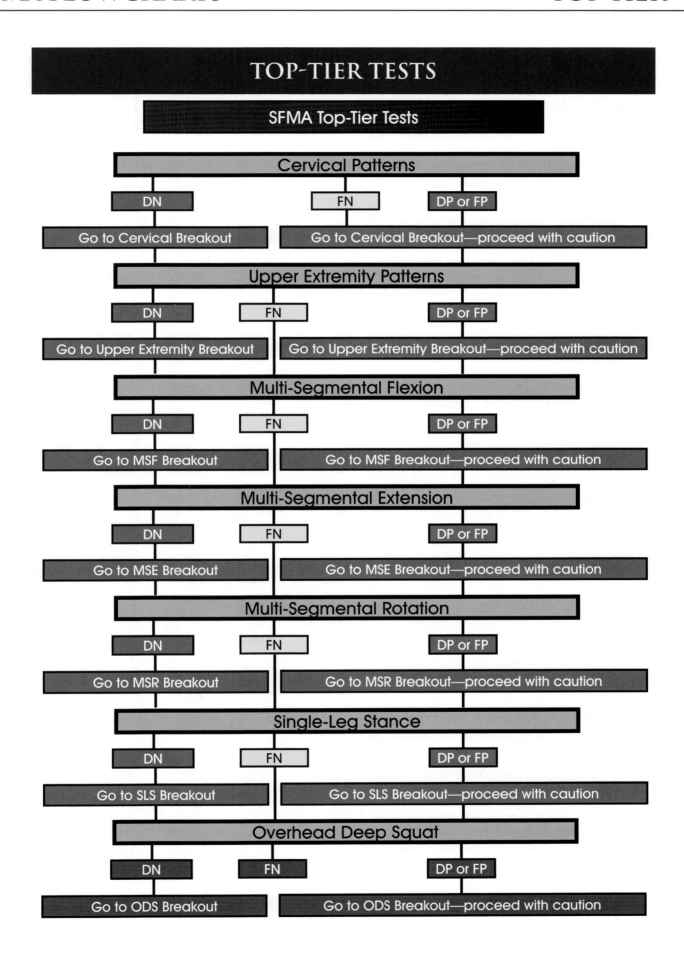

TOP-TIER TESTS

SFMA Top-Tier Tests

Cervical Patterns

| DN | FN | DP or FP |

Go to Cervical Breakout | Go to Cervical Breakout—proceed with caution

Upper Extremity Patterns

| DN | FN | DP or FP |

Go to Upper Extremity Breakout | Go to Upper Extremity Breakout—proceed with caution

Multi-Segmental Flexion

| DN | FN | DP or FP |

Go to MSF Breakout | Go to MSF Breakout—proceed with caution

Multi-Segmental Extension

| DN | FN | DP or FP |

Go to MSE Breakout | Go to MSE Breakout—proceed with caution

Multi-Segmental Rotation

| DN | FN | DP or FP |

Go to MSR Breakout | Go to MSR Breakout—proceed with caution

Single-Leg Stance

| DN | FN | DP or FP |

Go to SLS Breakout | Go to SLS Breakout—proceed with caution

Overhead Deep Squat

| DN | FN | DP or FP |

Go to ODS Breakout | Go to ODS Breakout—proceed with caution

SFMA

THE SELECTIVE FUNCTIONAL MOVEMENT ASSESSMENT

SFMA SCORING		FN	FP	DP	DN
Active Cervical Flexion		☐	☐	☐	☐
Active Cervical Extension		☐	☐	☐	☐
Cervical Rotation-Lateral Bend	L	☐	☐	☐	☐
	R	☐	☐	☐	☐
Upper Extremity Pattern 1 (MRE)	L	☐	☐	☐	☐
	R	☐	☐	☐	☐
Upper Extremity Pattern 2 (LRF)	L	☐	☐	☐	☐
	R	☐	☐	☐	☐
Multi-Segmental Flexion		☐	☐	☐	☐
Multi-Segmental Extension		☐	☐	☐	☐
Multi-Segmental Rotation	L	☐	☐	☐	☐
	R	☐	☐	☐	☐
Single-Leg Stance	L	☐	☐	☐	☐
	R	☐	☐	☐	☐
Overhead Deep Squat		☐	☐	☐	☐

PROVOCATION PATTERNS

		FN	FP	DP	DN
Impingement Sign	L	☐	☐	☐	☐
	R	☐	☐	☐	☐
Horizontal Adduction	L	☐	☐	☐	☐
	R	☐	☐	☐	☐

SFMA

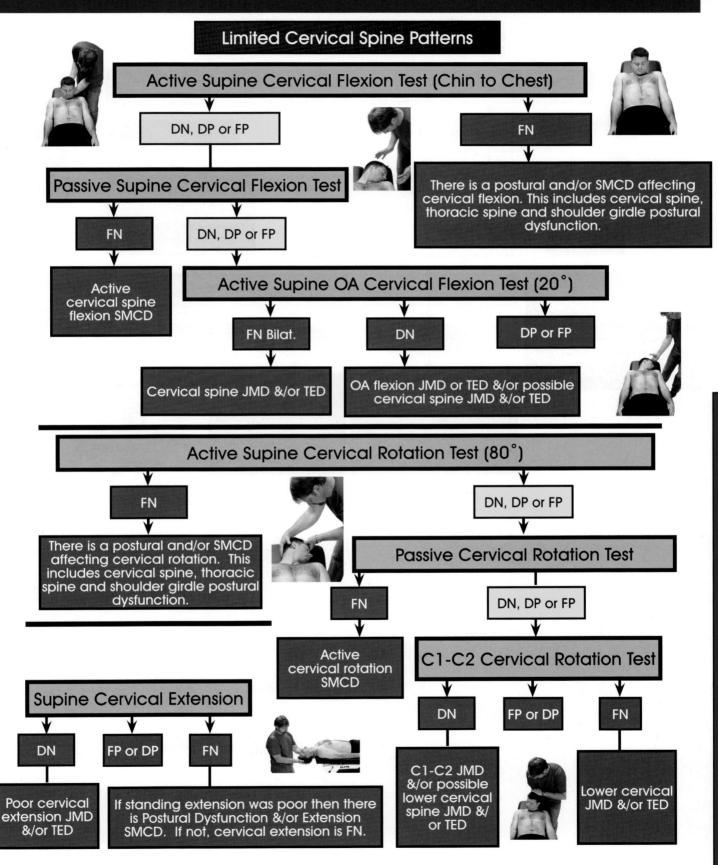

CERVICAL SPINE PATTERN BREAKOUTS

Limited Cervical Spine Patterns

Active Supine Cervical Flexion Test (Chin to Chest)

- DN, DP or FP
  - **Passive Supine Cervical Flexion Test**
    - FN
      - Active cervical spine flexion SMCD
    - DN, DP or FP
      - **Active Supine OA Cervical Flexion Test (20°)**
        - FN Bilat.
          - Cervical spine JMD &/or TED
        - DN
          - OA flexion JMD or TED &/or possible cervical spine JMD &/or TED
        - DP or FP
- FN
  - There is a postural and/or SMCD affecting cervical flexion. This includes cervical spine, thoracic spine and shoulder girdle postural dysfunction.

Active Supine Cervical Rotation Test (80°)

- FN
  - There is a postural and/or SMCD affecting cervical rotation. This includes cervical spine, thoracic spine and shoulder girdle postural dysfunction.
- DN, DP or FP
  - **Passive Cervical Rotation Test**
    - FN
      - Active cervical rotation SMCD
    - DN, DP or FP
      - **C1-C2 Cervical Rotation Test**
        - DN
          - C1-C2 JMD &/or possible lower cervical spine JMD &/or TED
        - FP or DP
        - FN
          - Lower cervical JMD &/or TED

Supine Cervical Extension

- DN
  - Poor cervical extension JMD &/or TED
- FP or DP
  - If standing extension was poor then there is Postural Dysfunction &/or Extension SMCD. If not, cervical extension is FN.
- FN

UPPER EXTREMITY PATTERN BREAKOUTS

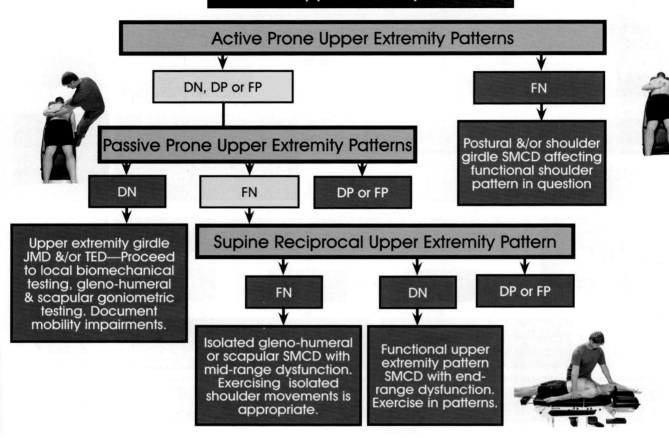

Limited Upper Extremity Patterns

Active Prone Upper Extremity Patterns

DN, DP or FP

FN

Passive Prone Upper Extremity Patterns

Postural &/or shoulder girdle SMCD affecting functional shoulder pattern in question

DN

FN

DP or FP

Upper extremity girdle JMD &/or TED—Proceed to local biomechanical testing, gleno-humeral & scapular goniometric testing. Document mobility impairments.

Supine Reciprocal Upper Extremity Pattern

FN

DN

DP or FP

Isolated gleno-humeral or scapular SMCD with mid-range dysfunction. Exercising isolated shoulder movements is appropriate.

Functional upper extremity pattern SMCD with end-range dysfunction. Exercise in patterns.

SFMA

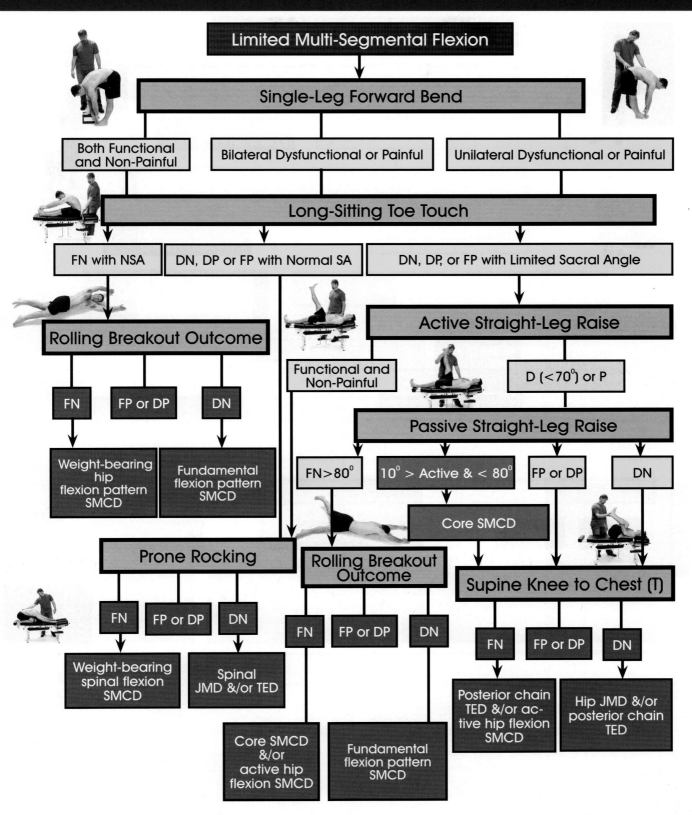

MULTI-SEGMENTAL FLEXION BREAKOUTS

Limited Multi-Segmental Flexion

Single-Leg Forward Bend

- Both Functional and Non-Painful
- Bilateral Dysfunctional or Painful
- Unilateral Dysfunctional or Painful

Long-Sitting Toe Touch

- FN with NSA
- DN, DP or FP with Normal SA
- DN, DP, or FP with Limited Sacral Angle

Rolling Breakout Outcome

- FN
- FP or DP
- DN

Weight-bearing hip flexion pattern SMCD

Fundamental flexion pattern SMCD

Active Straight-Leg Raise

- Functional and Non-Painful
- D ($<70°$) or P

Passive Straight-Leg Raise

- FN $>80°$
- $10° >$ Active & $< 80°$
- FP or DP
- DN

Core SMCD

Prone Rocking

- FN
- FP or DP
- DN

Weight-bearing spinal flexion SMCD

Spinal JMD &/or TED

Rolling Breakout Outcome

- FN
- FP or DP
- DN

Core SMCD &/or active hip flexion SMCD

Fundamental flexion pattern SMCD

Supine Knee to Chest (T)

- FN
- FP or DP
- DN

Posterior chain TED &/or active hip flexion SMCD

Hip JMD &/or posterior chain TED

SFMA

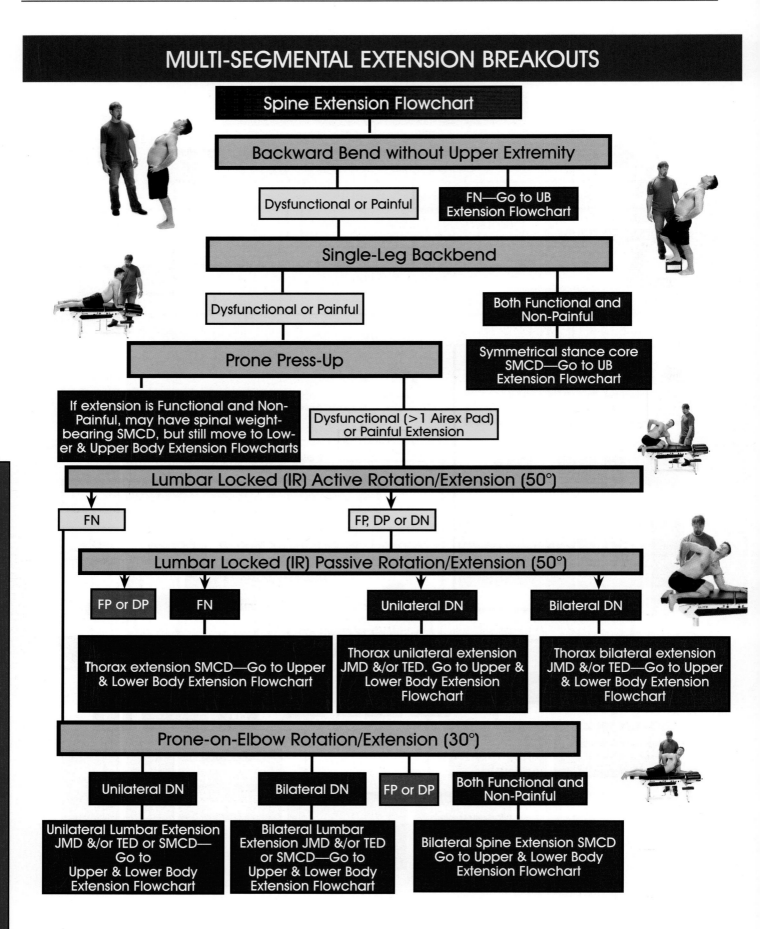

MULTI-SEGMENTAL EXTENSION BREAKOUTS

Spine Extension Flowchart

Backward Bend without Upper Extremity

Dysfunctional or Painful

FN—Go to UB Extension Flowchart

Single-Leg Backbend

Dysfunctional or Painful

Both Functional and Non-Painful

Symmetrical stance core SMCD—Go to UB Extension Flowchart

Prone Press-Up

If extension is Functional and Non-Painful, may have spinal weight-bearing SMCD, but still move to Lower & Upper Body Extension Flowcharts

Dysfunctional (>1 Airex Pad) or Painful Extension

Lumbar Locked (IR) Active Rotation/Extension (50°)

FN

FP, DP or DN

Lumbar Locked (IR) Passive Rotation/Extension (50°)

FP or DP

FN

Unilateral DN

Bilateral DN

Thorax extension SMCD—Go to Upper & Lower Body Extension Flowchart

Thorax unilateral extension JMD &/or TED. Go to Upper & Lower Body Extension Flowchart

Thorax bilateral extension JMD &/or TED—Go to Upper & Lower Body Extension Flowchart

Prone-on-Elbow Rotation/Extension (30°)

Unilateral DN

Bilateral DN

FP or DP

Both Functional and Non-Painful

Unilateral Lumbar Extension JMD &/or TED or SMCD— Go to Upper & Lower Body Extension Flowchart

Bilateral Lumbar Extension JMD &/or TED or SMCD—Go to Upper & Lower Body Extension Flowchart

Bilateral Spine Extension SMCD Go to Upper & Lower Body Extension Flowchart

SFMA

MULTI-SEGMENTAL EXTENSION BREAKOUTS

Lower Body Extension Flowchart

Standing Hip Extension

- > 10 degrees Extension Bilateral
- Dysfunctional or Painful

If there were previous signs of hip extension dysfunction, assume a weight bearing lower quarter SMCD &/or limited ankle dorsiflexion. If not, hip extension is normal. Check ODS & SLS.

Prone Active Hip Extension

- > or = 10 degrees Extension (FN)
- DP, FP, or DN

Rolling Pattern Outcomes

FN	FP or DP	DN
Spine weight-bearing hip extension SMCD		Fundamental extension pattern SMCD

Prone Passive Hip Extension

- DN or Painful
- If 25% > Active Hip Extension

Rolling Pattern Outcomes

FN	FP or DP	DN
Core SMCD &/or active hip extension SMCD		Fundamental extension pattern SMCD

FABER Test

FN	FP or DP	DN

Hip/SI JMD &/or TED &/or core SMCD—Perform local biomechanical testing of the hip

Modified Thomas Test

FN with knee straight	FN with hip abducted	FN with hip abducted & knee straight	DN	DP/FP	FN
Anterior chain TED	Lateral chain TED	Anterior and lateral chain TED	Hip JMD &/or TED and/or core SMCD. Perform local biomechanical testing of the hip.		Core SMCD

SFMA

MULTI-SEGMENTAL EXTENSION BREAKOUTS

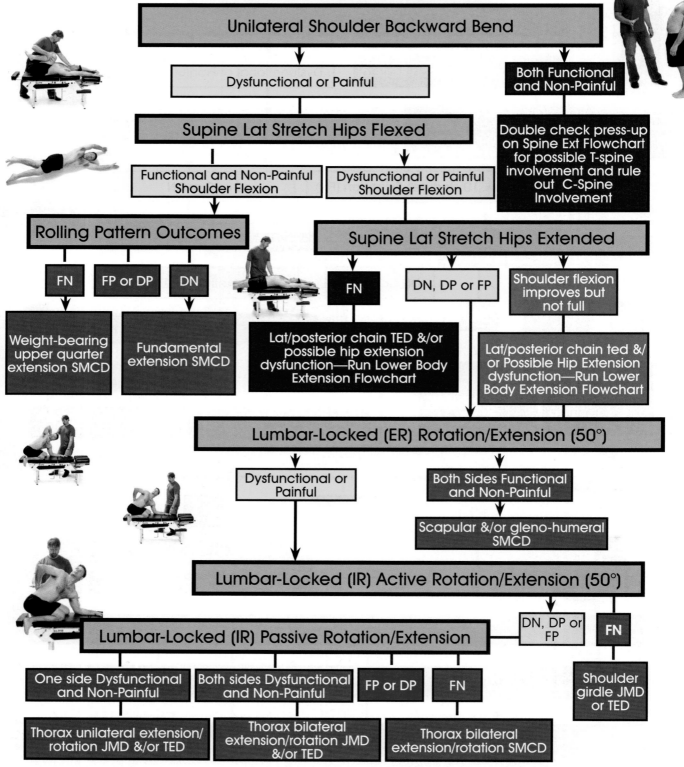

Upper Body Extension Flowchart

Unilateral Shoulder Backward Bend

Dysfunctional or Painful

Both Functional and Non-Painful

Supine Lat Stretch Hips Flexed

Double check press-up on Spine Ext Flowchart for possible T-spine involvement and rule out C-Spine Involvement

Functional and Non-Painful Shoulder Flexion

Dysfunctional or Painful Shoulder Flexion

Rolling Pattern Outcomes

FN | FP or DP | DN

Supine Lat Stretch Hips Extended

FN | DN, DP or FP | Shoulder flexion improves but not full

Weight-bearing upper quarter extension SMCD

Fundamental extension SMCD

Lat/posterior chain TED &/or possible hip extension dysfunction—Run Lower Body Extension Flowchart

Lat/posterior chain ted &/or Possible Hip Extension dysfunction—Run Lower Body Extension Flowchart

Lumbar-Locked (ER) Rotation/Extension (50°)

Dysfunctional or Painful

Both Sides Functional and Non-Painful

Scapular &/or gleno-humeral SMCD

Lumbar-Locked (IR) Active Rotation/Extension (50°)

DN, DP or FP | FN

Lumbar-Locked (IR) Passive Rotation/Extension

One side Dysfunctional and Non-Painful

Both sides Dysfunctional and Non-Painful

FP or DP | FN

Shoulder girdle JMD or TED

Thorax unilateral extension/rotation JMD &/or TED

Thorax bilateral extension/rotation JMD &/or TED

Thorax bilateral extension/rotation SMCD

SFMA

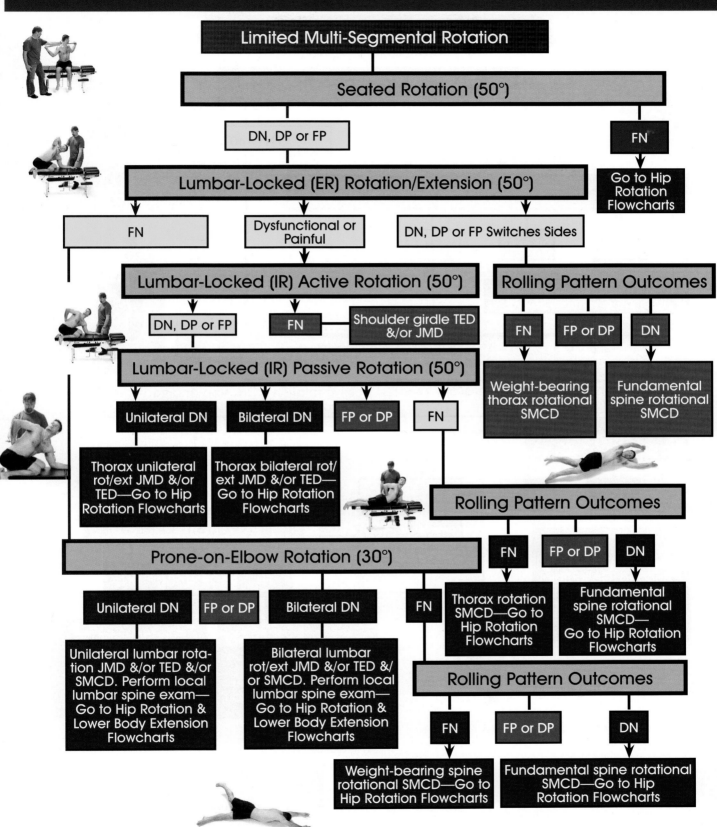

MULTI-SEGMENTAL ROTATION BREAKOUTS

Limited Multi-Segmental Rotation

Seated Rotation (50°)

DN, DP or FP → **Lumbar-Locked (ER) Rotation/Extension (50°)**

FN → **Go to Hip Rotation Flowcharts**

From Lumbar-Locked (ER) Rotation/Extension (50°):
- FN
- Dysfunctional or Painful
- DN, DP or FP Switches Sides

Dysfunctional or Painful → **Lumbar-Locked (IR) Active Rotation (50°)**
- DN, DP or FP
- FN → Shoulder girdle TED &/or JMD

DN, DP or FP Switches Sides → **Rolling Pattern Outcomes**
- FN → **Weight-bearing thorax rotational SMCD**
- FP or DP
- DN → **Fundamental spine rotational SMCD**

DN, DP or FP → **Lumbar-Locked (IR) Passive Rotation (50°)**
- **Unilateral DN** → **Thorax unilateral rot/ext JMD &/or TED—Go to Hip Rotation Flowcharts**
- **Bilateral DN** → **Thorax bilateral rot/ext JMD &/or TED—Go to Hip Rotation Flowcharts**
- **FP or DP**
- **FN**

Prone-on-Elbow Rotation (30°)
- **Unilateral DN** → **Unilateral lumbar rotation JMD &/or TED &/or SMCD. Perform local lumbar spine exam—Go to Hip Rotation & Lower Body Extension Flowcharts**
- **FP or DP**
- **Bilateral DN** → **Bilateral lumbar rot/ext JMD &/or TED &/or SMCD. Perform local lumbar spine exam—Go to Hip Rotation & Lower Body Extension Flowcharts**
- **FN**

Rolling Pattern Outcomes
- **FN** → **Thorax rotation SMCD—Go to Hip Rotation Flowcharts**
- **FP or DP**
- **DN** → **Fundamental spine rotational SMCD—Go to Hip Rotation Flowcharts**

Rolling Pattern Outcomes
- **FN** → **Weight-bearing spine rotational SMCD—Go to Hip Rotation Flowcharts**
- **FP or DP**
- **DN** → **Fundamental spine rotational SMCD—Go to Hip Rotation Flowcharts**

SFMA

MULTI-SEGMENTAL ROTATION BREAKOUTS

Hip Rotation Flowchart (Part 1)

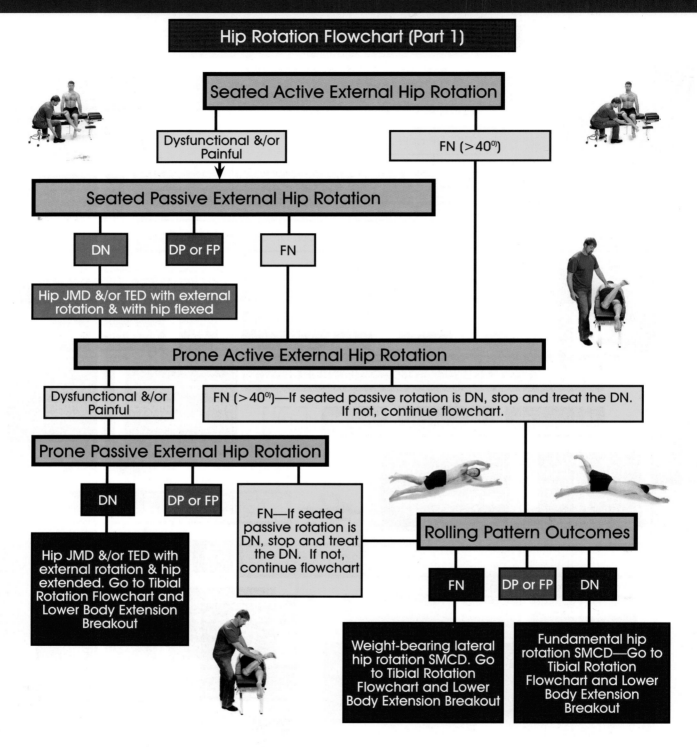

Seated Active External Hip Rotation

- Dysfunctional &/or Painful
- FN (>40⁰)

Seated Passive External Hip Rotation

- DN
- DP or FP
- FN

Hip JMD &/or TED with external rotation & with hip flexed

Prone Active External Hip Rotation

- Dysfunctional &/or Painful
- FN (>40⁰)—If seated passive rotation is DN, stop and treat the DN. If not, continue flowchart.

Prone Passive External Hip Rotation

- DN
- DP or FP
- FN—If seated passive rotation is DN, stop and treat the DN. If not, continue flowchart

Hip JMD &/or TED with external rotation & hip extended. Go to Tibial Rotation Flowchart and Lower Body Extension Breakout

Rolling Pattern Outcomes

- FN
- DP or FP
- DN

Weight-bearing lateral hip rotation SMCD. Go to Tibial Rotation Flowchart and Lower Body Extension Breakout

Fundamental hip rotation SMCD—Go to Tibial Rotation Flowchart and Lower Body Extension Breakout

SFMA

MULTI-SEGMENTAL ROTATION BREAKOUTS

Hip Rotation Flowchart (Part 2)

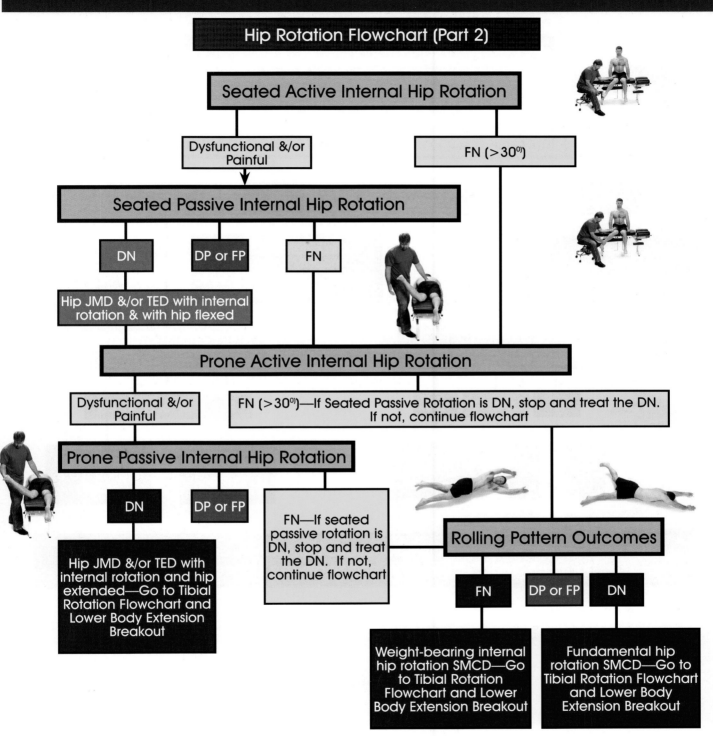

Seated Active Internal Hip Rotation

- Dysfunctional &/or Painful
- FN (>30⁰)

Seated Passive Internal Hip Rotation

- DN
- DP or FP
- FN

Hip JMD &/or TED with internal rotation & with hip flexed

Prone Active Internal Hip Rotation

- Dysfunctional &/or Painful
- FN (>30⁰)—If Seated Passive Rotation is DN, stop and treat the DN. If not, continue flowchart

Prone Passive Internal Hip Rotation

- DN
- DP or FP
- FN—If seated passive rotation is DN, stop and treat the DN. If not, continue flowchart

Hip JMD &/or TED with internal rotation and hip extended—Go to Tibial Rotation Flowchart and Lower Body Extension Breakout

Rolling Pattern Outcomes

- FN
- DP or FP
- DN

Weight-bearing internal hip rotation SMCD—Go to Tibial Rotation Flowchart and Lower Body Extension Breakout

Fundamental hip rotation SMCD—Go to Tibial Rotation Flowchart and Lower Body Extension Breakout

SFMA

MULTI-SEGMENTAL ROTATION BREAKOUTS

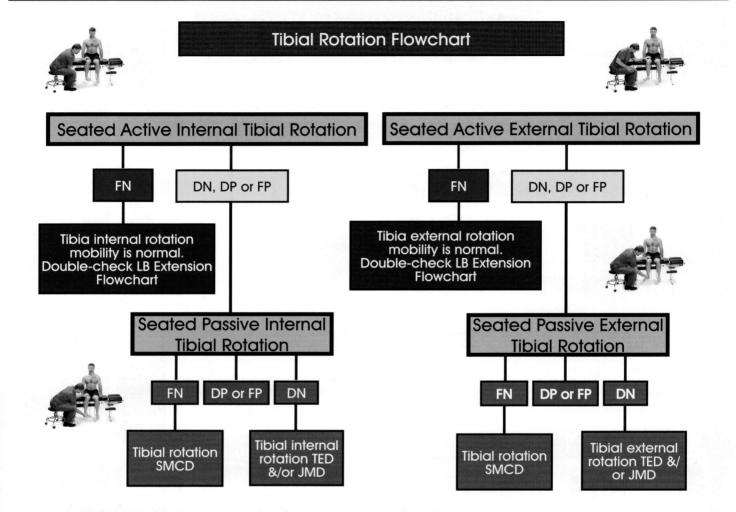

Tibial Rotation Flowchart

Seated Active Internal Tibial Rotation

FN

DN, DP or FP

Tibia internal rotation mobility is normal. Double-check LB Extension Flowchart

Seated Passive Internal Tibial Rotation

FN | **DP or FP** | **DN**

Tibial rotation SMCD

Tibial internal rotation TED &/or JMD

Seated Active External Tibial Rotation

FN

DN, DP or FP

Tibia external rotation mobility is normal. Double-check LB Extension Flowchart

Seated Passive External Tibial Rotation

FN | **DP or FP** | **DN**

Tibial rotation SMCD

Tibial external rotation TED &/or JMD

If spine, hips and tibia are all functional and non-painful, double-check rolling for spine SMCD, LB Extension and Single-Leg Stance Breakouts.

SFMA

SINGLE-LEG STANCE BREAKOUTS

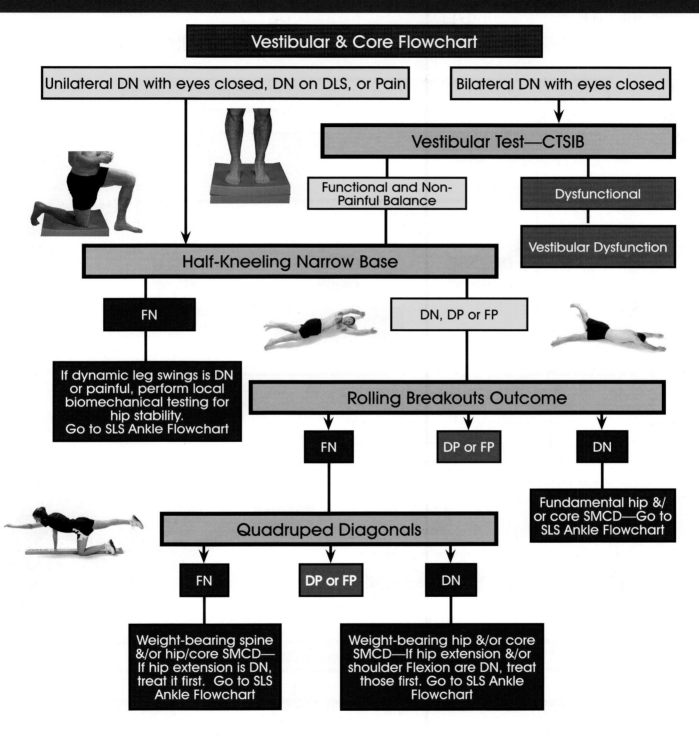

Vestibular & Core Flowchart

Unilateral DN with eyes closed, DN on DLS, or Pain

Bilateral DN with eyes closed

Vestibular Test—CTSIB

Functional and Non-Painful Balance

Dysfunctional

Vestibular Dysfunction

Half-Kneeling Narrow Base

FN

DN, DP or FP

If dynamic leg swings is DN or painful, perform local biomechanical testing for hip stability.
Go to SLS Ankle Flowchart

Rolling Breakouts Outcome

FN

DP or FP

DN

Fundamental hip &/or core SMCD—Go to SLS Ankle Flowchart

Quadruped Diagonals

FN

DP or FP

DN

Weight-bearing spine &/or hip/core SMCD—If hip extension is DN, treat it first. Go to SLS Ankle Flowchart

Weight-bearing hip &/or core SMCD—If hip extension &/or shoulder Flexion are DN, treat those first. Go to SLS Ankle Flowchart

SINGLE-LEG STANCE BREAKOUTS

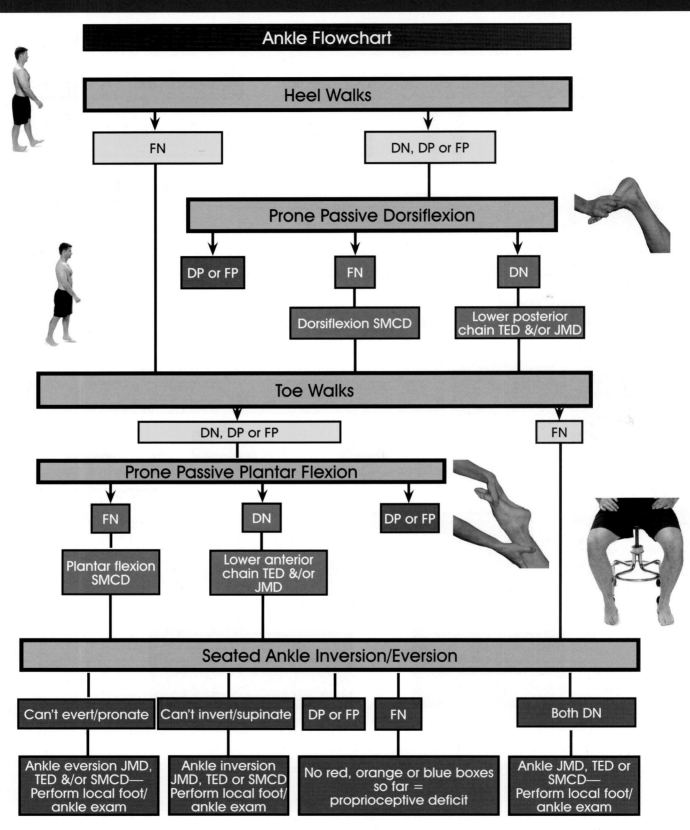

Ankle Flowchart

Heel Walks

FN

DN, DP or FP

Prone Passive Dorsiflexion

DP or FP

FN

Dorsiflexion SMCD

DN

Lower posterior chain TED &/or JMD

Toe Walks

DN, DP or FP

FN

Prone Passive Plantar Flexion

FN

Plantar flexion SMCD

DN

Lower anterior chain TED &/or JMD

DP or FP

Seated Ankle Inversion/Eversion

Can't evert/pronate

Can't invert/supinate

DP or FP

FN

Both DN

Ankle eversion JMD, TED &/or SMCD— Perform local foot/ankle exam

Ankle inversion JMD, TED or SMCD Perform local foot/ankle exam

No red, orange or blue boxes so far = proprioceptive deficit

Ankle JMD, TED or SMCD— Perform local foot/ankle exam

OVERHEAD DEEP SQUATTING PATTERN BREAKOUTS

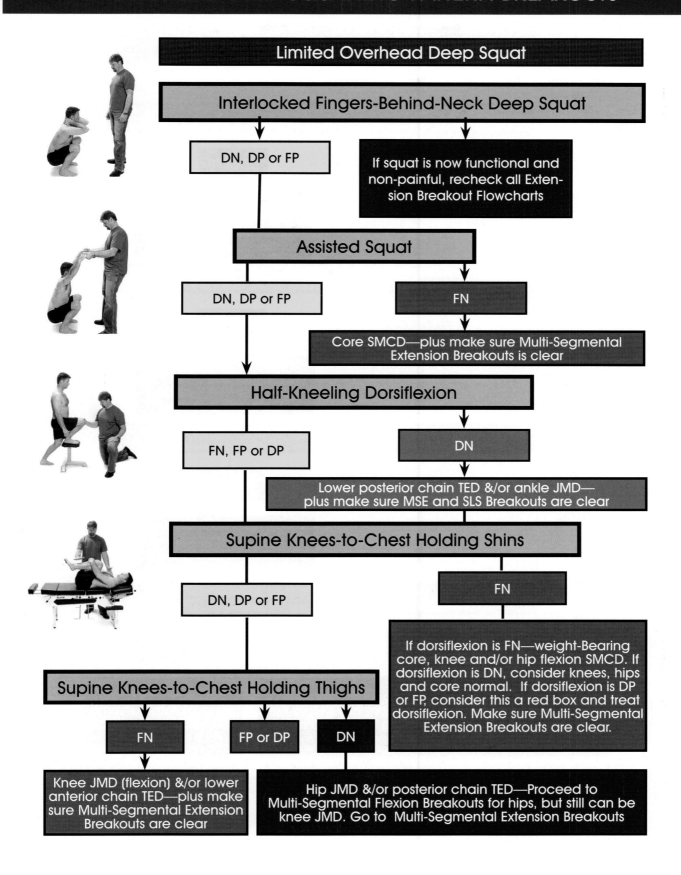

Limited Overhead Deep Squat

Interlocked Fingers-Behind-Neck Deep Squat

DN, DP or FP

If squat is now functional and non-painful, recheck all Extension Breakout Flowcharts

Assisted Squat

DN, DP or FP

FN

Core SMCD—plus make sure Multi-Segmental Extension Breakouts is clear

Half-Kneeling Dorsiflexion

FN, FP or DP

DN

Lower posterior chain TED &/or ankle JMD—plus make sure MSE and SLS Breakouts are clear

Supine Knees-to-Chest Holding Shins

DN, DP or FP

FN

If dorsiflexion is FN—weight-Bearing core, knee and/or hip flexion SMCD. If dorsiflexion is DN, consider knees, hips and core normal. If dorsiflexion is DP or FP, consider this a red box and treat dorsiflexion. Make sure Multi-Segmental Extension Breakouts are clear.

Supine Knees-to-Chest Holding Thighs

FN

FP or DP

DN

Knee JMD (flexion) &/or lower anterior chain TED—plus make sure Multi-Segmental Extension Breakouts are clear

Hip JMD &/or posterior chain TED—Proceed to Multi-Segmental Flexion Breakouts for hips, but still can be knee JMD. Go to Multi-Segmental Extension Breakouts

SFMA

ROLLING BREAKOUTS

Rolling Flowchart

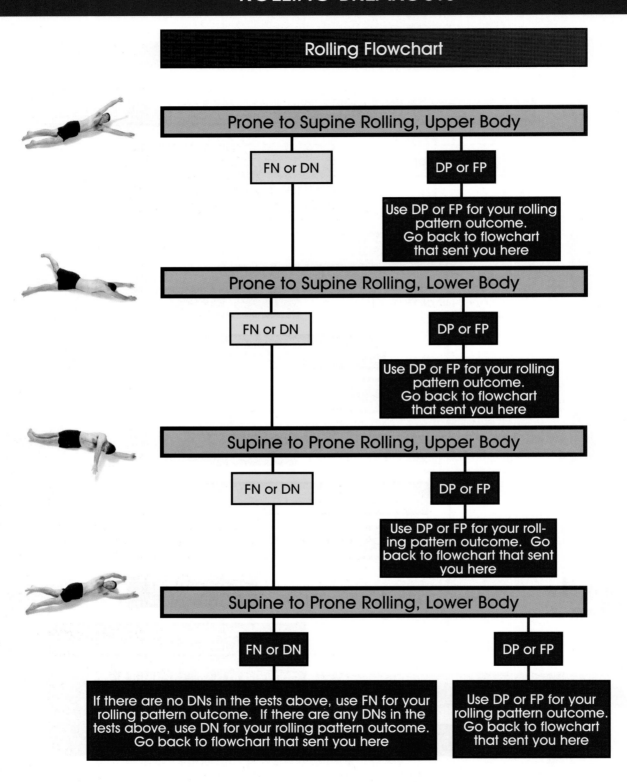

Prone to Supine Rolling, Upper Body

FN or DN

DP or FP

Use DP or FP for your rolling pattern outcome. Go back to flowchart that sent you here

Prone to Supine Rolling, Lower Body

FN or DN

DP or FP

Use DP or FP for your rolling pattern outcome. Go back to flowchart that sent you here

Supine to Prone Rolling, Upper Body

FN or DN

DP or FP

Use DP or FP for your rolling pattern outcome. Go back to flowchart that sent you here

Supine to Prone Rolling, Lower Body

FN or DN

DP or FP

If there are no DNs in the tests above, use FN for your rolling pattern outcome. If there are any DNs in the tests above, use DN for your rolling pattern outcome. Go back to flowchart that sent you here

Use DP or FP for your rolling pattern outcome. Go back to flowchart that sent you here

BREATHING

Laurie McLaughlin, PT, DSc, FCAMPT, CMAG

Healthy breathing is comprised of the mechanical aspect of breathing or ventilation (bringing air in and out of the lungs) and respiration. Respiration refers to gas exchange where oxygen is brought to the cells to fuel metabolism and the CO_2 produced through that metabolism is directed back to the lungs. While one might think the goal would be eliminate all CO_2 as a waste gas, that is not the case. Total elimination is not the goal; in fact about 85-88% of CO_2 is meant to be retained to balance pH and allow for proper allocation of oxygen. Optimal respiration occurs when CO_2 production matches elimination maintaining a baseline.

Breathing has both reflex and higher center control. Higher center control can be either conscious (e.g. talking, swimming) or unconscious. Pain, stress and fear are known ventilatory stimulants that can alter breathing. These are examples of unconscious higher center input leading to altered breathing. Under prolonged pain and stressful conditions, altered breathing can simply become a habit.

When breathing mechanics change, respiratory chemistry can change—specifically CO_2 levels. CO_2 levels are profoundly important in body system function, as the arterial CO_2 level represents the denominator of the Henderson–Hasselbach or pH equation (Levitsky 2003, Thomson et al 1997) and is totally determined by breathing. Any change in CO_2 levels will impact pH. Ideal partial pressure of CO_2 in alveoli and arterial blood is 40 mm Hg, with normal range being 35 to 45 mm Hg (Levitsky 2003).

When ventilation exceeds metabolic requirements by overbreathing (moving more air through the lungs than needed), excess CO_2 is exhaled, dropping arterial CO_2 levels below 35 mm Hg, which is a condition known as hypocapnia (Thomson et al 1997). This is the most common type of altered breathing in people who have normal heart and lung function. Overbreathing occurs by either breathing too quickly, with too large of a volume, or both.

When excess CO_2 is exhaled, the pH of blood, extracellular and cerebrospinal fluid become alkaline. This alteration in pH results in reduced blood flow, particularly to the brain (Eames et al 2004, Ito et al 2005), poorer oxygen delivery to the tissues (Thomson et al 1997), increased muscle tension (Thomson et al 1997) and increased nervous system excitability (Seyal et al 1998, Mogyoros et al 1997).

Overbreathing is difficult to diagnose through observation (Warburton and Jack 2006, Gardner 1996) and requires physiological testing for confirmation (Warburton and Jack 2006). The gold standard for CO_2 measurement is arterial blood gas measurement (Gardner 1996). Arterial blood gas measurement is difficult to obtain because it requires an arterial puncture, and blood gases give information about CO_2 levels only at that moment. Since CO_2 levels change from breath to breath (Levitsky 2003), testing one moment can limit detection of transient or situational hypocapnia (Gardner 1996). However, continuous values can be obtained using capnometry or capnography, which test CO_2 levels in exhaled air at the end of exhale, known as End Tidal CO_2.

CAPNOGRAPHY

Capnography is the measurement of CO_2 levels in exhaled air and is used as a proxy measure for the arterial CO_2 level. It is used in operating rooms to monitor ventilation status, in the emergency department to determine the success of invasive ventilation procedures and to assist in procedural sedation. It has been found to be an accurate, time-sensitive measure of CO_2 levels (Miner 2002).

The application outside of hospital settings is a relatively new and shows promise in being able to diagnose overbreathing, provide biofeedback to improve breathing, and potentially as an outcome measure following both specific breathing retraining and manual therapy interventions.

Testing can be performed in various postures and activities, during concentration, after continuous speaking, and at different breath rates and volumes. A variety of activities and body postures can be tested to determine if breathing deteriorates due to the activity or posture. Proper breathing can then be reinstated through education around breathing mechanics and the development of self-awareness using the feedback gained from capnography. This will help to determine which postures and movement patterns elicit poor breathing and to monitor changes in breathing throughout the rehabilitation process.

HEART RATE VARIABILITY

Although heart Rate Variability (HRV) sounds like it could be a bad thing, it is actually a good thing according to specialists. We often discuss heart rate in terms of a resting heart rate—like 70 beats per minute or in terms of a training range between 160-180 beats per minute. This approach, while representative of cardiac function, probably does not give the complete picture regarding cardiac physiology. HRV simply implies there is unevenness in the rhythmical representation of the heart's action. HRV could also be considered a reflection of the adaptability of the cardiac autonomic nervous system, which is vital for general physical fitness and overall health.

HRV actually measures the variability in intervals between the spikes, or R waves, represented in electrophysiological testing. It has been established as a non-invasive test for the assessment of cardiac and autonomic function.

Historically, the tests have been used to predict increased probability of sudden unexpected death. Since poor variability represents a system with poor adaptability and higher variability represents a system with higher adaptability, HRV is considered an important indicator. According to some reports, HRV has also been used as a screen for individuals in high stress jobs such as Russian Cosmonauts and submarine crews.

Many people assume that the heart rate—rhythm—is fixed and it simply gets faster or slower with activity. However, in healthy individuals the rate varies constantly, and it demonstrates a quality of adaptableness to internal and external stressors. It can be seen as a representation of fluid changes in the autonomic nervous system's ability to function on the stress continuum. These stressors can push us into our fight or flight responses driven by our sympathetic nervous system, and the removal of stressors can allow us to resume our resting and digesting state driven by the parasympathetic nervous system.

Like capnography discussed on the previous page, HRV seems to be emerging as a biomarker of nervous system quality. These qualities are closely associated with movement. But we don't need to make assumptions; we simply need to be observant of key variables that seem to represent a qualitative component of function.

The open mind can consider how poor nervous system quality can drive poor movement patterns, and how poor movement patterns can compromise nervous system quality. By monitoring both, we can develop better screening and correction, since it will ultimately help us modulate stress. If we over-stress the movement system or nervous system, we might compromise our corrective outcomes. If we avoid stress altogether, we will not create change or challenge the system's function and adaptability.

APPENDIX 6

THE FUNCTIONAL MOVEMENT SYSTEMS TEAM
A HANDFUL OF LEADERS

A blend of both art and science can be found in every noble profession. Periodically some professions unfortunately lean more in one direction than the other. When a profession moves too far in one direction, some quality of balance is lost. It takes hard work and honest self-appraisal to maintain perspective and be an advocate for balanced professional progress. I can think of no better insurance for continual progress than a respected group of professional friends, peers, coworkers, teachers and mentors. This brand of individual is essential if professional principles are to be maintained in the presence of modern progress and changing political climates. These individuals make up tribes and these tribes make things happen. Seth Godin in his book *Tribes* discusses how a tribe is any group of people, large or small, who are connected to one another by a theme, idea, principle or leader.

I have been fortunate to work with many individuals who are not intimidated by new ideas. Although they travel different professional paths, they are connected by a fundamental ideology that makes them special. They each demonstrate a dedication to principles and do not get swallowed in methodology. They embrace new perspectives and have an uncommon appreciation of the big picture. Their interest in improvement does not erode their appreciation for the lessons of the past. This book is based on the ideas that emerged from years of clinical practice, coaching, teaching, short discussions, long discussions, and way too many road trips with these great people.

With their helpful nudges, suggestions, hard work, support and dedication, this book is an attempt at a re-blend of art and science in exercise and rehabilitation. Screening and assessment may seem dry and organized on the surface, but these simple tools reveal a perspective that has been missing, a perspective provided by the observation of human movement patterns against other forms of movement data. This tribe appreciates the systematic platform that improves perspective and enhances understanding and communication.

When we look at human movement, we don't realize how much we see until it is missing. The best in computer animation still can't fool us—human movement has a unique signature. We recognize the active silhouette of a close friend or family member in low light by the way the person moves long before we can see the person's face. The systems in this book help us capture that signature and blend it with the other information so we can be as comprehensive and proactive as possible.

THE TEAM

My name is most often associated with functional movement screening and assessment and I am honored by the recognition. However, these functional movement systems are no more a solo endeavor than that of a successful NFL quarterback. I am part of a team, a great team, a network... a tribe. This team is made up of individuals who have helped shape the development of the systems and the way they are presented, performed and studied. Each have worked and contributed to create clarity, education and acceptance of these systems. Some have taken the systems to new places, while others have asked the hard questions and done the research. Each has contributed and developed to this work in some way, and I am glad to say each has left a fingerprint. Some have left a handprint and few have even left footprints. Together we have enjoyed making a small difference in the way our worlds look at movement.

We live in a digital world where technology and trends can overshadow skill and art.

We use the tools of movement screening and assessment to force ourselves to look at movement, to consider patterns, and to practice our science with a touch of art—or our art with a touch of science.

THE ORIGINAL TEAM

The original team has been here from the beginning. They have been chipping away at different parts of movement screening and assessment since day one. The functional movement systems in this book rest on the shoulders of the names below.

Gray Cook | Lee Burton | Kyle Kiesel | Brett Jones
Mike Voight | Keith Fields | Greg Rose

THE EXPANSION TEAM

The expansion team is made up of individuals who have provided support, feedback and a wide range of perspectives. They have each added something to the concept or acceptance of movement screening and movement assessment. They have taken the functional movement systems to many more places than we could have imagined.

Todd Arnold	Mike Lehr
Mike Boyle	Scott Livingston
Milo Bryant	Tim Maxey
Robert Butler	Stephanie Montgomery
Lisa Chase	Darcy Norman
Mark Cheng	Jeff O'Connor
Courtney Mizuhara-Cheng	Phil Plisky
Steve Conca	Thom Plummer
Mike Contreras	Chris Poirer
Geralyn Coopersmith	Jim Raynor
Alwyn Cosgrove	Anthony Renna
Rachael Cosgrove	Jay Shiner
Eric Dagati	Steve Smith
Pete Draovitch	Carla Sottovia
John Du Cane	Mike Strock
Sue Falsone	Nishin Tambay
Jeff Fish	Ed Thomas
Joe Gomes	Alan Tomczykowski
Paul Gorman	Jon Torine
Behnad Honarbakhsh	Pavel Tsatsouline
Rusty Jones	Joe VanAllen
Pat Kersey	Mark Verstegen
Thomas Knox	Charlie Weingroff

One additional team name must be added, and we want to make sure it is not lost in a list. Laree Draper our publisher, is the structure and organization behind this work. Her interest and attention to detail on the project goes far beyond the role of a typical publisher, and we are forever grateful. Her heroic efforts have forced us to present and discuss our material in the most organized and clear format to date.

The physical therapy education at the University of Miami prepared me to ponder movement and exercise from many different perspectives. My orthopedic education was straightforward and it applied the basic principles of kinesiology and biomechanics. My learning regarding the neurological system further broadened the scope of my understanding and reasoning as I started to consider movement and its many unique problems.

Studying proprioceptive neuromuscular facilitation (PNF) and developmental movement patterns triggered a recognition of how interconnected our dynamic functional patterns really are. I started to realize that conventional orthopedic rehabilitation did not incorporate neurological principles with the same weight it gave to basic biomechanics principles. Studying movement patterns provided perspective of the sequence of growth and development, and I became interested in movements like rolling, creeping, crawling, and kneeling and the way one movement pattern could serve as the stepping stone, an actual foundation for the next. I soon realized that attacking a functional pattern at a functional level might not produce the best possible result—I wanted to understand when it would be necessary to address a functional problem at a fundamental level.

As an exercise professional, I also noticed that general principles in fitness and athletic conditioning did not consider neurological principles with the same weight as those of exercise physiology. Neurological techniques were rarely employed in corrective exercise and orthopedic rehabilitation, but seemed reserved for mostly neurological problems. Fortunately I was starting to understand these techniques were not simply exercise and rehabilitation tools for the neurologically impaired. They could provide better perceptual opportunities for movement-pattern correction and facilitation than conventional exercises that focused on an iso-lated approach targeting the prime movers within a given pattern. The addition of neurological-based thinking could allow us to make perception, balance, timing, and muscle tone more appropriate and to facilitate the way fundamental patterns can create functional patterns.

Cerebral palsy, brain injuries and spinal cord injuries leave patients with increased muscle tone or spasticity. Other problems like Down's syndrome and paralysis leave patients with reduced muscle tone. Neurological techniques are basically ways to use manual interaction and movement to adjust the volume on input and output. Neurological methods were simply ways to make the best of a bad situation, tapping into the sensory motor system and using forms of stimulation to create more optimal environments for movement. PNF and other techniques use passive movement, assisting movements, tactile stimulation, body position, light resistance, breath control and other forms of subtle stimulation. We must simply embrace ways to incorporate these methods into corrective exercise even when the neurological system is considered normal by medical standards. However we must not simply apply these methods randomly. We must use some standard to gauge when these techniques would be most beneficial and when other methods might offer greater progress.

Based on natural perspectives of movement and movement control, many of these perspectives are so common we ignore them. We watch babies go through the progressive postures of growth and development in which they develop command of one mode of movement and then tinker with a more challenging pattern. We watch them use different parts of their bodies for locomotion, not realizing they are stimulating better support and movement with every point of weight bearing.

We watch sports and fitness movement without considering the many spiral and diagonal movements that go into each athletic form. We fail to note the subtle torso rotation or reciprocal arm action of an elite runner, but when these movements are absent in the less-polished runner, we immediately sense the awkwardness in the movement. We note the awkwardness and yet cannot identify what is lacking.

It's ironic that most people can identify awkward or dysfunctional movement, but usually have a problem exactly describing what is wrong. Since they cannot comment on the actual problem, they ignore the obvious awkwardness and subtle dysfunction, and awkward slowly become the norm.

This is why the movement screen is introduced as a non-diagnostic tool. The FMS identifies fundamental and functional movement problems in a rating and ranking system and first seeks agreement on what should be acceptable and what should not be acceptable before it attempts to suggest remedies or corrections.

PERSPECTIVE IS EARNED THROUGH HARD WORK AND MANY MISTAKES

The perspective that has just been discussed has grown and developed with my continual education. A few sparks accelerated my learning. As a young, overconfident physical therapy student, I was sure exercise and rehabilitation perspectives were oversimplified and ignorant of neurological fundamentals, and I wanted my final research project to reflect those impressions. I didn't know much, but I had a fair amount of experience in exercise prior to my physical therapy education. I felt that orthopedic rehabilitation and general exercise perspectives did not make full use of the knowledge that was routinely employed in neurological circumstances such as stroke rehabilitation and pediatric physical therapy. However, I didn't enter physical therapy school with this perspective. It grew through the challenges of a few professors who wanted me to appreciate the central nervous system as much as the musculoskeletal system. Their challenges made me notice situations where movement quality was prized equally with movement quantity; clinicians routinely employed neurological principles alongside basic exercise practices.

I slowly started to understand that these same neurological principles should be considered even when a neurological problem isn't present. There is no reason to assume the neurological system is functioning efficiently or optimally just because a neurological injury or disease process is not present. I had already observed orthopedic patients with poor static or dynamic stability and wondered, "Isn't this a neurological problem, too?" The orthopedic protocols seemed to treat poor stabilization as a strength problem, exercising muscles associated with poor stabilization waiting for strength to occur and assuming strengthening would somehow improve timing and coordination.

My professional development started alongside a move toward more purposeful movement in both conditioning and rehabilitation. Both Vern Gambetta in the conditioning world and Gary Gray in the rehabilitation world offered exercise examples more realistic and purposeful for normal activities. Their presentations and publications were influential in my early professional development, and they influenced my questioning of one-size-fits-all protocols and the popular isolation approaches of the 1980s and '90s. At that time, athletes were bodybuilding, thinking they were becoming more athletic, while physical therapists were obsessed with fixed-axis isokinetic devices that were technical and reliable, but not necessarily functional.

Both Vern and Gary helped develop the functional approach well accepted today. Unfortunately, many fitness and conditioning professionals still consider functional exercise as a soft alternative to "real training." They look at the elastic bands, medicine balls, lighter weights and balance boards, and question the efficacy when pitted against hard-nosed heavy strength training. They think of it as a replacement for strength and power work, instead of an adjunct to keep it real. Like many trends in exercise, the functional movement had a polarizing effect—people still refer to themselves as a strength guy or a functional guy, but never both.

Even though my experiences were limited, I had formulated some fundamental views. I had worked as a personal trainer, a student athletic trainer, a hospital orderly and as a wellness exercise instructor. My undergraduate degree was in exercise science with a minor in athletic training and psychology. Even though all my professional aspirations revolved around the exercise and rehabilitation occupations, I didn't think the most logical approaches were being applied.

I constantly wondered about efficiency and effectiveness, and considered how many exercises could be removed from any situation, while maintaining the same results. I pondered the potential of a single exercise performed at the right place and the right time to change movement in a measurable way—

- to demonstrate that movement performance could be enhanced by considering movement patterns and neurological facilitation

- to prove that facilitation of movement quality could have a benefit of improved quantity

- to demonstrate that if these neurological therapies could be of benefit against neurological problems, their effect on the, efficiency in an otherwise normal neurological system, could be profound

CURIOSITY

My idea was to take a movement pattern that was easy to measure and undisputed as a demonstration of athletic power and improve it in less than an hour. Could there be untapped power even in an otherwise normal system? Most exercise and performance programs assume the current level of physiology and metabolism limits power, but how about the neuromuscular system? Maybe it could be the limiting factor. Could each of us be walking around with untapped power?

Power could also be limited by the way we breathe, or by poor postural alignment. Our power could be limited by increased tone in muscles where we hold tension, or by reduced tone in muscles associated with patterns we fail to use. If I could prove power was equally associated with the neurological state and physical state, there would be a case for formulating a more practical exercise perspective.

I also wanted to calculate the result of continuous training and repetition on neurological programming and movement patterns that were inefficient. Would they spontaneously become more efficient, or would compensation and substi-

tution simply cover the problem and make it less obvious?

During my physical therapy education, plyometric exercise was gaining popularity in performance training, and some coaches were employing ballistic exercise packages into their programs. Plyometric training was used to improve power by refining reaction responses to loads producing quick stretch of muscle tendon structures. Plyometric training obviously fit the definition of neurological training with the benefits being improved power, coordination, dynamic stability and even stronger tendons from all the shock absorption activity.

Plyometric training relied on neurological reflexes and the elastic muscle tendon components to coordinate their efforts and produce improved levels of explosive power. One problem started to emerge, though. Plyometric training needed to be packaged and used appropriately. Was it part of conventional conditioning or was it facilitation?

If it was conventional conditioning, sets, reps, loads and the workout schedule would ultimately govern plyometric exercise packages. If, in contrast, it was used for facilitation, it would only be used to optimize neurological efficiency whenever it was not optimal. Obviously both benefits would be nice but to maximize the desired benefit one method should dictate the programming.

The neurological efficiency approach required a baseline of neurological efficiency. It would only juice a system that needed juice. Used as a form of facilitation, plyometrics would only be used when neural excitation was the most limiting factor. If flexibility, stability, endurance, posture, form or strength was a limiting factor, plyometric training might place unnecessary stress on other body systems. It might also fail to appropriately facilitative the neurological system if other body systems were unable to tolerate the plyometric load needed to improve neurological function.

It simply goes back to the weakest link scenario; training and conditioning focused on a system other than the weakest link will not likely improve the function of the chain. Once I realized plyometrics could only safely optimize neurological functioning when basic mobility and stability were at functional levels, it was obvious that screening should precede plyometrics or other forms of neurological training.

At that time, though, I did not ponder screening, assuming it existed somewhere. Little did I know that would be the next phase of my life. As a physical therapy student, I just wanted to prove that neurological training could be used more effectively in orthopedic rehabilitation, general fitness and performance enhancement.

MY RESEARCH PROJECT

For my research project, I chose to study the vertical leap. It was an unlikely topic for a physical therapy research project, but here is how I made my point. I wanted to pick normal, active individuals and use facilitation to see if I could influence a performance that was an undisputed representation of power. A demonstration of this would help build a case for facilitation exercises in all phases of rehabilitation, exercise and sports training.

Goals for the project—

- demonstrate that neuromuscular efficiency could be optimized with the right exercise choices

- demonstrate that power could improve in under an hour

- demonstrate improvement could occur without physiological adaptation or change in tissue

- make a case for accelerated movement-pattern learning

The vertical leap was a good choice because it is an unbiased activity; everyone can jump, and gravity affects all bodies equally, so body size is negated. I wouldn't use elite jumpers like college basketball players or volley players for the study, but instead wanted active individuals who were not elite jumpers and most likely had room to improve their jumping performance. This was not about designing a jump-training program to be performed over a few weeks or months, or even to train them for days. I wanted to see how one bout of exercise could affect the neuromuscular system, to see a response, not an adaptation.

I knew I needed to compare my facilitation exercise idea to a conventional form of jump training. To follow correct research form I also needed a control group. My research partners and I wanted to rule out the physiological effects of warm up, and make a case for movement pattern learning.

To develop our technique we used PNF principles in the form of facilitation exercise. PNF is historically hands-on, but this was to be a hands-off exercise technique. I also wanted create a non-visual and non-verbal movement learning situation.

With my advisor and my research partners, we put together the plan, including a way to provide light resistance to the entire jumping pattern. Popular jump training devices at the time used heavy elastic resistance anchored to a platform and to a shoulder and waist harness. The focus was mainly on the legs and the purpose was clearly to overload the lower body.

From advertisements and instructions, it looked like the device manufacturers expected a strengthening effect from the resistance and a plyometric effect from the harder landing imposed by the bands coupled with gravity. Some background research on jumping mechanics showed us however, that the biggest biomechanical difference in good jumpers and great jumpers was the contribution of the upper body and arms. This was not even addressed in the popular product, but the product looked good and fed into the basic strength mentality. I wanted to use subtle resistance to refine the jumping movement, and wanted to use the resistance to create proprioceptive input and facilitate coordination and timing. We didn't realize it at the time but we were putting our focus on perception instead of behavior. This would later become a key component to my understanding about corrective exercise.

Our movement pattern used light form of resistance and was held in the hands. The movement was performed by having the participant hold light resistance bands in their hands while they performed 10 jumps as high and as fast as possible.

The bands were attached to a platform about 24-30 inches outside of the jumpers' left and right feet. They provided some resistance, but mostly the bands exaggerated whatever mistake the jumpers exhibited when they jumped. If they hyper-extended their spines, the bands would pull them farther into extension. If the jumper leaned off to right, the bands would actually pull them even farther to the right. Each jumper received instantaneous proprioceptive feedback about his or her unique jumping mistake. If they over-corrected in the opposite direction, they would receive more feedback about the over-correction. If they jumped with great alignment and symmetry, they would not be pulled in any awkward direction; they would simply be pulled down to reset and jump again. No verbal or visual feedback or coaching was provided other than the initial instructions.

50 subjects would be chosen, 25 men and 25 women, all of college age and all actively participating in weekly exercise or recreational sports. No one in the group was injured or complained of pain with movement. Participants were randomly assigned to three groups, each performing 10 jumps.

The first group received the jumping facilitation technique we now call reactive neuromuscular training (RNT).

The second group received conventional jump training where heavy elastic resistance was attached to the platform and to a shoulder and waist harness.

The third group was the control group, who performed jumps as high and fast as possible, with no resistance.

Three tests were performed at three different times, and the average of three tests was recorded. The test measured both the height of the jump and the jump reaction time. Testing was done right before and after the 10-repetition jump training trail was performed. This allowed for us to look immediately at the effects of each technique.

We realized that fatigue would play a role in the second test series, since by that time each participant had performed 16 total jumps between a testing session, a training session and another testing session. Each participant rested 30 minutes and repeated the test a third time. The third test was specifically designed to negate the physiological effects of the 16 jumps. We wanted the third test to be absent of both fatigue and warm-up, and to specifically see if some motor learning effect could be captured. We would only capture a motor re-

sponse and not an adaptation, but we were satisfied to demonstrate a change in motor control, even if it was only temporary.

Our study revealed that we produced a positive effect on reaction time immediately following and 30 minutes after training when compared with conventional jump training and control. We also demonstrated an impressive improvement in jump height after 30 minutes of rest compared with conventional jump training and control. The bar graphs below demonstrate the initial effect (second test) and the carryover effect (third test).

See graphs at the end of the document

We surmised that by resisting the entire jumping pattern with a light load we created more opportunity for proprioceptive input and therefore more opportunity for motor control refinement. This improved motor efficiency was demonstrated immediately following the facilitation technique and although reduced, it lasted across the 30 minutes of rest. Our facilitation technique did not distinguish itself immediately, but the training effect was superior to both control and conventional jump training across 30 minutes of rest. We had demonstrated a two-fold improvement in power, since reaction time and jump height were both positively affected.

Our intention was not to create a jump training exercise, but to demonstrate that a simple 10-repetition exercise could have impressive motor effects if performed to maximize proprioceptive input. Our technique provided a light resistance to jumping that instantaneously magnified mistakes in the jumping activity. We wanted to demonstrate that PNF principles could be used to refine motor control even when the hands of a clinician did not provide the resistance.

LOOKING AT THINGS DIFFERENTLY

This study had a profound effect on me, one I didn't realize it at the time. I graduated from physical therapy school and worked while studying manual therapy for five years. I learned how to mobilize and manipulate the joints of the body, and how to address soft tissue problems with a command of modalities and manual skills.

Although I was receiving advanced training in all aspects of orthopedic and sports rehabilitation, I was unimpressed with the standardized prepackaged exercises that often followed brilliant manual treatment.

Corrective exercises were not correcting anything. These exercises simply rehearsed movements that were awkward or faulty with the hope that arbitrary resistance loads would somehow create strength, integrity and competency. Clinicians didn't know if the corrective exercises were designed to create a physiological demand or a neuromuscular demand. What was the weakest link? What was the most limited system?

Most of the corrective exercise targeted tissue physiology and not motor control. It was all highly coached verbal and visual two-dimensional movement, and did not fit my definition of function. We didn't make anyone react to anything; we didn't challenge the sensory motor system. We just rehearsed exercises that fit the simplest application of local kinesiology. We didn't design corrective exercises to enhance perception and provide the necessary mistakes for learning. Instead we provided instruction and told our patients the movement mistakes they should not make.

On the surface it seemed that some rehabilitation professionals provided activity at or around the dysfunctional region, and assumed motor control would spontaneously reset. These activities were not so much causing a reset, as they were creating greater opportunity for compensation behavior. We now know that pain affects motor control in unpredictable and inconsistent ways. This coupled with poorly planned and poorly reproduced exercises gave the average patient little chance of reestablishing authentic motor control. It might also speak to the research implying that previous injury was the most significant risk factor for a future injury.

We moved people around until their pain was gone or at least at a tolerable level, and think that we had done something. We didn't check function; we had no idea of how much compensation the patient had developed on the road to recovery. We concerned ourselves with removing pain, not restoring function. There was far more in my physical therapy discharge notes about pain resolution than functional movement pattern restoration.

As I started to refine my evaluation skills, I also started experimenting with drills that fit my definition of RNT, drills that use a light load to exaggerate a movement mistake. If I saw valgus collapse on a lunge pattern, I put an elastic band on the knee and pulled the knee inward. If I pulled too hard, the move was too difficult to complete. If I did not pull hard enough, the pattern wouldn't change, but if I pulled just enough, I would see a reactive countermeasure. The knee that was caving in would reset itself in a more functional position. The best resistance with respect to the RNT load is the one that causes the problem to correct itself with minimal verbal or visual feedback.

Physical therapists were classifying problems by a patient's diagnosis or by the site of the pain. Meanwhile, I was on a completely different path of choosing corrective exercise based on movement dysfunction, not pain or diagnosis. Previously I had provided treatments appropriate for the pain and dysfunction, but that had no bearing on movement pattern correction. Eventually, I decided on corrective exercise based on movement dysfunction, not pain or diagnosis.

Of course, I would not prescribe exercises that were contraindicated by the patient's condition. In many cases, I found myself using corrective exercise on regions of the body far from the site of pain. In this new model, it was possible for two patients with low back pain have completely different exercise programs. They might receive the same pain control treatments, but their movement dysfunctions may require completely different corrective exercise approaches.

This new approach to corrective exercise worked well and seemed to accelerate progress. However, two major rules became clear determinants of effectiveness. The first rule required consideration of movement patterns alongside other parameters like physical performance and diagnosis. These considerations soon became the basis for the FMS and SFMA.

The second rule was an acknowledgment of natural law. Mobility must precede stability. These new reactive drills were only effective if mobility was not compromised. This meant we must address mobility before expecting a new level of motor control. If there was no mobility problem, I could expect the RNT drills and exercises to improve motor control and improve movement patterns. If mobility was limited, I needed to address the mobility first.

Of course, it is unrealistic to expect to normalize mobility in all cases. However, it is also impractical assume that since it cannot become normal, no attempt should be made. In most cases, mobility has the potential to improve. With each measurable improvement, it is also likely that motor control can be addressed with a basic stabilization exercise or RNT drill.

Mobility problems are functional dysfunctions, probably the byproduct of inappropriate movement patterns. They could be the result of a poorly managed injury, physical stress, emotional stress, postural stress or inefficient stabilization. All these issues alone or in combination can reduce mobility in an attempt to provide function at some level. Those with a weak core might develop tightness in the shoulder girdle or neck musculature as a secondary attempt to continue to function at a desired level. Those with chronic low back pain may develop tightness in the hip flexors and hamstrings to function as secondary braces even if it reduces their mobility.

It became evident that nature had worked out a solution, and although the solution might compromise mobility in some regions, it afforded function at a desirable level… and that is survival.

As long as mobility is compromised, the stiffness and increased muscle tone are providing the requisite stability needed for function. If mobility is not addressed in any way, the system will not need or want a new level of motor control. However, if mobility is improved, a window of opportunity is opened. For a short amount of time, the body cannot rely on or lean on stiffness and inappropriate muscle tone.

Within this window, motor control exercises that engage both the sensory and motor systems will call on primary stabilizers to work, since tightness and stiffness are temporarily not options. Dosage is everything in this window. If exercise is too stressful, the individual will default back to old patterns, and if exercise does not challenge the primary stabilizers, they will not become reintegrated into posture and movement.

The system simply requires the user to improve mobility in a region where a limitation has been identified. The patient or client is put into a posture where they can be challenged, such as like rolling, quadruped, kneeling or half-kneeling. The person might perform movement or simply be challenged to hold the position. From the stable posture, he or she is then progressed to less stable postures, and then into dynamic movement patterns.

Babies enter the world with uncompromised mobility and follow this progression naturally. My best efforts in exercise and rehabilitation have attempted to replicate this gold standard when movement patterns are dysfunctional.

A NEW PARADIGM FOR MOVEMENT

As an attempt to apply what I've learned, I now map the dysfunctional movement patterns, noting asymmetry, limitation and inabilities, and then address the most fundamental movement pattern problem with specific attention to mobility issues. When measurable improvement is noted in mobility, I attempt the challenge the system without its crutch of stiffness and tightness.

I try to tap into natural reactions that maintain posture, balance or alignment at a level of stability the patient can handle, a level where he or she can demonstrate success and receive positive feedback.

- I attempt to avoid fatigue at all costs and minimize verbal instruction and visual feedback.

- I attempt to challenge the person, and get him or her to respond through feel.

- I always encourage people not to over-think or try too hard. Balance is automatic—balance is natural.

- I always make sure patients are not stress-breathing. If stress-breathing is noted, I stop the drill and try to get a laugh or perform breathing drills.

- As patients develop control, they're progressed, but I'm always mindful to not to overload or turn the motor control drill into conventional exercise.

- I end each session with a reappraisal of the dysfunctional pattern.

If I'm successful I know where to start next time, and if not, I know exactly where not to start. If successful, I recommend a small amount of corrective activity at home to maintain our gains. If I'm not successful, I only ask the patient to perform mobility exercises and maybe some breathing drills until the next session, since I have not yet established the best motor control exercise for that individual.

That's what I do…

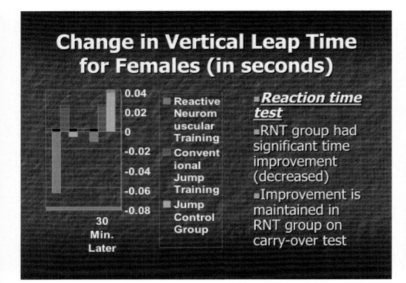

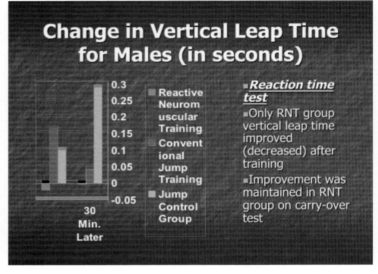

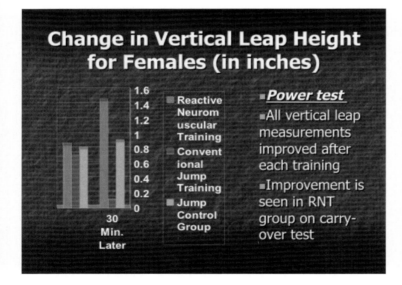

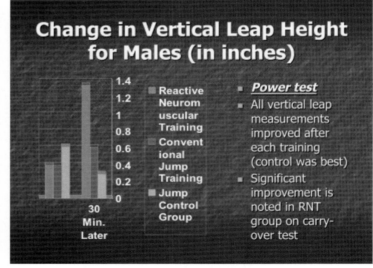

CORE TESTING and FUNCTIONAL GONIOMETRY

New Concepts in Core Testing—
Testing the Core in Quarters or Quadrants

Unfortunately, many attempts at core testing used in practice are isolated strength tests of muscles in the abdominal region. However, more and more professionals are recognizing that the geographic center of our strength has multiple dimensions. Gary Gray and other forward-thinking professionals have historically used more functional demonstrations to discuss core control and ability. Modern revisions have improved data collection and consistency without losing the essence of the quality and dynamics of true core control.

The best orthopedic manual therapists often discuss the body and its functional segments in groupings called quadrants or quarters. Semantics make it nearly impossible to discuss the shoulder without discussing the cervical spine and thoracic spine, and that is the point. Likewise, those skilled in orthopedic rehabilitation would not think of discussing the knee and its issues without also considering the foot, hip, pelvis and lumbar spine. This view represents the interdependence of segments.

By considering the core in functional quadrants, the rehabilitation and exercise professional can have a more comprehensive appraisal of function before specific or isolated testing is performed. By dividing the body into right and left sides, we can appreciate symmetry. By dividing the upper and lower body, we can appreciate core control from the bottom up or top down. This perspective is helpful because it demonstrates a wide-angle view of baseline function before reductions and impairments are investigated. This type of testing can complement screening and assessment as well as confirm progress with corrective exercise paradigms.

Y BALANCE TEST (YBT)— THE FUNCTIONAL GONIOMETER

Phil Plisky, PT, DSc, OCS, ATC, CSCS

As discussed in Chapter 3, a test gauges a person's ability and is a measurement that does not require interpretation. A comprehensive functional test would examine a client's ability across multiple domains and give a precise numerical rating that corresponds with aptitude in those domains. This is what the Y Balance Test does. It acts as a functional goniometer by allowing precise quantification of a person's body relative movement by simultaneously requiring strength, flexibility, neuromuscular control, core stability, range of motion, balance and proprioception.

UPPER QUARTER Y BALANCE TEST

Upper extremity closed kinetic chain trunk stability tests have been described in the literature—prone, supine and side bridge; the One-Arm Hop Test[1] and the Closed Kinetic Chain Upper Extremity Stability Test (CKCUEST).[2] The bridge tests are static tests that do not challenge dynamic stability. In the One-Arm Hop test, an individual assumes a one-arm pushup position on the floor and then uses his or her arm to hop onto a 10.2 cm step and back onto the floor. The time required to perform five repetitions as quickly as possible is recorded. The CKCUEST begins in a traditional pushup position with the hands placed 36 inches apart on strips of athletic tape. The person then reaches with alternating hands across the body to touch the piece of tape under the opposing hand; the number of cross-body touches performed in 15 seconds is recorded.

Although these tests place the individual in a closed kinetic chain position, they measure stability in a limited range, and do not take into account mobility or end-range stability. Further, these tests are performed within a person's comfortable base of support and therefore do not challenge beyond the point of stability. None of the tests described adequately assess other essential aspects of natural movement such as thoracic and scapular mobility.

The Upper Quarter Y Balance Test (YBT-UQ) is a body relative quantitative analysis of a person's ability to reach with the free upper limb while maintaining single-limb weight-bearing on the contralateral upper limb. The YBT-UQ can be used to test dynamic upper extremity and trunk stability. To perform the YBT-UQ, the individual assumes a starting position *(Figure 1)* with the hand of the limb being tested on the YBT platform with the thumb adducted. The thumb is aligned along the red starting line, with the hand on the label side of the line. While maintaining a pushup position with feet shoulder width apart, the person is asked to reach with the free hand in the medial, inferolateral, and superolateral directions in relation to the stance hand *(Figures 2-4)*. Unlike the Lower Quarter YBT, all three reach directions are performed sequentially without touching down or resting between directions. The person may touch down and rest prior to performing the next trial. Shoes are not worn during the performance of the test.

In order to compare the performance to normative data or to other team members, upper limb length should be measured. Upper limb length is measured from the C7 spinous process to the tip of the longest finger with the shoulder elevated to 90 degrees in the sagittal plane. To calculate normalized composite reach distance, the sum of the greatest reach in the three reach directions is divided by three times upper limb length, and then multiplied by 100.

The YBT-UQ attempts to address some of the limitations of the previously developed tests. First, mobility and stability are both maximally challenged during the test. Stability of the trunk and loaded arm is challenged at the same time mobility of the thorax and reach arm is challenged. During each reach, components of scapular stability, mobility, thoracic rotation and core stability are combined as the person is encouraged to reach as far as possible without loss of balance. Reaching as far as possible outside of a narrow base of support requires balance, proprioception, strength and full range of motion. Most healthy people can perform the test without much training or cueing.

STANDARDIZED YBT-UQ TESTING INSTRUCTIONS

This is intended to test how well a person can balance on one hand. The goal is to balance on one hand in a pushup position in the center of a Y Balance Test, and reach with the free hand as far to the side, under and behind, and above and across while maintaining single-hand balance. The person is allowed to practice two times on each hand before the test begins.

You start in a push up position with feet shoulder width apart. The examiner will tell you which hand to lift off the ground first. You will then reach in each of the three directions pushing the target as far as you can and then return the reach hand to the starting position. The reach will be repeated if you are unable to maintain your balance on one hand, lift or move the balance hand from the platform, touchdown with the reach hand, fail to return the reach hand to the starting point, or do not maintain contact with the target when it comes to rest (e.g. shove the target). This process will be repeated until you have performed three trials in each direction on each hand.

Figure 1. YBT-UQ Start Position

Figure 2. YBT-UQ Medial Reach

Figure 3. YBT-UQ Inferolateral Reach

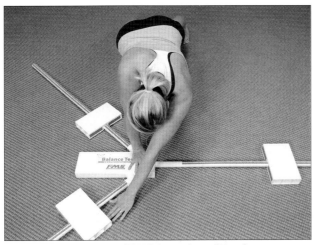

Figure 4. YBT-UQ Superolateral Reach

Athlete Model Beth Ross

Photos by Keith Leonhardt

LOWER QUARTER Y BALANCE TEST

The Lower Quarter Y Balance Test (YBT-LQ) is a dynamic test that requires stability, strength, flexibility and proprioception of the lower quadrant of the body. The goal of the YBT-LQ is to maintain single-leg stance on one leg while reaching as far as possible with the contralateral leg.[3,4] This dynamic task requires the person to perform at the limit of stability. [5-8] Gary Gray first described a similar test called the Balance Reach test; it has been subsequently modified for research and clinical purposes and is known as the Star Excursion Balance Test. Since there are many common sources of error and method variation for administration of the Star Excursion Balance Test, Plisky et al improved the repeatability of measurement and standardized performance using the Y Balance Test protocol.[9] Since the Y Balance Test is comprised of two comprehensive assessments of the upper and lower quarter, the lower quarter contains the LQ designation.

The importance of symmetry on this test has been well established in identifying chronic ankle instability, ACL deficiency, and injury prediction.[7,10-14] Researchers first demonstrated that the test can reliably identify individuals with chronic ankle instability. The test was then modified to make it more efficient by including only three of eight directions.[7] Plisky et al found that performance on the three directions as well as asymmetry was able to predict lower extremity injury in high school basketball players.[7]

Since there is a learning effect that occurs after four to six trials, it is recommended that the client practice four to six trials on each leg in each of the three reach directions prior to formal testing.[9,15-16] Shoes are not worn during the performance of the test. The client stands on one leg on the center of the platform with the most distal aspect of the toes at the starting line. While maintaining single-leg stance, the client is asked to push the reach indicator in the red target area with the free limb in the anterior *(Figure 5)*, posteromedial *(Figure 6)*, and posterolateral *(Figure 7)* directions in relation to the stance foot. The testing order is three trials standing on the right foot reaching in the anterior direction (right anterior reach) followed by three trials standing on the left foot reaching in the anterior direction repeating this procedure for the posteromedial and then the posterolateral reach directions. The specific testing order is right anterior, left anterior, right posteromedial, left posteromedial, right posterolateral and left posterolateral.

The client is allowed to touch down with the reach foot between trials and the stance foot heel may come off the ground. The hands and arms can be in position of comfort for the client. The maximal reach distance is measured by reading the tape measure at the edge of the reach indicator, at the point where the most distal part of the foot reached. The trial is discarded and repeated if the client: **1)** fails to maintain unilateral stance on the platform (e.g. touches down to the floor with the reach foot or falls off the stance platform), **2)** fails to maintain reach foot contact with the reach indicator on the target area while it is in motion (e.g. kicks the reach indicator), **3)** uses the reach indicator for stance support (e.g. places foot on top of reach indicator), or **4)** fails to return the reach foot to the starting position under control. The starting position for the reach foot is defined by the area immediately between the standing platform and the pipe opposite the stance foot. The greatest successful reach for each direction on each leg is used for the client's score. Also, the greatest reach distance from each direction is summed to yield a composite reach distance for analysis of overall performance on the test.

Researchers have shown that performance on the Lower Quarter Y Balance Test is dependent on gender, competition level and sport. In order to compare the client's performance to normative data or other team members, lower limb length should be measured. While the client is in hooklying, the hips are lifted off the table and returned to the starting position. Then, the legs are passively straightened to equalize the pelvis. The client's right limb length is then measured in centimeters from the most inferior aspect of the anterior superior iliac spine to the most distal portion of the medial malleolus with a cloth tape measure. To calculate normalized composite reach distance, the sum of the greatest reach in the three reach directions is divided by three times limb length, and then multiplied by 100.

WHAT DO WE LOOK FOR ON THE YBT-LQ?

Researchers indicate that there should not be a greater than 4 centimeter right/left reach distance difference in the anterior reach direction. It is hypothesized that there should not be greater than a 6 cm reach distance difference in the posteromedial and posterolateral directions. Also, the composite score (sum of three reach directions is divided by three times limb length, and then multiplied by 100) should not be less than the cut points that are specific for the age, gender, and sport of the client.

STANDARDIZED YBT-LQ INSTRUCTIONS

You will practice six trials in three reach directions before the test begins. You will stand on one leg in the center of the YBT platform and reach with the free leg pushing the target as far forward, to the side, and backward while maintaining single leg stance. The maximal reach distance will be measured by observing the point where the target stopped.

The trial will be considered incomplete and repeated if you—

- fail to maintain unilateral stance on the platform (e.g. touch down to the floor with the reach foot)

- fail to maintain reach foot contact with the reach indicator on the target area while it is in motion (e.g. kick the reach indicator)

- use the reach indicator for stance support (e.g. place foot on top of reach indicator)

- fail to return the reach foot to the starting position under control

You will repeat this process until you have performed three trials in each direction on each leg.

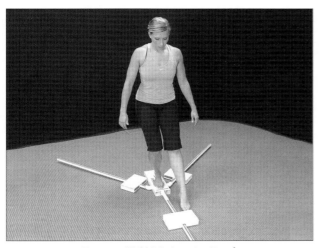

Figure 5. YBT-LQ Anterior Reach

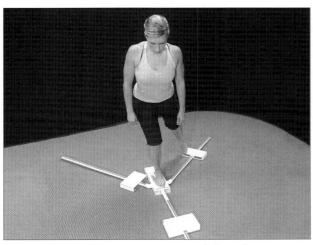

Figure 6. YBT-LQ Posteromedial Reach

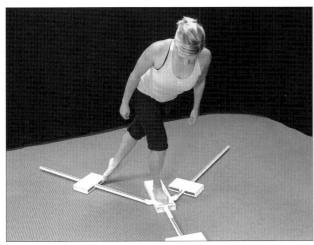

Figure 7. YBT-LQ Posterolateral Reach

HOW TO USE THE FUNCTIONAL GONIOMETER

Since the YBT measures across all domains of movement (range of motion, strength, proprioception, core stability, etc.), one faulty component of any of these systems will cause a positive test. The YBT requires a unique harmony of the entire neuromusculoskeletal system that is not seen in many other tests. This might bring up the question: Why not use the YBT instead of the Functional Movement Screen? First of all, the YBT measures a different construct of movement. Rather than breaking movement patterns down into components, it puts many of them together. This makes the YBT powerful as it measures comprehensive movement harmony, but limiting in the fact that you are unable to determine a client's weak link. The YBT helps determine risk and identify that there is a problem, but you would be limited prescribing corrective exercise for a poor Y Balance Test score. You could not narrow the numerous possibilities that can cause a positive test. This prescriptive ability is one of the many utilities of the FMS.

Since the test requires the person to perform at the limit of stability, it can be used to discriminate neuromuscular control abilities at a more demanding level, which is required for most clients. Further, the YBT can be used in the rehabilitation setting as a marker of ability at the beginning of rehabilitation and as a return-to-activity criterion near discharge. Additionally, the YBT can be used in the pre-participation physical. If a client has a previous injury, the YBT can be used to quickly identify those clients who have not fully rehabilitated or normalized their dynamic neuromuscular control after an injury.

Y BALANCE REFERENCES

1 Falsone SA, Gross MT, Guskiewicz KM, Schneider RA. One-arm hop test: reliability and effects of arm dominance. *J Orthop Sports Phys Ther.* Mar 2002;32(3):98-103.

2 Roush JR, Kitamura J, Waits MC. Reference values for the closed kinetic chain upper extremity stability test (CKCUEST) for collegiate baseball players. *NAJSPT.* August 2007;2(3):159-163.

3 Kinzey S, Armstrong C. The reliability of the star-excursion test in assessing dynamic balance. *J Orthop Sports Phys Ther.* 1998;27(5):356-360.

4 Gray G. Lower Extremity Functional Profile. Adrian, MI: Wynn Marketing, Inc; 1995.

5 English T, Howe K. The effect of Pilates exercise on trunk and postural stability and throwing velocity in college baseball pitchers: single subject design. *NAJSPT.* 2007;2(1):8-19.

6 Lanning CL, Uhl TL, Ingram CL, Mattacola CG, English T, Newsom S. Baseline values of trunk endurance and hip strength in collegiate athletes. *J Athl Train.* Oct-Dec 2006;41(4):427-434.

7 Plisky PJ, Rauh MJ, Kaminski TW, Underwood FB. Star Excursion Balance Test as a predictor of lower extremity injury in high school basketball players. *J Orthop Sports Phys Ther.* Dec 2006;36(12):911-919.

8 Hale SA, Hertel J, Olmsted-Kramer LC. The effect of a 4-week comprehensive rehabilitation program on postural control and lower extremity function in individuals with chronic ankle instability. *J Orthop Sports Phys Ther.* Jun 2007;37(6):303-311.

9 Plisky PJ, Gorman P, Kiesel K, Butler R, Underwood F, Elkins B. The reliability of an instrumented device for measuring components of the Star Excursion Balance Test. .*NAJSPT.* 2009;4(2):92-99.

10 Herrington L, Hatcher J, Hatcher A, McNicholas M. A comparison of Star Excursion Balance Test reach distances between ACL deficient patients and asymptomatic controls. Knee 2009;16(2):149-52.

11 Hertel J, Braham R, Hale S, Olmsted-Kramer L. Simplifying the star excursion balance test: analyses of subjects with and without chronic ankle instability. Journal of Orthopaedic and Sports Physical Therapy 2006;36(3):131-7.

12 Gribble P, Hertel J, Denegar C, Buckley W. The effects of fatigue and chronic ankle instability on dynamic postural control. Journal of Athletic Training 2004;39(4):321-9.

13 Hubbard TJ, Kramer LC, Denegar CR, Hertel J. Contributing factors to chronic ankle instability. *Foot Ankle Int.* Mar 2007;28(3):343-354.

14 Olmsted L, Carcia C, Hertel J, Shultz S. Efficacy of the Star Excursion Balance Tests in detecting reach deficits in subjects with chronic ankle instability. *J Athl Train.* 2002;37(4):501-506.

15 Hertel J, Miller S, Denegar C. Intratester and intertester reliability during the Star Excursion Balance Tests. *J Sport Rehabil.* 2000;9:104-116.

16 Robinson RH, Gribble PA. Support for a reduction in the number of trials needed for the star excursion balance test. *Arch Phys Med Rehabil.* Feb 2008;89(2):364-370.

FMS SCORING CRITERIA

DEEP SQUAT

3

Upper torso is parallel with tibia or toward vertical | Femur below horizontal
Knees are aligned over feet | Dowel aligned over feet

2

Upper torso is parallel with tibia or toward vertical | Femur is below horizontal
Knees are aligned over feet | Dowel is aligned over feet | Heels are elevated

1

Tibia and upper torso are not parallel | Femur is not below horizontal
Knees are not aligned over feet | Lumbar flexion is noted

The athlete receives a score of zero if pain is associated with any portion of this test.
A medical professional should perform a thorough evaluation of the painful area.

HURDLE STEP

3

Hips, knees and ankles remain aligned in the sagittal plane
Minimal to no movement is noted in lumbar spine | Dowel and hurdle remain parallel

2

Alignment is lost between hips, knees and ankles | Movement is noted in lumbar spine
Dowel and hurdle do not remain parallel

1

Contact between foot and hurdle occurs | Loss of balance is noted

The athlete receives a score of zero if pain is associated with any portion of this test.
A medical professional should perform a thorough evaluation of the painful area.

INLINE LUNGE

3

Dowel contacts maintained | Dowel remains vertical | No torso movement noted
Dowel and feet remain in sagittal plane | Knee touches board behind heel of front foot

2

Dowel contacts not maintained | Dowel does not remain vertical | Movement noted in torso
Dowel and feet do not remain in sagittal plane | Knee does not touch behind heel of front foot

1

Loss of balance is noted

The athlete receives a score of zero if pain is associated with any portion of this test.
A medical professional should perform a thorough evaluation of the painful area.

SHOULDER MOBILITY

3

Fists are within one hand length

2

Fists are within one-and-a-half hand lengths

1

Fists are not within one and half hand lengths

The athlete will receive a score of zero if pain is associated with any portion of this test. A medical professional should perform a thorough evaluation of the painful area.

CLEARING TEST

Perform this clearing test bilaterally. If the individual does receive a positive score, document both scores for future reference. If there is pain associated with this movement, give a score of zero and perform a thorough evaluation of the shoulder or refer out.

FMS

ACTIVE STRAIGHT-LEG RAISE

3

Vertical line of the malleolus resides between mid-thigh and ASIS
The non-moving limb remains in neutral position

2

Vertical line of the malleolus resides between mid-thigh and joint line
The non-moving limb remains in neutral position

1

Vertical line of the malleolus resides below joint line
The non-moving limb remains in neutral position

The athlete will receive a score of zero if pain is associated with any portion of this test.
A medical professional should perform a thorough evaluation of the painful area.

TRUNK STABILITY PUSHUP

3

The body lifts as a unit with no lag in the spine

Men perform a repetition with thumbs aligned with the top of the head
Women perform a repetition with thumbs aligned with the chin

2

The body lifts as a unit with no lag in the spine
Men perform a repetition with thumbs aligned with the chin | Women with thumbs aligned with the clavicle

1

Men are unable to perform a repetition
with hands aligned with the chin

Women unable with thumbs aligned with the clavicle

The athlete receives a score of zero if pain is associated with any portion of this test.
A medical professional should perform a thorough evaluation of the painful area.

SPINAL EXTENSION CLEARING TEST

Spinal extension is cleared by performing a press-up in the pushup position. If there is pain associated with this motion, give a zero and perform a more thorough evaluation or refer out. If the individual does receive a positive score, document both scores for future reference.

ROTARY STABILITY

 **3**

Performs a correct unilateral repetition

 **2**

Performs a correct diagonal repetition

 **1**

Inability to perform a diagonal repetition

The athlete receives a score of zero if pain is associated with any portion of this test.
A medical professional should perform a thorough evaluation of the painful area.

SPINAL FLEXION CLEARING TEST

Spinal flexion can be cleared by first assuming a quadruped position, then rocking back and touching the buttocks to the heels and the chest to the thighs. The hands should remain in front of the body, reaching out as far as possible. If there is pain associated with this motion, give a zero and perform a more thorough evaluation or refer out. If the individual receives a positive score, document both scores for future reference.

THE FUNCTIONAL MOVEMENT SCREEN

SCORING SHEET

NAME _____ DATE _____ DOB _____

ADDRESS _____

CITY, STATE, ZIP _____ PHONE _____

SCHOOL/AFFILIATION _____

SSN _____ HEIGHT _____ WEIGHT _____ AGE _____ GENDER _____

PRIMARY SPORT _____ PRIMARY POSITION _____

HAND/LEG DOMINANCE _____ PREVIOUS TEST SCORE _____

TEST		RAW SCORE	FINAL SCORE	COMMENTS
DEEP SQUAT				
HURDLE STEP	L			
	R			
INLINE LUNGE	L			
	R			
SHOULDER MOBILITY	L			
	R			
IMPINGEMENT CLEARING TEST	L			
	R			
ACTIVE STRAIGHT-LEG RAISE	L			
	R			
TRUNK STABILITY PUSHUP				
PRESS-UP CLEARING TEST				
ROTARY STABILITY	L			
	R			
POSTERIOR ROCKING CLEARING TEST				
TOTAL				

   Raw Score: This score is used to denote right and left side scoring. The right and left sides are scored in five of the seven tests and both are documented in this space.

   Final Score: This score is used to denote the overall score for the test. The lowest score for the raw score (each side) is carried over to give a final score for the test. A person who scores a three on the right and a two on the left would receive a final score of two. The final score is then summarized and used as a total score.

APPENDIX 10

VERBAL INSTRUCTIONS FOR
THE FUNCTIONAL MOVEMENT SCREEN

The following is a script to use while administering the FMS. For consistency throughout all screens, this script should be used during each screen. The bold words represent what you should say to the client.

Please let me know if there is any pain while performing any of the following movements.

DEEP SQUAT

EQUIPMENT NEEDED: DOWEL

INSTRUCTIONS

- **Stand tall with your feet approximately shoulder width apart and toes pointing forward.**

- **Grasp the dowel in both hands and place it horizontally on top of your head so your shoulders and elbows are at 90 degrees.**

- **Press the dowel so that it is directly above your head.**

- **While maintaining an upright torso, and keeping your heels and the dowel in position, descend as deep as possible.**

- **Hold the descended position for a count of one, then return to the starting position.**

- **Do you understand the instructions?**

Score the movement.
The client can perform the move up to three times total if necessary.
If a score of three is not achieved, repeat above instructions using the 2 x 6 under the client's heels.

HURDLE STEP

EQUIPMENT NEEDED: DOWEL, HURDLE

INSTRUCTIONS

- **Stand tall with your feet together and toes touching the test kit.**

- **Grasp the dowel with both hands and place it behind your neck and across the shoulders.**

- **While maintaining an upright posture, raise the right leg and step over the hurdle, making sure to raise the foot towards the shin and maintaining foot alignment with the ankle, knee and hip.**

- **Touch the floor with the heel and return to the starting position while maintaining foot alignment with the ankle, knee and hip.**

- **Do you understand these instructions?**

Score the moving leg.
Repeat the test on the other side.
Repeat two times per side if necessary.

INLINE LUNGE

EQUIPMENT NEEDED: DOWEL, 2X6

INSTRUCTIONS

- **Place the dowel along the spine so it touches the back of your head, your upper back and the middle of the buttocks.**

- **While grasping the dowel, your right hand should be against the back of your neck, and the left hand should be against your lower back.**

- **Step onto the 2x6 with a flat right foot and your toe on the zero mark.**

- **The left heel should be placed at _____ mark.** *This is the tibial measurement marker.*

- **Both toes must be pointing forward, with feet flat.**

- **Maintaining an upright posture so the dowel stays in contact with your head, upper back and top of the buttocks, descend into a lunge position so the right knee touches the 2x6 behind your left heel.**

- **Return to the starting position.**

- **Do you understand these instructions?**

Score the movement.
Repeat the test on the other side.
Repeat two times per side if necessary.

SHOULDER MOBILITY

EQUIPMENT NEEDED: MEASURING DEVICE

INSTRUCTIONS

- **Stand tall with your feet together and arms hanging comfortably.**

- **Make a fist so your fingers are around your thumbs.**

- **In one motion, place the right fist over head and down your back as far as possible while simultaneously taking your left fist up your back as far as possible.**

- **Do not "creep" your hands closer after their initial placement.**

- **Do you understand these instructions?**

Measure the distance between the two closest points of each fist.
Score the movement.
Repeat the test on the other side.

ACTIVE SCAPULAR STABILITY (SHOULDER CLEARING)

INSTRUCTIONS

- **Stand tall with your feet together and arms hanging comfortably.**

- **Place your right palm on the front of your left shoulder.**

- **While maintaining palm placement, raise your right elbow as high as possible.**

- **Do you feel any pain?**

Repeat the test on the other side.

ACTIVE STRAIGHT-LEG RAISE

EQUIPMENT NEEDED: DOWEL, MEASURING DEVICE, 2X6

INSTRUCTIONS

- **Lay flat with the back of your knees against the 2x6 with your toes pointing up.**

- **Place both arms next to your body with the palms facing up.**

- **Pull the toes of your right foot toward your shin.**

- **With the right leg remaining straight and the back of your left knee maintaining contact with the 2x6, raise your right foot as high as possible.**

- **Do you understand these instructions?**

Score the movement.
Repeat the test on the other side.

TRUNK STABILITY PUSHUP

EQUIPMENT NEEDED: NONE

INSTRUCTIONS

- **Lie face down with your arms extended overhead and your hands shoulder width apart.**

- **Pull your thumbs down in line with the ___ (forehead for men, chin for women).**

- **With your legs together, pull your toes toward the shins and lift your knees and elbows off the ground.**

- **While maintaining a rigid torso, push your body as one unit into a pushup position.**

- **Do you understand these instructions?**

Score the movement.
Repeat two times if necessary.
Repeat the instructions with appropriate hand placement if necessary.

SPINAL EXTENSION CLEARING

INSTRUCTIONS

- **While lying on your stomach, place your hands, palms down, under your shoulders.**

- **With no lower body movement, press your chest off the surface as much as possible by straightening your elbows.**

- **Do you understand these instructions?**

- **Do you feel any pain?**

Rotary Stability

Equipment needed: 2 x 6

Instructions

- **Get on your hands and knees over the 2x6 so your hands are under your shoulders and your knees are under your hips.**

- **The thumbs, knees and toes must contact the sides of the 2x6, and the toes must be pulled toward the shins.**

- **At the same time, reach your right hand forward and right leg backward, like you are flying.**

- **Then without touching down, touch your right elbow to your right knee directly over the 2x6.**

- **Return to the extended position.**

- **Return to the start position.**

- **Do you understand these instructions?**

Score the movement.
Repeat the test on the other side.
If necessary, instruct the client to use a diagonal pattern of right arm and left leg.
Repeat the diagonal pattern with left arm and right leg.
Score the movement.

Spinal Flexion Clearing

Instructions

- **Get on all fours, and rock your hips toward your heels.**

- **Lower your chest to your knees, and reach your hands in front of your body as far as possible.**

- **Do you understand these instructions?**

- **Do you feel any pain?**

EXAMPLE OF A CONVENTIONAL DEEP SQUAT EVALUATION PROCESS

On the following page you'll find a case of how a movement pattern can fall under premature specific analysis and potentially hinder corrective exercise choices. The example represents a typical approach to movement-pattern evaluation seen in the literature, introduced here to create an opportunity to discuss the differences between evaluation and screening.

In professional circles, the word evaluation carries more weight and seems to be much more scientific and thorough than screening, but this can be a logical mistake. I want our readers to be able to discuss screening and evaluation intelligently. Each has its place, with benefits and limitations, and it is important that we embrace both tools.

Screening should initially direct focus and attention to the most limiting factor in a given situation. In contrast, evaluation should identify specific information within the most limiting predetermined variable.

When specific evaluation is placed before generalized screening, it can potentially cause assumptions and neglect the systematic logic necessary to rate and rank movement problems into a manageable hierarchy. Evaluation without screening is a typical example of reductionist science and it produces limited and oversimplified corrective solutions. The premature evaluation might seem systematic, but in reality it produces a narrow-minded outcome.

Note that the remedy for each problem in the example below is introduced as a tightness or weakness of a particular muscle group. The unwitting professional using this movement-pattern evaluation method could potentially follow all the instructions for corrections and observe no change in movement-pattern quality. This evaluation follows a basic kinesiological framework without considering motor control and developmental models of movement pattern acquisition, and it can greatly limit the potential successful outcome.

WHAT TO LOOK FOR:

FOOT & ANKLE

- Foot pronation: Y/N
- Externally rotation: Y/N

KNEES

- Valgus collapse: Y/N
- Varus: Y/N

LUMBO-PELVIC-HIP COMPLEX

- Asymmetrical weight shift: Y/N
- Lumbar lordosis: Y/N
- Hip adduction: Y/N
- Hip internal rotation: Y/N

WHAT TO DO WITH FINDINGS:

FOOT PRONATION & EXTERNAL ROTATION

- Tightness: Soleus, lateral gastrocnemius, biceps femoris, peroneals, piriformis

KNEE VALGUS & INTERNAL ROTATION

- Tightness: Gastrocnemius/soleus, adductors, IT band
- Weakness: Gluteus medius

LUMBAR LORDOSIS

- Tightness: Erector spinae & psoas
- Weakness: Transverse abdominis, internal obliques

HIP ADDUCTION

- Tightness: Hip adductors
- Weakness: Gluteus medius

HIP INTERNAL ROTATION

- Weakness: Gluteus maximus, hip external rotators

William Prentice, Rehabilitation Techniques for Sports Medicine and Athletic Training, 4th Edition, 2004, McGraw-Hill. Reprinted with publisher's permission.

The squat evaluation on the opposite page appeared in the literature following the introduction of the FMS and may have been an attempt to improve analysis of the squatting pattern. To many professionals, this view of the deep squat seems to be more thorough, but actually it is not because it can potentially be misleading.

First, it is referred to as a deep squat evaluation when the squat movement is stopped at an arbitrary point instead of being allowed to complete a full movement pattern. This is like saying you've evaluated a golfer's swing when in reality you decided to impose the limits of the backswing and follow-through. It would be incorrect to imply you have performed an evolution of movement pattern if you arbitrarily set limits and verbally impose range limits into the pattern. It is more logical to ask that a full pattern be performed with a standardized setup and allow natural proficiency or deficiency to present itself.

Second, in this example the deep squat movement pattern is put through a typical mechanical assessment. Although the assessment seems to demonstrate a comprehensive checklist of potential problems, it does not consider the problems over multiple movement patterns, and that is where perspective can be lost. The problems identified could represent a more fundamental problem or be only limited to the squatting pattern, however the evaluation produces no indication of which problem is present.

By not looking at multiple movement patterns, the issues in the squat could potentially be managed as independent problems within the squat, neglecting more fundamental issues of mobility, motor control and movement pattern acquisition. In contrast, screening simply introduces multiple patterns and rates and ranks the most dysfunctional pattern. The most dysfunctional pattern is broken down by the introduction of corrective strategies targeting basic mobility and basic motor control, followed by movement pattern retraining. The screening system is designed to reintroduce movement patterns following developmental requirements that complement learning. The evaluation of the deep squat seems to be a checklist of imperfections, whereas the screening system identifies those patterns that fall below a minimum level of quality.

Third, this evaluation suggests solutions to problems without considering the multiple causes of movement-pattern dysfunction. The loss of alignment might actually be a compensation strategy in a normal segment for a dysfunction elsewhere. In the deep squat evaluation model, the loss of alignment always implicates tightness and weakness. Fundamental problems with mobility and stability can create compensatory balance strategies that appear incorrect on the surface when in reality they are the only option when movement patterns are imposed on a dysfunctional base.

Let's consider three individuals with deep squat imperfections. For the sake of simplicity, these imperfections are so obvious they produce a FMS deep squat score of one in each of our three subjects.

Subject one's deep squat score is a one; all other scores on the FMS are threes.

Subject two's deep squat score is a one, with a one/two asymmetry noted on both the active straight-leg raise and shoulder mobility tests. All other scores are symmetrical twos.

Subject three's deep squat score is a one, with a score of one on the pushup. All other scores are symmetrical twos.

The FMS has identified three completely different problems that would all seem similar, if not identical, under the deep squat evaluation perspective. In the evaluation model, each individual would receive the exact same stretches and strengthening exercise to improve the deficient squat pattern. The FMS would only focus on the squat pattern corrections for subject one. Subjects two and three have obvious fundamental issues that should be managed before the squat pattern is considered according to the FMS corrective hierarchy.

Ironically, under the evaluation model, subject one would receive stretching and strengthening work as a corrective solution for the squat pattern. This would not be the case in the FMS model. It should be noted that six movement screen tests reveal no significant mobility or stability problems—the only problem is the squat pattern itself. The other FMS tests use nearly the same mobility and stability requirements of the deep squat—remember… the intentional redundancy. In this situation, subject one would probably have all the necessary mobility and stability to perform the deep squat. The problem is more likely a timing or motor control problem with squat movement pattern.

This example is provided to demonstrate the natural tendency of professionals to identify a list of imperfections in the evaluation process. A systematic approach starting with screening provides an appraisal of the most significant deviations from acceptable movement-pattern standards.

PATIENT SELF-EVALUATION FORMS

Modified Low Back Pain Disability Questionnaire[a]

This questionnaire has been designed to give your therapist information as to how your back pain has affected your ability to manage in everyday life. Please answer every question by placing a mark in the **one** box that best describes your condition today. We realize you may feel that 2 of the statements may describe your condition, but **please mark only the box that most closely describes your current condition.**

Pain Intensity
❑ I can tolerate the pain I have without having to use pain medication.
❑ The pain is bad, but I can manage without having to take pain medication.
❑ Pain medication provides me with complete relief from pain.
❑ Pain medication provides me with moderate relief from pain.
❑ Pain medication provides me with little relief from pain.
❑ Pain medication provides has no effect on my pain.

Personal Care (eg, Washing, Dressing)
❑ I can take care of myself normally without causing increased pain.
❑ I can take care of myself normally, but it increases my pain.
❑ It is painful to take care of myself, and I am slow and careful.
❑ I need help, but I am able to manage most of my personal care.
❑ I need help everyday in most aspects of my care.
❑ I do not get dressed, wash with difficulty, and stay in bed.

Lifting
❑ I can lift heavy weights without increased pain.
❑ I can lift heavy weights, but it causes increased pain.
❑ Pain prevents me from lifting heavy weights off the floor, but I can manage if the weights are conveniently positioned (eg, on a table).
❑ Pain prevents me from lifting heavy weights, but I can manage light to medium weights if they are conveniently positioned.
❑ I can lift only very light weights.
❑ I cannot lift or carry anything at all.

Walking
❑ Pain does not prevent me from walking any distance.
❑ Pain prevents me from walking more than 1 mile.[b]
❑ Pain prevents me from walking more than ½ mile.
❑ Pain prevents me from walking more than ¼ mile.
❑ I can only walk with crutches or a cane.
❑ I am in bed most of the time and have to crawl to the toilet.

Sitting
❑ I can sit in any chair as long as I like.
❑ I can only sit in my favorite chair as long as I like.
❑ Pain prevents me from sitting for more than 1 hour.
❑ Pain prevents me from sitting for more than ½ hour.
❑ Pain prevents me from sitting for more than 10 minutes.
❑ Pain prevents me from sitting at all.

Standing
❑ I can stand as long as I want without increased pain.
❑ I can stand as long as I want, but it increases my pain.
❑ Pain prevents me from standing more than 1 hour.
❑ Pain prevents me from standing more than ½ hour.
❑ Pain prevents me from standing more than 10 minutes.
❑ Pain prevents me from standing at all.

Sleeping
❑ Pain does not prevent me from sleeping well.
❑ I can sleep well only by using pain medication.
❑ Even when I take pain medication, I sleep less than 6 hours.
❑ Even when I take pain medication, I sleep less than 4 hours.
❑ Even when I take pain medication, I sleep less than 2 hours.
❑ Pain prevents me from sleeping at all.

Social Life
❑ My social life is normal and does not increase my pain.
❑ My social life is normal, but it increases my level of pain.
❑ Pain prevents me from participating in more energetic activities (eg, sports dancing).
❑ Pain prevents me from going out very often.
❑ Pain has restricted my social life to my home.
❑ I have hardly any social life because of my pain.

Traveling
❑ I can travel anywhere without increased pain.
❑ I can travel anywhere, but it increases my pain.
❑ My pain restricts my travel over 2 hours.
❑ My pain restricts my travel over 1 hour.
❑ My pain restricts my travel to short necessary journeys under ½ hour.
❑ My pain prevents all travel except for visits to the physician/therapist or hospital.

Employment/Homemaking
❑ My normal homemaking/job activities do not cause pain.
❑ My normal homemaking/job activities increase my pain, but I can still perform all that is required of me.
❑ I can perform most of my homemaking/job duties, but pain prevents me from performing more physically stressful activities (eg, lifting, vacuuming).
❑ Pain prevents me from doing anything but light duties.
❑ Pain prevents me from doing even light duties.
❑ Pain prevents me from performing any job or homemaking chores.

Reprinted from Fritz, Julie A and Irrgang, James J, A Comparison of a Modified Oswestry Low Back Pain Disability Questionnaire and the Quebec Back Pain Disability Scale, Vol. 81, No. 2, February 2001, pp. 776-788, with permission of the American Physical Therapy Association.

Neck Disability Index

This questionnaire has been designed to give the doctor information as to how your neck pain has affected your ability to manage in everyday life. Please answer every section and mark in each section only the ONE box that applies to you. We realize you may consider that two of the statements in any one section relate to you, but please just mark the ONE box that most closely describes your problem.

1. PAIN INTENSITY
☐ I have no pain at the moment.
☐ The pain is very mild at the moment.
☐ The pain is moderate at the moment.
☐ The pain is fairly severe at the moment.
☐ The pain is very severe at the moment.
☐ The pain is the worst imaginable at the moment.

6. CONCENTRATION
☐ I can concentrate fully when I want to with no difficulty.
☐ I can concentrate fully when I want to with slight difficulty.
☐ I have a fair degree of difficulty in concentrating when I want to.
☐ I have a lot of difficulty in concentrating when I want to.
☐ I have a great deal of difficulty in concentrating when I want to.
☐ I cannot concentrate at all.

2. PERSONAL CARE (washing, dressing, etc.)
☐ I can look after myself normally without causing extra pain.
☐ I can look after myself normally but it causes extra pain.
☐ It is painful to look after myself; I am slow and careful.
☐ I need some help but manage most of my personal care.
☐ I need help every day in most aspects of self-care.
☐ I don't get dressed; I wash with difficulty and stay in bed.

7. WORK
☐ I can do as much as I want to.
☐ I can do only my usual work, but no no more.
☐ I can do most of my usual work, but no more.
☐ I cannot do my usual work.
☐ I can hardly do any work at all.
☐ I can't do any work at all.

3. LIFTING
☐ I can lift heavy weights without extra pain.
☐ I can lift heavy weights but it gives me extra pain.
☐ Pain prevents me from lifting heavy weights off the floor, but I can manage if they are conveniently positioned, for example on a table.
☐ Pain prevents me from lifting heavy weights, but I can manage light to medium weights if they are conveniently positioned.
☐ I can lift very light weights.
☐ I cannot lift or carry anything at all.

8. DRIVING
☐ I can drive my car without any neck pain.
☐ I can drive my car as long as I want with slight pain in my neck.
☐ I can drive my car as long as I want with moderate pain in my neck.
☐ I can't drive my car as long as I want because of moderate pain in my neck.
☐ I can hardly drive at all because of severe pain in my neck.
☐ I can't drive my car at all.

4. READING
☐ I can read as much as I want to with no pain in my neck.
☐ I can read as much as I want to with slight pain in my neck.
☐ I can read as much as I want to with moderate pain in my neck.
☐ I can't read as much as I want because of moderate neck pain.
☐ I can hardly read at all because of severe pain in my neck.
☐ I cannot read at all.

9. SLEEPING
☐ I have no trouble sleeping.
☐ My sleep is slightly disturbed (less than 1 hr. sleepless).
☐ My sleep is mildly disturbed (1-2 hrs. sleepless).
☐ My sleep is moderately disturbed (2-3 hrs. sleepless).
☐ My sleep is greatly disturbed (3-5 hrs. sleepless).
☐ My sleep is completely disturbed (5-7 hrs. sleepless).

5. HEADACHES
☐ I have no headaches at all.
☐ I have slight headaches that come infrequently.
☐ I have moderate headaches that come infrequently.
☐ I have moderate headaches that come frequently.
☐ I have severe headaches that come frequently.
☐ I have headaches almost all the time.

10. RECREATION
☐ I am able to engage in all my recreation activities with no neck pain at all.
☐ I am able to engage in all my recreational activities, with some pain in my neck.
☐ I am able to engage in most, but not all my usual recreation activities because of pain in my neck.
☐ I am able to engage in few of my ususal recreation activities because of pain in my neck.
☐ I can hardly do any recreation activities because of pain in my neck.
☐ I can't do any recreation activities at all.

Please rate your pain, based on how you feel today, on the following scale:

Pain   Worst Imaginable Pain

0   1   2   3   4   5   6   7   8   9   10

Name: _____   Account Number: _____

Date: _____   Diagnosis: _____   Score: _____ %   Post-Op   Hip   Conservative   Knee

THE LOWER EXTREMITY FUNCTIONAL SCALE

We are interested in knowing whether you are having any difficulty at all with the activities listed below because of your lower limb problem for which you are currently seeking attention.

Please provide an answer for **each** activity.

	Activities	Extreme Difficulty or Unable to Perform Activity	Quite a Bit of Difficulty	Moderate Difficulty	A Little Bit of Difficulty	No Difficulty
1	Any of your usual work, housework or school activities.	0	1	2	3	4
2	Your usual hobbies, recreationor sporting activities.	0	1	2	3	4
3	Getting into or out of the bath.	0	1	2	3	4
4	Walking between rooms.	0	1	2	3	4
5	Putting your shoes or socks on.	0	1	2	3	4
6	Squatting.	0	1	2	3	4
7	Lifting an object, like a bag of groceries, from the floor.	0	1	2	3	4
8	Performing light activities around your home.	0	1	2	3	4
9	Performing heavy activities around your home.	0	1	2	3	4
10	Getting into or out of a car.	0	1	2	3	4
11	Walking 2 blocks.	0	1	2	3	4
12	Walking a mile.	0	1	2	3	4
13	Going up or down 10 stairs (about 1 flight of stairs).	0	1	2	3	4
14	Standing for 1 hour.	0	1	2	3	4
15	Sitting for 1 hour.	0	1	2	3	4
16	Running on even ground.	0	1	2	3	4
17	Running on uneven ground.	0	1	2	3	4
18	Making sharp turns while running fast.	0	1	2	3	4
19	Hopping.	0	1	2	3	4
20	Rolling over in bed.	0	1	2	3	4
	Column Totals:					

Minimum Level of Detectable Change (90% Confidence): 9 points

SCORE: _____ /80

Please rate your pain, based on how you feel today, on the following scale:

No Pain
0   1   2   3   4   5   6   7   8   9   10
                                  Worst Imaginable Pain

GLOBAL RATE OF CHANGE

Please rate the overall condition of your lower limb from the time you began treatment until now (check only one):

____ A very great deal better
____ A great deal better
____ Quite a bit better
____ Moderately better
____ Somewhat better
____ A little bit better
____ A tiny bit better (almost the same)

____ About the same

____ A very great deal worse
____ A great deal worse
____ Quite a bit worse
____ Moderately worse
____ Somewhat worse
____ A little bit worse
____ A tiny bit worse (almost the same)

Name: _____ Account Number: _____ Post-Op   Conservative

Date: _____ Therapist: _____ Score: _____ %

Foot and Ankle Ability Measure (FAAM)
Activities of Daily Living Subscale

Please answer **every question** with **one response** that most closely describes your condition within the past week.
If the activity in question is limited by something other than your foot or ankle mark not applicable (N/A).

Because of your foot and ankle how much difficulty do you have with:	No Difficulty	Slight Difficulty	Moderate Difficulty	Extreme Difficulty	Unable To Do	N/A
Standing	4	3	2	1	0	N/A
Walking on even ground	4	3	2	1	0	N/A
Walking on even ground without shoes	4	3	2	1	0	N/A
Walking up hills	4	3	2	1	0	N/A
Walking down hills	4	3	2	1	0	N/A
Going up stairs	4	3	2	1	0	N/A
Going down stairs	4	3	2	1	0	N/A
Walking on uneven ground	4	3	2	1	0	N/A
Stepping up and down curbs	4	3	2	1	0	N/A
Squatting	4	3	2	1	0	N/A
Coming up on your toes	4	3	2	1	0	N/A
Walking initially	4	3	2	1	0	N/A
Walking 5 minutes or less	4	3	2	1	0	N/A
Walking approximately 10 minutes	4	3	2	1	0	N/A
Walking 15 minutes or greater	4	3	2	1	0	N/A
Home responsibilities	4	3	2	1	0	N/A
Activities of daily living	4	3	2	1	0	N/A
Personal care	4	3	2	1	0	N/A
Light to moderate work (standing, walking)	4	3	2	1	0	N/A
Heavy work (push/pulling, climbing, carrying)	4	3	2	1	0	N/A
Recreational activities	4	3	2	1	0	N/A

Please rate your pain, based on how you feel today, on the following scale:

Pain

0   1   2   3   4   5   6   7   8   9   10

                                        Worst Imaginable Pain

Overall, how would you rate your current level of function?

□ Normal   □ Nearly Normal   □ Abnormal   □ Severely Abnormal

* Scored from 0-4 with 0 = unable and 4 = no difficulty *
Take the total score and divide by highest total possible (84) then x 100
A lower score represents a greater level of disability

ICC = 0.89   SEM = 2.1
MDC$_{95}$ = 5.7   MCID = 8

LIST OF ILLUSTRATIONS

1 Brushoj C, Larsen K, Albrecht-Beste E, Nielsen MB, Loye F, Holmich P. Prevention of overuse injuries by a concurrent exercise program in subjects exposed to an increase in training load: a randomized controlled trial of 1020 army recruits. *Am J Sports Med.* Apr 2008;36(4):663-670.

2 Shrier I. Stretching before exercise does not reduce the risk of local muscle injury: a critical review of the clinical and basic science literature. *Clin J Sport Med.* Oct 1999;9(4):221-227.

3 Kiesel K, Plisky P, Kersey P. Functional Movement Test Score as a Predictor of Time-loss during a Professional Football Team's Pre-season Paper presented at: American College of Sports Medicine Annual Conference 2008; Indianapolis, IN.

4 Kiesel K, Plisky PJ, Voight M. Can serious injury in professional football be predicted by a preseason Functional Movement Screen? *North Am J Sports Phys Ther.* August 2007;2(3):147-158.

5 Peate WF, Bates G, Lunda K, Francis S, Bellamy K. Core strength: a new model for injury prediction and prevention. *J Occup Med Toxicol.* 2007;2:3.

6 Meghan F, McFadden D, Deuster P, et al. Functional Movement Screening: A Novel Tool for Injury Risk Stratification of Warfighters. American College of Sports Medicine Annual Conference. Baltimore Maryland 2010.

7 Kiesel K, Plisky P, Butler R. Functional movement test scores improve following a standardized off-season intervention program in professional football players. *Scand J Med Sci Sports.* Dec 18 2009.

8 Wainner RS, Whitman JM, Cleland JA, Flynn TW. Regional interdependence: a musculoskeletal examination model whose time has come. *J Orthop Sports Phys Ther.* Nov 2007;37(11):658-660.

9 Cook, G., Burton, L., Fields, K., Kiesel, K., & Van Allen, J. (1999). Functional Movement Screening: Upper and Lower Quarter Applications. Paper presented at the Mid-America Athletic Trainer's Annual Symposium, Sioux Falls, South Dakota.

10 Cook, G., Burton,L.& Van Allen, J., (1999). Functional Movement Screening. Presented at the National Athletic Trainer's Association Annual Symposium, Kansas City, Kansas.

11 Kiesel K, Plisky PJ, Voight M. Can serious injury in professional football be predicted by a preseason Functional Movement Screen? *North Am J Sports Phys Ther.* August 2007;2(3):147-158.

12 Plisky PJ, Rauh MJ, Kaminski TW, Underwood FB. Star Excursion Balance Test as a predictor of lower extremity injury in high school basketball players. *J. Orthop Sports Phys Ther* Dec 2006;36(12):911-919.

13 Nadler SF, Moley P, Malanga GA, Rubbani M, Prybicien M, Feinberg JH. Functional deficits in athletes with a history of low back pain: a pilot study. *Arch Phys Med Rehabil.* Dec 2002;83(12):1753-1758.

14 Nadler SF, Malanga GA, Feinberg JH, Rubanni M, Moley P, Foye P. Functional performance deficits in athletes with previous lower extremity injury. *Clin J Sport Med.* Mar 2002;12(2):73-78.

15 Nadler SF, Malanga GA, Bartoli LA, Feinberg JH, Prybicien M, Deprince M. Hip muscle imbalance and low back pain in athletes: influence of core strengthening. *Med Sci Sports Exerc.* Jan 2002;34(1):9-16.

16 Cichanowski HR, Schmitt JS, Johnson RJ, Niemuth PE. Hip strength in collegiate female athletes with patellofemoral pain. *Med Sci Sports Exerc.* Aug 2007;39(8):1227-1232.

17 Lehance C, Binet J, Bury T, Croisier JL. Muscular strength, functional performances and injury risk in professional and junior elite soccer players. *Scand J Med Sci Sports.* Mar 31 2008.

18 Lombard, W.P. & Abbott, F.M. (1907). The mechanical effects produced by the contraction of individual muscles of the thigh of the frog. *Am J Physiol.* 20, 1-60

19 Dekker JM, Crow RS, Folsom AR, et al. Low heart rate variability in a 2-minute rhythm strip predicts risk of coronary heart disease and mortality from several causes: the ARIC Study. *Atherosclerosis Risk In Communities. Circulation.* Sep 12 2000;102(11):1239-1244.

20 van Dieen JH, Selen LP, Cholewicki J. Trunk muscle activation in low-back pain patients, an analysis of the literature. *J Electromyogr Kinesiol.* Aug 2003;13(4):333-351.

21 Flor H. Cortical reorganisation and chronic pain: implications for rehabilitation. *J Rehabil Med.* May 2003(41 Suppl):66-72.

22 Richardson C, Hodges P, Hides J. *Therapeutic Exercise for Lumbopelvic Stabilization: A Motor Control Approach for the Treatment and Prevention of Low Back Pain 2nd ed*: Churchill Livingstone; 2004.

23 Fleming DW, Binder S, eds. National Center for Injury Prevention and Control. CDC Injury Research Agenda. Atlanta (GA): Centers for Disease Control and Prevention; 2002.

24 Peate WF, Bates G, Lunda K, Francis S, Bellamy K. Core strength: a new model for injury prediction and prevention. *J Occup Med Toxicol.* 2007;2:3.

25 Emery CA. Injury prevention and future research. Med Sport Sci. 2005;48:179-200.

26 Cholewicki J, Silfies SP, Shah RA, et al. Delayed trunk muscle reflex responses increase the risk of low back injuries. *Spine.* Dec 1 2005;30(23):2614-2620.

27 Faude O, Junge A, Kindermann W, Dvorak J. Risk factors for injuries in elite female soccer players. *Br J Sports Med.* Sep 2006;40(9):785-790.

28 McHugh MP, Tyler TF, Tetro DT, Mullaney MJ, Nicholas SJ. Risk factors for noncontact ankle sprains in high school athletes: the role of hip strength and balance ability. *Am J Sports Med.* Mar 2006;34(3):464-470.

29 McKay GD, Goldie PA, Payne WR, Oakes BW. Ankle injuries in basketball: injury rate and risk factors. *Br J Sports Med.* Apr 2001;35(2):103-108.

30 Turbeville SD, Cowan LD, Owen WL, Asal NR, Anderson MA. Risk factors for injury in high school football players. *Am J Sports Med.* Nov-Dec 2003;31(6):974-980.

31 Tyler TF, McHugh MP, Mirabella MR, Mullaney MJ, Nicholas SJ. Risk factors for noncontact ankle sprains in high school football players: the role of previous ankle sprains and body mass index. *Am J Sports Med.* Mar 2006;34(3):471-475.

32 Zazulak BT, Hewett TE, Reeves NP, Goldberg B, Cholewicki J. Deficits in neuromuscular control of the trunk predict knee injury risk: a prospective biomechanical-epidemiologic study. *Am J Sports Med.* Jul 2007;35(7):1123-1130.

33 Cholewicki J, Panjabi MM, Khachatryan A. Stabilizing function of trunk flexor-extensor muscles around a neutral spine posture. *Spine* (Phila Pa 1976). Oct 1 1997;22(19):2207-2212.

34 Nadler SF, Moley P, Malanga GA, Rubbani M, Prybicien M, Feinberg JH. Functional deficits in athletes with a history of low back pain: a pilot study. *Arch Phys Med Rehabil.* Dec 2002;83(12):1753-1758.

35 Bullock-Saxton JE, Janda V, Bullock MI. The influence of ankle sprain injury on muscle activation during hip extension. *Int J Sports Med.* Aug 1994;15(6):330-334.

36 Choudhry NK, Fletcher RH, Soumerai SB. Systematic review: the relationship between clinical experience and quality of healthcare. *Ann Intern Med.* Feb 15 2005;142(4):260-273.

37 Hickey J, Barrett B, Butler R, Kiesel K, Plisky P. Reliability of the Functional Movement Screen Using a 100-point Grading Scale. Paper presented at: American College of Sports Medicine Annual Meeting 2010; Baltimore, MD.

38 Minick KI, Kiesel KB, Burton L, Taylor A, Plisky P, Butler RJ. Interrater reliability of the functional movement screen. *J Strength Cond Res.* Feb 2010;24(2):479-486.

39 Cook G, Burton L, Fields K, Kiesel K, Van Allen J. Functional Movement Screening: Upper and Lower Quarter Applications. Paper presented at: Mid-America Athletic Trainer's Annual Symposium 1999; Sioux Falls, South Dakota.

40 Deyo RA, Mirza SK, Martin BI. Back pain prevalence and visit rates: estimates from U.S. national surveys, 2002. *Spine* (Phila Pa 1976). Nov 1 2006;31(23):2724-2727.

41 Zedka M, Prochazka A, Knight B, Gillard D, Gauthier M. Voluntary and reflex control of human back muscles during induced pain. *J Physiol.* Oct 15 1999;520 Pt 2:591-604.

42 Lund JP, Donga R, Widmer CG, Stohler CS. The pain-adaptation model: a discussion of the relationship between chronic musculoskeletal pain and motor activity. *Can J Physiol Pharmacol.* May 1991;69(5):683-694.

43 Richardson C, Hodges P, Hides J. *Therapeutic Exercise for Lumbopelvic Stabilization: A Motor Control Approach for the Treatment and Prevention of Low Back Pain 2nd ed:* Churchill Livingstone; 2004.

44 Kiesel K BR, Duckworth A, Underwood, F. Experimentally induced pain alters the EMG activity of the lumbar multifidus in asymptomatic subjects. 6th Interdisciplinary World Congress on Low Back & Pelvic Pain. (Platform Presentation) Barcelona Spain; 2007.

45 van Dieen JH, Selen LP, Cholewicki J. Trunk muscle activation in low-back pain patients, an analysis of the literature. *J Electromyogr Kinesiol.* Aug 2003;13(4):333-351.

46 Clark M, Russell A. Optimum Performance Training for the Performance Enhancement Specialist. Calabasas, CA: *National Academy of Sports Medicine.* 2001.

47 Cholewicki J, Greene HS, Polzhofer GK, Galloway MT, Shah RA, Radebold A. Neuromuscular function in athletes following recovery from a recent acute low back injury. *J Orthop Sports Phys Ther.* Nov 2002;32(11):568-575.

48 Pirouzi S, Hides J, Richardson C, Darnell R, Toppenberg R. Low back pain patients demonstrate increased hip extensor muscle activity during standardized submaximal rotation efforts. *Spine.* Dec 15 2006;31(26):E999-E1005.

49 van Dieen JH, Selen LP, Cholewicki J. Trunk muscle activation in low-back pain patients, an analysis of the literature. *J Electromyogr Kinesiol.* Aug 2003;13(4):333-351.

50 Kiesel K, Plisky P, Butler R. Functional movement test scores improve following a standardized off-season intervention program in professional football players. *Scand J Med Sci Sports.* Dec 18 2009.

51 Hodges P, Richardson C, Jull G. Evaluation of the relationship between laboratory and clinical tests of transversus abdominis function. *Physiother Res Int.* 1996;1(1):30-40.

52 Hodges PW, Moseley GL, Gabrielsson A, Gandevia SC. Experimental muscle pain changes feedforward postural responses of the trunk muscles. *Exp Brain Res.* Jul 2003;151(2):262-271.

53 Hodges PW, Richardson CA. Delayed postural contraction of transversus abdominis in low back pain associated with movement of the lower limb. *J Spinal Disord.* Feb 1998;11(1):46-56.

54 Hodges PW, Richardson CA. Altered trunk muscle recruitment in people with low back pain with upper limb movement at different speeds. *Arch Phys Med Rehabil.* Sep 1999;80(9):1005-1012.

55 Cowan SM, Schache AG, Brukner P, et al. Delayed onset of transversus abdominus in long-standing groin pain. *Med Sci Sports Exerc.* Dec 2004;36(12):2040-2045.

56 Yamamoto K, Kawano H, Gando Y, et al. Poor trunk flexibility is associated with arterial stiffening. *Am J Physiol Heart Circ Physiol.* Oct 2009;297(4):H1314-1318.

57 Richter RR, VanSant AF, Newton RA. Description of adult rolling movements and hypothesis of developmental sequences. *Phys Ther.* Jan 1989;69(1):63-71.

58 Meghan F, McFadden D, Deuster P, et al. Functional Movement Screening: A Novel Tool for Injury Risk Stratification of Warfighters. American College of Sports Medicine Annual Conference. Baltimore Maryland 2010.

59 Peate WF, Bates G, Lunda K, Francis S, Bellamy K. Core strength: a new model for injury prediction and prevention. *J Occup Med Toxicol.* 2007;2:3.